En vertu de l'article 1er de la Convention signée le 14 décembre 1960, à Paris, et entrée en vigueur le 30 septembre 1961, l'Organisation de Coopération et de Développement Économiques (OCDE) a pour objectif de promouvoir des politiques visant :

- à réaliser la plus forte expansion de l'économie et de l'emploi et une progression du niveau de vie dans les pays Membres, tout en maintenant la stabilité financière, et à contribuer ainsi au développement de l'économie mondiale ;
- à contribuer à une saine expansion économique dans les pays Membres, ainsi que non membres, en voie de développement économique ;
- à contribuer à l'expansion du commerce mondial sur une base multilatérale et non discriminatoire conformément aux obligations internationales.

Les signataires de la Convention relative à l'OCDE sont : la République Fédérale d'Allemagne, l'Autriche, la Belgique, le Canada, le Danemark, l'Espagne, les Etats-Unis, la France, la Grèce, l'Irlande, l'Islande, l'Italie, le Luxembourg, la Norvège, les Pays-Bas, le Portugal, le Royaume-Uni, la Suède, la Suisse et la Turquie. Les pays suivants ont adhéré ultérieurement à cette Convention (les dates sont celles du dépôt des instruments d'adhésion) : le Japon (28 avril 1964), la Finlande (28 janvier 1969), l'Australie (7 juin 1971) et la Nouvelle-Zélande (29 mai 1973).

La République socialiste fédérative de Yougoslavie prend part à certains travaux de l'OCDE (accord du 28 octobre 1961).

L'Agence de l'OCDE pour l'Énergie Nucléaire (AEN) a été créée le 20 avril 1972, en remplacement de l'Agence Européenne pour l'Énergie Nucléaire de l'OCDE (ENEA) lors de l'adhésion du Japon à titre de Membre de plein exercice.

L'AEN groupe désormais tous les pays Membres européens de l'OCDE ainsi que l'Australie, le Canada, les États-Unis et le Japon. La Commission des Communautés Européennes participe à ses travaux.

L'AEN a pour principaux objectifs de promouvoir, entre les gouvernements qui en sont Membres, la coopération dans le domaine de la sécurité et de la réglementation nucléaires, ainsi que l'évaluation de la contribution de l'énergie nucléaire au progrès économique.

Pour atteindre ces objectifs, l'AEN :

- *encourage l'harmonisation des politiques et pratiques réglementaires dans le domaine nucléaire, en ce qui concerne notamment la sûreté des installations nucléaires, la protection de l'homme contre les radiations ionisantes et la préservation de l'environnement, la gestion des déchets radioactifs, ainsi que la responsabilité civile et les assurances en matière nucléaire ;*
- *examine régulièrement les aspects économiques et techniques de la croissance de l'énergie nucléaire et du cycle du combustible nucléaire, et évalue la demande et les capacités disponibles pour les différentes phases du cycle du combustible nucléaire, ainsi que le rôle que l'énergie nucléaire jouera dans l'avenir pour satisfaire la demande énergétique totale ;*
- *développe les échanges d'informations scientifiques et techniques concernant l'énergie nucléaire, notamment par l'intermédiaire de services communs ;*
- *met sur pied des programmes internationaux de recherche et développement, ainsi que des activités organisées et gérées en commun par les pays de l'OCDE.*

Pour ces activités, ainsi que pour d'autres travaux connexes, l'AEN collabore étroitement avec l'Agence Internationale de l'Énergie Atomique de Vienne, avec laquelle elle a conclu un Accord de coopération, ainsi qu'avec d'autres organisations internationales opérant dans le domaine nucléaire.

The NEA Committee on the Safety of Nuclear Installations (CSNI) is an international committee made up of scientists and engineers who have responsibilities for nuclear safety research and nuclear licensing. The Committee was set up in 1973 to develop and co-ordinate the Nuclear Energy Agency's work in nuclear safety matters, replacing the former Committee on Reactor Safety Technology (CREST) with its more limited scope.

The Committee's purpose is to foster international co-operation in nuclear safety amongst the OECD Member countries. This is done in a number of ways. Full use is made of the traditional methods of co-operation, such as information exchanges, establishment of working groups, and organisation of conferences. Some of these arrangements are of immediate benefit to Member countries, for example by enriching the data base available to national regulatory authorities and to the scientific community at large. Other questions may be taken up by the Committee itself with the aim of achieving an international consensus wherever possible. The traditional approach to co-operation is increasingly being reinforced by the creating of co-operative (international) research projects, such as PISC and LOFT, and by a novel form of collaboration known as the international standard problem exercise, for testing the performance of computer codes, test methods, etc. used in safety assessments. These exercises are now being conducted in most sectors of the nuclear safety programme.

The greater part of the CSNI co-operative programme is concerned with safety technology for water reactors. The principal areas covered are operating experience and the human factor, reactor system response during abnormal transients, various aspects of primary circuit integrity, the phenomenology of radioactive releases in reactor accidents, and risk assessment. The Committee also studies the safety of the fuel cycle, conducts periodic surveys of reactor safety research programmes and operates an international mechanism for exchanging reports on power plant incidents.

The Committee has set up a sub-Committee on Licensing which examines a variety of nuclear regulatory problems, provides a forum for the free discussion of licensing questions and reviews the regulatory impact of the conclusions reached by CSNI.

Le Comité de l'AEN sur la sûreté des installations nucléaires (CSIN) est un Comité international composé de scientifiques et d'ingénieurs qui ont des responsabilités dans les domaines de la recherche sur la sûreté nucléaire et de l'autorisation des installations nucléaires. Ce Comité a été créé en 1973 en vue de développer et coordonner les travaux de l'Agence pour l'Energie Nucléaire en matière de sûreté nucléaire ; il a remplacé le précédent Comité sur les techniques de sécurité des réacteurs (CREST) dont le mandat était plus limité.

Le Comité a pour but d'encourager la coopération internationale en matière de sûreté nucléaire entre les pays Membres de l'OCDE, et ce, par divers moyens. A cet effet, il utilise pleinement les méthodes classiques de coopération, telles que les échanges d'informations, la mise en place de groupes de travail et l'organisation de conférences. Certains de ces arrangements sont d'un intérêt immédiat pour les pays Membres, en ce sens notamment qu'ils permettent aux autorités nationales chargées de la réglementation et à la communauté scientifique dans son ensemble d'enrichir la base de donnés dont elles disposent. Le Comité lui-même peut être amené à aborder d'autres questions en vue de parvenir à un consensus international chaque fois que cela est possible. Cette démarche classique à l'égard de la coopération est de plus en plus renforcée par la création de projets de recherche (internationaux) en coopération, comme le Programme d'inspection des composants en acier (PISC) et le projet fondé sur l'installation d'essais de perte de fluide caloporteur (LOFT), et par une forme inédite de collaboration, à savoir l'exécution de problèmes standards internationaux, visant à tester le fonctionnement des programmes de calcul, les méthodes d'essai, etc., utilisés dans les évaluations de sûreté. De tels travaux sont désormais menés dans la plupart des domaines couverts par le programme de sûreté nucléaire.

Le programme en coopération du Comité sur la sûreté des installations nucléaires (CSIN) a essentiellement trait à la technolgie de la sûreté des réacteurs à eau. Les principaux domaines qu'il couvre sont les suivants : expérience en matière d'exploitation et facteurs humains, réponse des systèmes de réacteurs au cours de transitoires anormaux, aspects divers de l'intégrité du circuit primaire, phénoménologie des rejets de radioactivité lors d'accidents affectant des réacteurs et évaluation des risques. Le Comité étudie également la sûreté du cycle du combustible, procède périodiquement à des enquêtes sur les programmes de recherche en matière de sûreté des réacteurs et exploite un mécanisme international pour l'échange de rapports sur les incidents survenus dans des centrales.

Ce Comité a créé un Sous-Comité chargé des questions d'autorisation, qui examine toute une gamme de problèmes liés à la réglementation nucléaire, offre un cadre permettant de débattre librement des questions d'autorisation et fait le point des incidences, sur le plan réglementaire, des conclusions découlant des travaux du CSIN.

PREFACE

The Workshop on Test Methods for Ductile Fracture was conceived by the CSNI Principal Working Group 3 (Primary Circuit Integrity). The topic of ductile fracture has been investigated from both the theoretical and experimental viewpoints for many years, and recently test procedures for ductile fracture have been developed by laboratories in Europe, USA and Japan. A critical review of these procedures was considered necessary to provide the basis for standard test methods. Most OECD countries have organisations responsible for standards, and some test methods for elastic-plastic fracture have been evolved. Nevertheless, it is necessary to promote an interaction among investigators worldwide in order to develop rational test procedures that have a relationship to structural integrity. An informal workshop was believed to be more conducive to this goal than a symposium.

One objective of the CSNI Workshop was to provide a forum for a free exchange of views, based upon informal presentations, covering advances in test methods and techniques for experimental characterization of elastic-plastic fracture. Another aim was to critically assess the problems and limitations associated with these methods to better define the current status and to highlight areas requiring further investigation. While the CSNI is involved exclusively with nuclear structures, the Workshop was not restricted to this area since elastic-plastic test methods are applicable to a number of structures, both nuclear and non-nuclear. With its multinational membership, the CSNI forms a logical focus for this type of activity.

The Workshop provided a unique opportunity for representatives of twelve countries to make presentations and discuss views with the opportunity for critique from the participants. Fifty-two persons were in attendance, all of whom were invited on the basis of demonstrated experience in ductile fracture testing. Each participant was asked to prepare a short presentation focussed on a particular topic to serve as the basis for discussion. These presentations were to emphasise experimental procedures as opposed to data trends, theory or applications. A total of thirty-nine presentations were made and these are assembled as the Proceedings of the Workshop. Summaries of the individual technical sessions are also included.

The contribution of all the participants in making this a highly successful meeting are gratefully acknowledged. Future workshops of this type can better establish progress in ductile test methods by providing a cross-fertilization of ideas on an international basis, often years before journal publication of this research.

F. J. Loss
Chairman

AVANT-PROPOS

La séance de travail sur les méthodes d'essai relatives à la rupture ductile a été organisée à l'initiative du Groupe de travail principal No. 3 du CSIN (sur l'intégrité du circuit primaire). Voici de nombreuses années que la question de la rupture ductile est étudiée du point de vue tant théorique qu'expérimental; or, des procédures d'essai y afférentes ont été récemment mises au point par des laboratoires situés en Europe, aux Etats-Unis et au Japon. On a jugé qu'un examen critique de ces procédures s'imposait afin de jeter les fondements de méthodes d'essai normalisées. Dans la plupart des pays de l'OCDE, il existe des organismes de normalisation et certaines méthodes d'essai ont été élaborées pour la rupture élastoplastique. Néanmoins, il est nécessaire d'inciter les chercheurs du monde entier à collaborer à la mise au point de procédures d'essai rationnelles qui soient en corrélation avec l'intégrité structurelle. On a estimé qu'une séance de travail informelle contribuerait davantage à la réalisation de cet objectif qu'un symposium.

La séance de travail du CSIN avait notamment pour objet de servir de cadre à un libre échange de vues reposant sur des exposés présentés à titre officieux et portant sur les progrès accomplis dans les méthodes d'essai et techniques applicables à la caractérisation expérimentale de la rupture élastoplastique. En outre, on se proposait d'évaluer d'un point de vue critique les problèmes et limitations liés à ces méthodes afin de mieux définir l'état d'avancement des travaux et de mettre en relief les domaines appelant un complément de recherches. Bien que le CSIN s'occupe uniquement des structures nucléaires, la séance de travail ne se limitait pas à ce domaine car les méthodes d'essai relatives à la rupture elastoplastique peuvent être appliquées à un certain nombre de structures aussi bien nucléaires que non nucléaires. Etant donné sa composition multinationale, le CSIN constitue un mécanisme de coordination logique pour des activités de ce type.

Cette séance de travail a offert une occasion unique à des représentants de douze pays de faire des exposés et de débattre des diverses thèses en présence, tout en donnant aux participants la possibilité d'émettre des critiques. Elle a rassemblé cinquante-deux personnes qui étaient toutes invitées en raison de leur expérience confirmée en matière d'essais relatifs à la rupture ductile. Chaque participant a été prié d'établir un bref exposé axé sur un sujet particulier et destiné à servir de base aux débats. Ces exposés devaient mettre l'accent sur les procédures expérimentales par opposition à l'évolution des données, à la théorie ou aux applications dans ce domaine. Les trente-neuf exposés au total qui ont été présentés à cette occasion sont regroupés dans le compte rendu de la séance de travail, qui contient également des résumés des différentes séances techniques.

Nous tenons à exprimer notre vive reconnaissance à tous les participants qui ont contribué à la pleine réussite de cette réunion. Les futures séances de travail de ce type pourront permettre de mieux démontrer les progrès réalisés dans les méthodes d'essai relatives à la rupture ductile en suscitant un brassage d'idées fécond à l'échelon international, souvent plusieurs années avant qu'il ne soit fait état de ces recherches dans la presse.

F.J. Loss
Président

SUMMARY

The Workshop highlighted a considerable activity in ductile fracture test methods now under way in Europe, USA and Japan. Sessions were held on J-R curve test techniques, experience and problems with existing techniques, testing of structural elements and new or improved test techniques. Areas of agreement as well as disagreement on certain test methods surfaced during the meeting. There appeared to be agreement on the use of the J-R curve as a useful method to characterize elastic-plastic fracture toughness. Since the Workshop did not emphasize theory, the interest in J does not necessarily imply agreement that this is the best tool to characterize elastic-plastic fractures. For example, there was little discussion on the CTOD procedure even though this is widely used in structural integrity assessment. The fact that CTOD is not a standard in the nuclear industry may explain the small emphasis given to this topic. Nevertheless, it is suggested that the CTOD should be included in future meetings of this type.

Major emphasis was given to methods of measuring the J-R curve and to assessing the magnitude of J at the initiation of crack extension, J_0. Generally, three methods are used to measure the R curve: multiple specimen (heat tint), single specimen compliance and single specimen potential drop (AC or DC). In addition, stretch zone width (SZW) measurements are often used to determine J_0. Problem areas with these methods are described further in the summaries of the individual sessions.

There was considerable discussion on the ASTM E 813 procedure for J_{Ic}. It was agreed that J_{Ic} is meant to be an engineering term and that this quantity generally overestimates J at the true initiation point. In addition, the E 813 procedure does not correctly represent the non-linear R curve behaviour for structural steels, nor does it accurately reflect the actual blunting behaviour of certain steels. Nevertheless, the need exists for an engineering critical value of J such as J_{Ic}.

In addition to the method of E 813 for initiation toughness, the ASTM has proposed a method for J-R curve determination which is based on the single specimen compliance technique. A similar organised standardisation activity is required for the J-R curve determined by the single specimen potential drop technique. Since the PD methods have been more actively investigated in Europe than the USA, it is possible that such an activity could be led by the European Group on Fracture (EGF).

As advances are made in elastic-plastic fracture test methods, the need remains to analyse older specimens in terms of the "modern" technology. This applies especially to reactor surveillance specimens. The Workshop highlighted means by which J-R curves can be obtained from Charpy-V and X-type WOL specimens as well as ASTM E 399 compact specimens. It now appears clear that the E 399 compact specimen can be used to define the full toughness range from LFM to elastic plastic.

One session was held on the testing of structural elements, including tubes, pipes and plates. This session served to focus discussion on the utility of test methods, such as those based on compact or bend specimen data, in the prediction of structural performance. Out of the discussion came a need for analysis of specimens other than the foregoing. For example, a surface flaw is a common structural defect, yet little specimen fracture testing is done with flaws of this type. Basically, test methods are sometimes developed with no structure in mind. Therefore, discussions on the development of test methods should include the applications of these methods.

The discussion of structural elements also highlighted the area of specimen size dependence. No agreement was reached on whether small or large specimen tests are better to predict structural behaviour. Opinions were neutral as to whether the compact specimen provides a conservative estimate of fracture toughness (with or without the use of side grooves). A bend-type specimen such as the compact is generally considered to give a conservative estimate of fracture toughness compared with a flat plate in tension. Nevertheless, some R curve tests with pipes produced more conservative R curves than those defined by compact specimen tests. Finally, there is a need to further explore the applicability of an initiation parameter, such as J_0 or J_{Ic}, in structural integrity analysis vis-à-vis the potential conservatism of this approach in the presence of elastic plastic behaviour.

The Workshop provided a forum for a valuable exchange of ideas among leading scientists in elastic plastic test methods. The discussions served as a sounding board for new ideas and a means to focus on problems with existing methods. This opportunity to exchange ideas helped to develop a perspective that individual investigators may not have achieved otherwise. Scientists in Europe, USA and Japan share common problem areas. These areas could be resolved in a more expeditious manner if lines of communication between investigators were kept open through informal meetings and cooperative programmes. Consequently, it is considered worthwhile to continue these workshops on a regular basis.

TABLE OF CONTENTS
TABLE DES MATIERES

Session I - Séance I

J-R CURVE TEST TECHNIQUES
METHODES D'ESSAI POUR DETERMINER LES COURBES J-R

Chairman - Président : Mr. B. MARANDET (France)

11

Session II – Séance II

J-R CURVE TEST TECHNIQUES
METHODES D'ESSAI POUR DETERMINER LES COURBES J-R

Chairman – Président : Mr. K.H. SCHWALBE (F.R. of Germany)

Session III – Séance III

EXPERIENCE AND PROBLEMS WITH EXISTING TECHNIQUES
METHODES COURANTES: EXPERIENCE ET DIFFICULTES

Chairman – Président : Mr. J.D. LANDES (United States)

Session IV - Séance IV

TESTING OF STRUCTURAL ELEMENTS
ESSAIS SUR LES ELEMENTS DE STRUCTURE

Chairman - Président : Mr. L. CRESWELL (United Kingdom)

Session V – Séance V

NEW AND IMPROVED TEST TECHNIQUES
METHODES NOUVELLES ET AMELIOREES

Chairman – Président : Mr. M.I. DeVRIES (Netherlands)

Session I

J-R CURVE TEST TECHNIQUES

Chairman - Président

B. MARANDET

(France)

Séance I

METHODES D'ESSAI POUR DETERMINER LES COURBES J-R

SESSIONS NO. 1 AND 2: J-R CURVE TEST TECHNIQUES

Summary by the Chairmen: K.-H. Schwalbe and M. Marandet

Of the many aspects discussed during the first two sessions, some points common to most of the presentations are summarised.

Of the many aspects discussed during the first two sessions, some points common to most of the presentations are summarised.

The indirect crack growth measurement techniques discussed were the partial unloading (PUL) technique and the potential drop (PD) method in the form of AC or DC. By calibration both of these techniques should be equally well suitable for quantitative determination of crack growth. However, problems can be caused by crack front tunnelling since both techniques are affected differently by tunnelling as compared to saw-cut calibration. Two theoretical presentations (Prandtl and Prij) showed how the tunnelling can be considered with respect to the specimen's compliance.

A very complicated subject is the determination of the intitiation of growth of a pre-existing fatigue crack. Since high resolution is required to detect initiation, the PD methods seem to be superior to the PUL technique. However, the various PD techniques presented use different criteria to define initiation:

- minimum of PD-clip gauge record (Sarno et al., de Roo et al., Schwalbe et al.),

- deviation from linearity of PD-clip gauge record (Hollstein et al.,Wilkowski, Berger),

- deviation of initial R-curve from blunting line (Hollstein et al, Schwalbe et al.).

The existence of a minimum in the ACPD versus displacement diagram can be explained by the competition between the magnetostrictive effect in ferritic steels and the crack growth, whereas in the DCPD method, the minimum appears if the specimen is not insulated and could therefore be related to a loss of current at the loading pins. Whether the ACPD minimum coincides with the first steps of crack growth or not was said to depend on the current frequency.

It was not clear from the presentation whether these criteria were checked by fractographic observations. Since several experimental parameters affect the PD response (particularly near initiation) future efforts should be directed towards systematic investigation of these parameters. In particular, it is recomended to take into account the following points:

- contact locations of current input and potential measurement,

- effect of frequency of the ACPD method,

- insulation of loading pins from CT specimen,

- sensitivity of PD methods to crack blunting and/or crack growth,

- type of test record used to detect initiation,

- criterion for point on test record believed to be the initiation point (physical background!),

- verification of initiation criterion by fractography,

- effect of crack front tunnelling.

A number of the technical problems can be overcome:

- The negative Δa (PUL) seems to be due to friction between the loading pins and the grips and can be avoided by using hard inserts in the grips holes (Voss, Klausnitzer).

- The thermal EMF voltages (DCPD) can be eliminated by keeping the temperature constant at the two connections or by remoting these connections in a thermally insulated device (Wilkowski).

- The effect of electrical resistivity changes with temperature during an experiment (ACPD and DCPD) can be minimized simply by correction or by using a reference specimen (Wilkowski, De Roo).

- Side grooving improves the results of these methods (Voss).

Another subject of interest was the comparison of J_{Ic} as determined by ASTM E 813 and the true initiation point, J_0. There was agreement that J_{Ic} is not a mesure of the true initiation, as J_0 is usually smaller than J_{Ic}; in addition, there does not seem to be a constant relationship between J_0 and J_{Ic}. The blunting line equation $J_0 = 2\sigma_y \Delta a$ is not always adequate. Moreover, the method of constructing the straight line for the J_{Ic} determination can be a source of uncertainty. On the other hand, there is a need for an engineering critical value of J such as J_{Ic} and it seems to be possible to improve E 813 by certain changes of the evaluation procedure in order to derive a quantity which is physically more relevant than the present J_{Ic}. Thus, the vast amount of experience behind E 813 can be rescued. This should be discussed in more depth at meetings of ASTM Committee E24.

Measurement of Stable Crack Growth Including Detection of Initiation of Growth Using the DC Potential Drop and the Partial Unloading Methods

K.-H. Schwalbe,* D. Hellmann,** J. Heerens,** J. Knaack,**
and J. Müller-Roos,**
GKSS Research Center, 2054 Geesthacht,
Federal Republic of Germany

SUMMARY

The ability of the DC potential drop method and the partial
unloading technique to measure crack growth and to detect
initiation of crack growth was investigated using a number
of steels and aluminum alloys. It was found that within the
range of parameters investigated both of these methods can
be recommended for the determination of the R-curve; however,
since at small amounts of crack growth the DC potential drop
method gave more consistent results it is therefore considered
to be superior. Two contacting arrangements of the DC potential
drop method were checked for initiation detection: one indicates
initiation by a potential minimum, the other by the point
of 0.1mm apparent crack growth or by the intersection of
the R-curve with the blunting line. The initiation values
of J were compared with J_{Ic} as obtained by current practice.
It was found that J_{Ic} is poorly related to initiation or
to a specific amount of crack growth.

KEY WORDS

DC potential drop method, partial unloading technique, initiation
of crack growth, crack growth, fracture toughness J_{Ic}, multiple-
specimen method.

---

* Head of department   ** Engineer

INTRODUCTION

For monotonically increasing loads two kinds of fracture
mechanics properties are determined experimentally:
- The initiation of growth of a pre-existing crack in terms
  of stress intensity, K_o, J-integral, J_o, or crack tip opening
  displacement, σ_o;
- the resistance against crack growth as a function of crack
  length increase, Δa, in terms of K_R, J_R, and σR.

The determination of these properties requires the

accurate measurement of the actual crack length during the
course of loading. Although numerous techniques for crack
length measurements have been developped (a compilation is
given in Ref./1/) only two methods are regularly being used
in laboratory practice: the compliance method and the potential
drop (PD) method using alternating current (AC) or direct
current (DC). If these indirect methods work reliably the
multiple specimen method can be avoided and needs to be used
for calibration purposes only.
The simplest technique is used in the ASTM Standard Test
Method E399-81 /2/ for the determination of K_{Ic}. A five percent
change of the specimen's compliance indicates nominally two
percent increase of crack length. The pre-crack length is
measured microscopically on the fracture surface and no absolute
crack length value is required from the compliance measurement.
A shortcoming of this simple technique is that plastic deformations affect the result by an unknown amount. In addition,
depending on the specimen size Δa = 2% means different locations
on the R-curve and hence K_{Ic} can be size dependent if there
is a rising R-curve. The former effect can be avoided by
partially unloading the specimen in certain intervals. Guidance
for carrying out the partial unloading procedure (PUL) is
given by Refs /3-5/. Since unloading occurs elastically even
after comprehensive plastic deformation the unloading trace
is a measure of the elastic compliance of the specimen and
hence of the actual crack length. But unlike the K_{Ic} procedure
the change of crack length is determined quantitatively which
requires very accurate measurement of the compliance. Particularly, any hysteresis effects due to friction at the loading
pin and at the knife edges for the clip gage must be carefully
avoided. Thus, although in principle identical instrumentation
is used as for a K_{Ic} test (which may make the method attractive)
much higher requirements concerning accuracy are needed.

Whereas the partial unloading technique is discontinous in
nature continous crack length monitoring is provided by the
potential method, which requires, however, additional expensive
instrumentation. Several techniques are in use: AC with varying
frequency /6-8/ and DC with different positions for current
input and potential measurement /1,7,9-13/, resulting in
diffent resolution and sensitivity.
By experimental or theoretical calibration both the partial
unloading and the PD methods are capable of generating crack
growth data.

A particular problem arises when the initiation of growth has to be detected.
During the course of the present investigation it turned out that "initiation" is very much a matter of definition. An operational definition of initiation has to comply with the true physical events and with the needs for reproducible experimental determination. A tentative definition will be given in a following section. In current practice three ways of detecting initiation are being used:

a) Measure crack growth, Δa, as a function of a crack field parameter, K, J or δ, and extrapolate back to vanishing crack growth to obtain K_o, J_o or δ_o, Fig. 1a. The British COD test method /14/ determines the initiation value of δ this way.

b) Since crack growth is often measured including crack tip blunting (particularly if the crack length is measured by an indirect method) the point of incipient cracking from the blunted crack tip can be found by introducing a blunting line and intersecting it with the extrapolated J-Δa relationship, Fig. 1b. This is the procedure adopted by the J_{Ic} test standard /3/. In both of these cases the J-Δa or δ-Δa data are represented by a straight line relationship.

This ignores the fact that at its very beginning the R-curve is by no means a straight line, Fig. 1b, Fig. 2. Therefore, it is more appropriate to represent the R-curve data by a power law /15,16/. The intersection of this power law crack growth curve with either the crack extension line

Eq(1)

$$\Delta a = \frac{3}{4} \frac{J}{\sigma_F'}$$

(1)

or the 0.15 mm exclusion line - whichever yields the highest value - was defined by Ref./15/ as the J_{Ic} value and was found to coincide closely with the J_{Ic} value determined after the standard test procedure /3/, Fig. 2a. This proposal has three special properties:

- It is based on the actual growth data, which can deviate significantly from the straight line construction of E813.

- It is aimed at defining a critical J at a certain amount of crack growth to avoid the sometimes very large data scatter at the very initiation of growth.

- It is capable of determining a well defined J_{Ic} value even if instability occurs so closely to the 0.15 mm exclusion line that no valid J_{Ic} can be derived according to E813.

However, it should be kept in mind that the statements made in this proposal are based on indirect crack length measurements (unloading compliance technique) and on a single material (A533-B weld deposit).

In order to examine the physical events close to crack growth
initiation it is necessary to make direct fracture surface
observations by means of light or electron microscopy. This
is the basis of a further proposal /16/ which is aimed at
finding J_o, the true initiation value of J, i. e. the value
of J for $\Delta a \rightarrow 0$. According to Ref. /16/ this can be done by
intersecting the power law crack growth curve with the blunting
line, Fig. 2b. The thus determined J value coincides closely
with the initiation value, J_o, determined by fracture surface
observations done with a low-power microscope and is appreciably
below the standard J_{Ic} value, Fig. 2b. The first and third
of the statements above apply for the second proposal as
well. This investigation /16/ was done on steel ASTM A533,
Grade B, Clas 1, HSST plate 03.

c) The third method consists of searching for a characteristic
point on the test record which indicates unequivocally the
onset of crack growth. On a potential drop-clip gage displacement
$(U-v_g)$ record such a point can often be detected: it is either
a change in slope /9/ or a minimum on the $U-v_g$ record
/6,8,10,13/, Fig. 1c.

The work described by the present paper was aimed at further
investigation of the indirect detection of crack growth initiation (single specimen method); a particular point of interest
was the relation between J_o (the initiation value of J) and
the standard J_{Ic}.
More specifically, the ability of the
- DC PD method and the
- partial unloading technique

to determine

- initiation of crack growth and
- crack growth
was investigated. Another motivation for this work was that
the DCPD method as described in /12/ and earlier reports
has been in use in the authors' lab since a decade; on the
other hand, in the meantime the partial unloading technique
appeared and additional PD techniques were developped. Thus,
the question arose whether our technique is still competitive.
The main emphasis, however, was on the DCPD method.

MATERIALS AND TEST PROCEDURE

Table 1 contains the materials investigated and their tensile
properties. All materials were tested in the form of CT specimens, 20--25 mm thick; specimen width was 50 or 100 mm. The
planview geometry of the specimens corresponded to the recommendations of E813.
All tests satisfied the conditions for valid J_{Ic} values
as specified by Ref. /3/. A number of 25mm thick WOL-X specimens
were made of material D; this is also indicated on the respective
diagram.

Three types of tests were performed:
- Multiple specimen tests to determine J_Q by means of scanning electron microscopical (SEM) fractur surface observations.

Table 1   Tensile properties of the materials investigated

	Material	$\sigma 0.2$, MPa	UTS, MPa
A	7075-T7351 aluminum	421	504
B	2024-T351 aluminum	317	440
C	2024-FC aluminum [1]	75	217
D	A533-B1,1 steel [2]	465	614
E	20MnMoNi55 steel [3]	478	612
F	A542 steel [4]	599	715

Crack growth (excluding the stretch zone) was measured at equally spaced locations along the crack front whereby at each location seven individual measurements were made as shown in Fig. 3.

 - Computer controlled partial unloading tests were done on materials C, E, and F only. Data acquisition and data reduction were done in a fully automated manner (including J_{Ic} determination and check of the criteria of validity) by the computer (HP 9826) which controlled the testing machine.

---

[1] furnace cooled condition

[2] HSST plate 03

[3] heat KS 11 of the German FKS program, corresponds to A533-B

[4] EGF round robin steel

For the DC PD tests the two contacting geometrics shown
in Fig. 4 were used. Similar contact arrangements were used
for the WOL specimens (see Fig. 13 a). The current loads
were fixed to the specimens using screws, the potential was
picked up via small interference pins. A Simac Electronica
power supply served as current source, the potential drop
was measured by a Keithley 140 nanovoltmeter. The current
was adjusted such that a nominal resolution with respect
to crack length of roughly 10 - 20 μm was obtained in all
tests.
Normally the CT specimens were insulated from the testing
machine, i.e. insulated loading pins were used and the specimens
side faces were protected by thin teflon foils. This practice
was applied for the contacting geometry A in Fig. 4. In addition,
the clip gages were insulated so that no short circuit could
occur. The CT specimens with contacting geometry B had
non-insulated loading pins; all WOL specimens were insulated
with respect to the loading pin and at the side faces whereas
the loading screw was not insulated (for further details
of the WOL tests see Ref. /13/.

DATA EVALUATION

J integral for CT specimens

The clip gage displacement values were corrected for rotation
/18/ and the J integral was determined using the relationship
given in Ref. /3/ which accounts for crack growth.

J integral for WOL specimens

The J integral for the WOL specimens was calculated according
to the recommendations in Ref. /19/:

$$J = \frac{1.8U}{B(W-a)} \qquad (2)$$

for $\qquad 0.5 \leq a/W \leq 0.6$

where U is the area under the load $-v_{LL}$ curve; since v_{LL}
cannot be measured on a WOL-X-specimen it was determined
by converting the displacement, v_g, measured at the specimen's
front face /10/:

$$\frac{v_{LL}}{v_g} = 0.40\frac{a}{W} + 0.44 \qquad (3)$$

No crack growth correction was applied to Eq. (2) since only
initiation values were determined on the WOL specimens.

Crack Length Determination by Unloading Compliance

As mentioned above the evaluation of the unloading compliance tests was carried out fully automatically. For the conversion of the compliance data to crack length the table in Ref./ 3 / was used. In addition, since there are always slight deviations between the measured and calculated crack lengths for the starting point, i.e. a = a_o an effective modulus of elasticity was calculated to give coincidence between both values. This effective modulus was then used for data evaluation, a practice which we have employed to all our fracture mechanics tests including compliance measurements irrespective of the specimen type and which is also refered to in the literature /15/.
At the end of each test the final crack extension was marked by further fatigue loading and compared with the predicted value.

Crack Length Determination by DCPD

As was demonstrated in Ref. /12/ the tests on CT specimens with contacting geometry A (see Fig. 4) can be evaluated using the relationship

$$a = \frac{2W}{\pi}\cos^{-1}\frac{\cosh(\pi y/2W)}{\cosh\{(U/U_0)\cosh^{-1}[\cosh(\pi y/2W)/\cos(\pi a_0/2W)]\}} \tag{4}$$

which we use as a standard calibration for all our fracture mechanics tests with DCPD measurements. (For the meaning of the symbols see the nomenclature). This calibration formula has a number of advantages:

- reasonable resolution and sensitivity,
- little effect of misplacement of potential probes (see Ref. /1/),
- application convenient since Eq(4) serves as calibration function,
- due to its normalised form (U/U_o) Eq(4) is independent of
  - material,
  - test temperature,
  - current,
  - size and shape of specimen (see Ref. /12/).

This last point is of particular convenience when a laboratory has to investigate specimens of various shapes and sizes. Again, as with the partial unloading technique, the final crack extension obtained in each test is compared with the prediction. Fig. 5 gives a compilation of this data. All data points fall into an error band of $\pm$10%.

During the tests the potential signal was recorded as a function
of load (F-U diagram) and of load line displacement (U-v_{LL}
diagram), Fig. 6. For contacting geometry A the crack length
was determined from the F-U record as shown in Fig. 7 using
the initial linear portion as the U_o baseline; U_o is thus
defined as the potential drop occurring under load but without
crack growth. In some cases the F-U record exhibited only
a weekly pronounced linear portion. It was then necessary
to <u>define</u> the U_o base line as well as possible; an example
is shown by Fig. 8. The error in Δa introduced thereby was
estimated at about 50 μm.

RESULTS

Crack Growth

Fig. 9a shows J_R-curves obtained on two specimens, whereby
the crack length was determined by the partial unloading
and by the DCPD techniques, respectively. The same situation
is depicted by Fig. 9b for two side-grooved specimens made
from the same material as those of Fig. 9a. As was already
expected from Fig. 5 both methods yield identical results.
The same results will be shown again below at a different
Δa scale (see Fig. 12a-d).

Crack Initiation

In Tables 2 through 8 the information relevant for the
determination of initiation of crack growth by means of the
potential method is compiled. In particular, it is shown
at what point of the test record the initiation point was
taken. In addition, the specimens tested by the partial unloading
technique are listed; these tests yielded J_{Ic} values. From
the types of test records shown in Tables 2 through 8 it
can be concluded that the contact geometry A leads to F-U
records with a more or less linear initial portion. The
transition to the nonlinear portion was tentatively taken
as the initiation point. No specific point which could indicate
an event like initiation was found on the U-v_{LL} records
(Exception: specimens 7/5 and 7/8 in Table 5 with negative
slope change). The experiments with contact geometry B are
characterized by potential minima in both kinds of test records.
This is true for the CT specimens as well as for the WOL
specimens the contact geometry of which was somewhat different
(see Fig. 13a). In case of contact geometry B the potential
minimum was taken as the initiation point.

The true initiation of crack growth was determined by the
multiple specimen method; the specimens were loaded to different
amounts of Δa, unloaded and subsequently re-fatigued. Crack
growth was then measured in the scanning electron microscope
as shown in Fig. 3. The results are compiled in Fig. 10 through
15. The scatter bands for J_o indicate the uncertainty in
the back-extrapolation procedure.
Some common observations can be made as follows:

- The value J_o as determined for $\Delta a \to o$ is always lower
  than J_{Ic}.

- For small amounts of crack growth the partial unloading
  technique yields more scatter than the DCPD method
  (see Fig. 12b, d and 15).

- In the range of microscopical crack growth the DCPD method
  measures larger amounts of crack growth than is observed
  on the fracture surface (It should be noted that a was
  measured
  without the stretch zone).

- A corresponding statement for the partial unloading technique
  is not possible since not enough results are available
  to rule out the scatter observed at small amounts of crack
  growth.

- Side grooving lowers J_{Ic} but not J_o (compare Figs. 12a
  through d). The two fractographical data points of Fig.12b
  (side-grooved) fall into the scatterband of the fractographical   results of Fig. 12a (non side-grooved).
  It can also be seen from Figs. 12c and d that the R-curves
  of the side-grooved and non side-grooved specimens tend
  to emerge from the same origin whereas beyond the first
  exclusion line they are already splitt off.

- The J_{Ic} determination procedure according to E813 yields
  well reproducible values, however, the ideal behaviour
  of
  crack growth just after initiation ($\Delta a \stackrel{\sim}{<} 0.2mm$) is completely
  ignored.

DISCUSSION

Crack Growth (DCPD Versus Partial Unloading)

According to our results both of the crack length measurement
techniques produce the same macroscopic R-curve, see Fig.9.
Moreover, the calibration data of Fig. 5 is within a $\pm10\%$
error band and is thus better than the 15% accuracy required
in E813 for the partial unloading technique. However, for
small amounts of crack growth the scatter of the partial
unloading technique (including the scattering of the data
points obtained in a single test around the average curve)
is bigger than that of the DCPD method (see Figs. 12b, 12c,
12d, and 15).

One of the reasons for this observation may be that the DCPD
method is continuous in nature whereas the data of a partial
unloading test are obtained in a discontinuous process which
may produce itself variations of the quantity to be measured.
Another point is the compliance measurement. Minute amounts
of friction (in the loading and measurement arrangements)
and of misplacement of the displacement measuring device
can introduce substantial amounts of errors. Thus, although
the compliance method is more sensitive than the potential
method (see Fig. 8 in /1/) its resolution problems (sensitivity
to errors of the measurement of the compliance) tend to make
it inferior to the electric potential method when small amounts
of crack growth are considered. These arguments apply even
more in the case of the stiff center cracked tensile specimens
where much smaller compliances are to be measured /19/. Further
efforts in friction avoidance (e.g. needle bearings in the
loading clevises) may improve the situation for the partial
unloading method.
But even so, apart from some exceptions the potential method
in the form described above will remain the standard method
in our lab since it is more universal, particularly due to
the fact that a single calibration curve (Eq(4)) is needed
for the commonly used specimen geometries. A further point
of interest is the concern that the plastic deformation of
the specimen's ligament may affect the material's resistivity
in such a way that no reliable conversion of potential drop
signal to crack length can be achieved /20/. However, no
significant difference between "as tested specimen calibration
points" as in Fig. 5 and saw cut calibrations (see /12/)
can be observed. Thus, we conclude that the order of magnitude
of resistivity change has no impact on the crack length
calibration.

In the form described in the present paper the DCPD method
is part of a recommended practice for the measurement of
R-curves on center cracked tension specimens /21, 22/.

Determination of Initiation from Fractography

As indicated by Fig. 1a existing (BS 5762 /14/) and drafted
(ASTM Test Method for CTOD Testing) test methods impose a
straight line on the crack growth data and determine the
initiation point by extrapolation of the straight line to
$\Delta a = 0$. The determination of J_{Ic} - which is supposed to be
a measure of initiation toughness - employs a similar procedure
(see Fig. 1b). The findings of the present paper show that
close to initiation the R-curve is anything but a straight
line. Moreover, straight line extrapolations lead to over-
estimations of initiation. Thus, in the present paper initiation
was determined by
- measuring crack growth down to very small amounts and by
- approximating the data by a smooth curve (which in some
cases was linear within a very limited range of crack growth).

The J_o values thus obtained are listed in Table 9.
It should be mentioned that no distinction was made between
side-grooved and smooth specimens since the data fell into
a common scatterband.

It is sometimes argued that extrapolation to $\Delta a = 0$ leads
to unduly conservative values and initiation should be defined
by the achievement of a fully developed crack front. The
latter definition may make sense in case of homogenous materials
whereas for inhomogenous materials local crack growth may
be large while at the major part of the crack front the crack
has not yet propagated.

Detection of Initiation by the DCPD Method

Potential Minimum

The potential minimum obtained with contacting geometry B
was taken to calculate J_{min} which is listed in Table 9. Although
the occurance of the minimum is not fully understood (it
is assumed that the load dependend contact area between pin
and specimen provides a less resistant current path with
increasing load and that initiation of growth overbalances
this effect) it is obvious from Table 9 that the potential
minimum is a good indicator of initiation.

End of Linear Portion of F-U Record

The J-values at the end of the linear portion of the load-
potential records for configuration A were designated as
J_{lin}. For the high strength aluminium alloy 7075-T7351 J_{lin}
corresponds approximately to the initiation value, J_o (see
Table 9); i.e. in this case the potential drop is a function
of crack length only. For all other materials (which are
characterized by relatively high ductility) J_{lin} is lower
than J_o. The reason is that prior to initiation due to the
high resolution the potential drop responds to crack tip
blunting and thus feigns crack growth. Therefore in Figs.
12a and 12b the DCPD measures somewhat more crack growth
than can be observed on the fracture surface. This difference
is only noticeable on a microscopic scale, of course; it
becomes meaningless for the macroscopic R-curve.

Alternative Methods

In Table 9 two additional J values are listed which are regarded
as candidate initiation values based on the very initial
portion of the R-curve: $J_{o.1}$, the J value at o.1mm of crack
growth as calculated from a sensitive indirect method like
the DCPD method and J_{int}, the J at the intersection of the
blunting line with the potential drop based R-curve. The
latter one could not be established safely for all materials.

Comparing the various J values of Table 9 it can be concluded that J_{IC} is very poorly related to initiation. Depending on the material the J_{IC}/J_o ratios range from 1.28 to 3.6 (For this calculation the average values for J_o in Table 9 were used). Thus, in terms of crack initiation and very early growth J_{IC} is poorly defined and can't be regarded as an appropriate material characterisation. The true initiation toughness is physically much better defined. According to Table 9 it can be determined as follows:

- By fractography (multiple specimen method),

- By the DCPD method with contacting geometry B and taking the potential minimum as the initiation point,

- By a calibrated indirect measurement technique like the DCPD method with contacting geometry A and taking that point at which 0.1mm average crack growth is indicated (Any other technique with a resolution of at least 50 µm can be used of course). This point roughly coincides with the true initiation point.

- A physically more appropriate method is the determination of the intersection of the initial part of the R-curve with the blunting line.

An abvious disadvantage for application is that J at initiation can be much lower than J_{IC} we are now accustomed with. But in the present paper no attempt is made to discuss possible effects on design philosophy, only some facts are presented. A possible solution to this problem may be the allowance for a specified amount of crack growth.

CONCLUSIONS

- For macroscopic crack growth the partial unloading and the DCPD techniques yield the same results, i.e. both methods are equally well suitable to determine an R-curve.

- The partial unloading technique is inferior to the DCPD method when only small amounts of crack growth are considered, i.e. just after initiation.

- The DCPD method in the form described in the present paper has a number of advantages the most important of which is the independance of its calibration with respect to test variables, such as material, temperature, and specimen size and shape.

- By special contacting arrangement (contacting geometry B in Fig. 4) the DCPD method is capable of detecting initiation of crack growth which can be related to a minimum of the clip gage-potential drop record.

- Contacting geometry A in Fig. 4 is more suitable for the quantitative determination of crack growth; initiation can be determined by taking the point at which 0.1mm crack growth is indicated by the potential signal or alternatively by determining the intersection of the initial R-curve with the blunting line.

- The reason that the DCPD method with contacting geometry A indicates a finite amount of growth at initiation is the response of the potential drop to crack tip blunting.

- J_o, the J value at initiation of crack growth as determined by SEM fractography was found to be 1.28 to 3.6 times lower than J_{Ic} as determined by E813.

NOMENCLATURE

a	crack length
Δa	crack length increment
Δa_{end}	Δa at end of test
B	specimen thickness
F	applied load
J_{end}	J at end of test
J_{int}	J at intersection of initial R-curve with blunting line
J_{min}	J at minimum of F-v_g record
J_o	J at initiation of crack growth
$J_{0.1}$	J at apparent $\Delta a = 0.1$ as indicated by the DCPD method
J_{Ic}	fracture toughness determined according to ASTM E813
K	stress intensity factor
K_o	K at initiation of crack growth
U	actual potential drop
U_o	U before crack growth
v_g	displacement measured by clip gage (which is in some cases identical with v_{LL})
v_{LL}	load line displacement
W	specimen width (half with for center cracked tensile specimen)
y	half gage span over which U is measured
δ	crack tip opening displacement
δ_o	δ at initiation of crack growth

TABLE 2 : Compilation of Test Results Obtained on Aluminum 7075-T7351 (Material-A of Table 1)

Specimen No.	B, mm	W, mm	a/W	POTENTIAL CONTACTING	TYPE OF TEST RECORD	J_{Ic} kJ/m²	J_0 kJ/m²	J_{end} kJ/m²	Δa_{end} (μm)	REMARKS
13	24.5	50	0.51	A	F	./.	19.2	24	7	
21	24.5	50	0.53	A	F	./.	21.8	34.8	260	
22	24.5	50	0.52	A	F	33.	23.6		7790	
30	24.6	50	0.52	./.	./.	34.8	./.	./.	20500	PARTIAL UN-LOADING
38	24.7	50	0.51	B	U (v_{LL})	36	26.3	./.	7540	
54	24.7	50	0.52	./.	./.	37	./.	./.	14700	PARTIAL UN-LOADING

TABLE 3: Compilation of Test Results Obtained on Aluminum 2024-T351 (Material B of Table 1)

Specimen No.	B, mm	W, mm	a/W	POTENTIAL CONTACTING	TYPE OF TEST RECORD	J_{IC}, kJ/m²	J_o kJ/m²	J_{end} kJ/m²	Δa_{end} (μm)	REMARKS
A1-23	20	100	0.71	A	F⌐⌐ U	34.5	18.2	162.6	19000	
L1-23	20	50	0.75	A	F⌐⌐ U	·/·	14.6	77.4	4500	
L6-23	B_{net} 15	50	0.54	A	F⌐⌐ U	26.3	19	36.5	·/·	SIDE GROOVED
N1-23	B_{net} 15	100	0.51	·/·	·/·	30.2	·/·	62.4	·16000	SIDE GROOVED PARTIAL UN-LOADING

TABLE 4 : Compilation of Test Results Obtained on Aluminum 2024-FC (Material C of Table 1)

Specimen No.	B, mm	W, mm	a/W	POTENTIAL CONTACTING	TYPE OF TEST RECORD	J_{Ic}, kJ/m²	J_0 kJ/m²	J_{end} kJ/m²	Δa_{end} (µm)	REMARKS
L2-23	20	50	0.52	A	[F-U graph]	·/·	7.5	215.8	9.0	
T1-23	20	50	0.72	A	[F-U graph]	·/·	6.1	142.0	4.7	
M1-23	20	100	0.72	A	[F-U graph]	·/·	6.8	194.5	6.2	
P1-23	15	50	0.53	A	[F-U graph]	·/·	9.6	162.2	8.6	SIDE GROOVED
P2-23	20	50	0.53	·/·	·/·	35.3	·/·	178.8	5.7	PARTIAL UN-LOADING
T2-23	15	50	0.53	·/·	·/·	17.9	·/·	161.9	8.4	SIDE GROOVED PARTIAL UN-LOADING

TABLE 5 : Compilation of Test Results Obtained on WOL-X-Specimens of HSST Plate 03 (Material D of Table 1) The Potential Contacting Deviates Somewhat from that of the CT-Specimens; it is Shown in Fig. 13a

Specimen No.	B, mm	W, mm	a/W	POTENTIAL CONTACTING	TYPE OF TEST RECORD	$J_{I\zeta}$ kJ/m²	J_0 kJ/m²	J_{end} kJ/m²	Δa_{end} (µm)	REMARKS
2	25.4	28.6	0.58	B		·/·	46.9	444	650	
3	25.4	28.6	0.72	A		·/·	27.7	347	330	
6	25.4	28.6	0.64	B		·/·	43.2	116	42.5	
7	25.4	28.6	0.57	B		·/·	≥31.8	31.8	2.4 1)	POTENTIAL MINIMUM NOT PASSED
8	25.4	28.6	0.57	B		·/·	45.5	72.3	12	

TABLE 5   Compilation of Test Results Obtained on WOL-X-Specimens of HSST Plate 03 (Material D of Table 1). The Potential Contacting Deviates Somewhat from that of the CT Specimens; it is Shown in Fig. 13a.

Specimen No.	B, mm	W, mm	a/W	POTENTIAL CONTACTING	TYPE OF TEST RECORD	J_{Ic} kJ/m²	J_o kJ/m²	J_{end} kJ/m²	Δa_{cnd} (µm)	REMARKS
9	25.4	28.6	0.58	B		·/·	47.1	61.7	7.6	
11	25.4	28.6	0.57	B		·/·	47.6	62.7	4	
7/1	25.4	28.6	0.55	B		·/·	52.7	168.6	62.5	
7/2	25.4	28.6	0.54	B		·/·	≥ 54.2	54.2	6.7 1)	POTENTIAL MINIMUM NOT PASSED
7/3	25.4	28.6	0.55	B		·/·	43.4	99.4	22	

TABLE 5 Compilation of Test Results Obtained on WOL-X-Specimens of HSST Plate 03 (Material D of Table 1). The Potential Contacting Deviates Somewhat from that of the CT Specimens; it is Shown in Fig. 13a.

Specimen No.	B, mm	W, mm	a/W	POTENTIAL CONTACTING	TYPE OF TEST RECORD	J_{Ic}, kJ/m²	J_0 kJ/m²	J_{end} kJ/m²	Δa_{end} (µm)	REMARKS
7/5	25.4	28.6	0.54	A		·/.	33.6	58.7	9	
7/6	25.4	28.6	0.51	A		·/.	≥ 30.5	30.5	0[1]	
7/8	25.4	28.6	0.54	A		·/.	25.4	69.4	10.2	
7/10	25.4	28.6	0.52	B		·/.	≥ 33.1	33.1	0[1]	

[1] Crack front not fully developped; isolated dimples only.

TABLE 6   Compilation of Test Results Obtained on CT Specimens of HSST Plate 03
(Material D of Table 1)

Specimen No.	B, mm	W, mm	a/W	POTENTIAL CONTACTING	TYPE OF TEST RECORD	J_{Ic}, kJ/m²	J_o kJ/m²	J_{end} kJ/m²	Δa_{end} (mm)	REMARKS
D7/1	25	50	0.62	A		·/·	33.1	288	370	UNSTABLE FRACTURE
D7/2	10	50	0.54	A		150	20.1	785	2500	
D7/3	25	50	0.61	A		·/·	16	127	80	UNSTABLE FRACTURE

TABLE 7

Specimen No.	B, mm	W, mm	a/W	POTENTIAL CONTACTING	TYPE OF TEST RECORD	J_{Ic}, kJ/m²	J_0 kJ/m²	J_{end} kJ/m²	Δa_{end} (µm)	REMARKS
A1	25	50	0.56	B	U	·/·	150	211	?	
A2	25	50	0.56	A	F	·/·	35	55.5	0	
A8	25	50	0.55	A	F	·/·	23.5	35.6	0	
A10	25	50	0.55	A	F	135	30.5	604.5	1500	
A11	25	50	0.51	A	F	125	57	522	1040	
A14	25	50	0.61	A	F	255	29.4	1249	6860	

TABLE 8 — Compilation of Test Results Obtained on Steel A542 (Material F of Table 1)

Specimen No.	B, mm	W, mm	a/W	POTENTIAL CONTACTING	TYPE OF TEST RECORD	J_{IC} kJ/m²	J_0 kJ/m²	J_{end} kJ/m²	Δa_{end} (µm)	REMARKS
ECN 35	25	50	0.62	B		./.	./.	75.7	≈ 1µm, some isolated dimples	
ECN 36	25	50	0.77	A		./.	./.	97.21	22	
ECN 34	25	50	0.69	B		130	71.5	463	2970	PARTIAL UNLOADING
ECN 37	25	50	0.71	A		130	./.	273.6	1300	
ECN 38	25	50	0.70	A		./.	./.	138.5	400	
ECN 39	25	50	0.695	./.		./.	./.	139.5	48	PARTIAL UN. LOADING, UN STABLE FRACTURE

TABLE 9   Compilation of initiation data with various definitions, explanations see text; J values in kJ/m^2

Material	J_{Ic}	J_o	$\dfrac{J_{Ic}}{J_o}$ [2]	J_{min}	J_{lin}	$J_{0.1}$	J_{int}
A	33 - 37	25 - 30	1.28	26.3	19.2-26.3	28 - 32	23 - 28
B	26.3-34.5	21.5-28	1.22	./.	14.6-19	24 - 28	18 - 22
C	22 -35.4	8 - 12	2.87	./.	6.1- 9.6	15 - 17	8 - 10
C[1]	17.9	8 - 12	1.79	./.	9.6	19	15
D	150	28 - 55	3.6	43.2-52.7	16 - 33.1	60 - 75	35 - 50
E	125 - 255	3)	./.	150[4]	23.5-57	70 -115	50 -115
F	130	∿90	1.44	71.5	./.	120	105

1) side grooved

2) average values used

3) due to the lack of very small Δa values J_o could not be safely determined; it is estimated at 100-150 kJ/m^2

4) two additional tests with ACPD (provided by F. Schmelzer, GKSS) resulted in J_{min} = 125 and 135 kJ/m^2.

REFERENCES

/1/ Schwalbe, K.-H. "Test Techniques for Fracture Mechanics Testing" in Advances in Fracture Research. Proceedings of the 5th International Conference on Fracture (ICFS), D. Francois, Ed., Pergamon Press, Oxford, 1982, pp. 1421-1446.

/2/ E399-81 Standard Test Method for Plane-Strain Fracture Toughness of Metallic Materials, Annual Book of ASTM Standards, 1981.

/3/ E813-81 Standard Test for J_{Ic}, a Measure of Fracture Toughness, ibid.

/4/ Clarke, G.A., W.R. Andrews, J.A. Begley, J.K. Donald, G.T. Embley, J.D. Landes, D.E. McCabe, and J.H. Underwood, "A Procedure for the Determination of Ductile Fracture Toughness Values Using J Integral Techniques", JTEVA, Vol. 7, No. 1, 1979, pp. 49-56.

/5/ Mayville, R.A., and J.G. Blauel, "Die Methode der partiellen Entlastung zur Ermittlung von Rißwiderstandskurven", Proceedings 12th Meeting of Arbeitskreis "Bruchvorgänge", Deutscher Verband für Materialprüfung, Berling, 1981.

/6/ Marandet, B., J.C. Devaux, and A. Pellissier-Tanon, "Correlation Between General Yielding and Tear Initiation in CT Specimens", CNSI-Specialist Meeting on Tear Instabilities, St.-Louis, 1979.

/7/ Ingham, T., and E. Morland, "The Measurement of Ductile Crack Initiation: a Comparison of Data from Multiple and Single Specimen Methods and Some Considerations of Size Effects", ibid.

/8/ Okumura, N., T.V. Venkatasubramanian, B.A. Unvala, and T.J. Baker, "Application of the AC Potential Drop Technique to the Determination of R-Curves of Tough Ferritic Steels" Eng. Fracture Mech., Vol. 14, 1981, pp. 617-625.

/9/ McIntyre, P., and D. Elliott, "A Technique for Monitoring Crack Extension During C.O.D. Measurement", British Steel Corporation, Report MG/15/72, 1972.

/10/ Hollstein, T. and J.A. Blauel, "Vergleichende Ermittlung von Rißzähigkeitskennwerten mit Hilfe von Klein- und Großproben des Typs WOL-X", Bericht VII/80 des Fraunhofer Instituts für Werkstoffmechanik, Freiburg, 1980.

/11/ Ritchie, R.O., and K.J. Bathe, "On the Calibration of the Eelctrical Potential Techniquefor Monitoring Crack Growth Using Finite Element Methods", Int. Journ. of Fracture, Vol. 15, Nr. 1, 1979, pp. 47-55.

/12/ Schwalbe, K.-H. and D. Hellmann, "Application of the Electrical Potential Method to Crack Length Measurements Using Johnson's Formula", JTEVA, Vol. 9, Nr.3, 1981. pp. 218-221.

/13/ Schwalbe, K.-H., J. Heerens, and D. Hellmann, "Vorberei- tende Arbeiten zur Untersuchung bestrahlter WOL-Proben", GKSS-report 82/E/2, 1982.

/14/ Methods for Crack Opening Displacement (COD) Testing, BS 5762:1979, British Standards Institution, London,1979.

/15/ Loss, F.J. Editor, "Structural Integrity of Water Reactor Pressure Boundary Components", NUREG/CR-1128, NRL Memoran- dum Report 4122, 1979.

/16/ Carlson, K.W., and J.A. Williams, "A More Basic Approach of the Analysis of Multiple Specimen R-Curves for Deter- mination of J_c", ASTM STP 743, 1981, pp. 503-524.

/17/ Schwalbe, K.-H., "Crack Propagation in AlZnMgCuO5 During Static Loading, Engineering Fracture Mechanics, Vol. 6, 1974, pp. 415-434.

/18/ Hawthorne, J.R., Editor, "The NRL-EPRI Research Program (RP 886-2), Evaluation and Prediction of Neutron Embrittle- ment in Reactor Pressure Vessel Materials, Annual Progress Report for CY 1978", NRL Report 8 327, 1979.

/19/ Schwalbe, K.-H., and W. Setz, "R-Curve and Fracture Toughness of Thin Sheet Materials", JTEVA, Vol. 9, No. 4, 1981, pp. 182-194.

/20/ Curry, D.A., and I. Milne, "The Detection and Measurement of Crack Growth During Ductile Fracture", CERL Report RD/L/N64/79, CERL Leatherhead, 1979.

/21/ Schwalbe, K.-H., W. Setz, L. Schwarmann, W. Geier, C. Wheeler, D. Rooke, A.U. deKoning, and J. Eastabrook, "Application of Fracture Mechanics to Thin-Walled Struc- tures" (in German), Fortschr. - Ber. VDI-Z. Reihe 18, Nr. 9, 1980, 114 pages.

/22/ Wheeler, C., J. Easterbrook, D.P. Rooke, K.-H. Schwalbe, W. Setz, and A.U. deKoning, "Recommendations for the Measurement of R-Curves Using Centre-Cracked Panels", Journal of Strain Analysis, Vol. 17, No. 4, 1982, pp. 205-213.

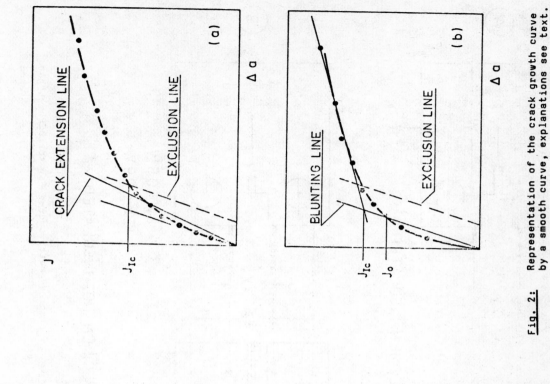

Fig. 2: Representation of the crack growth curve by a smooth curve, explanations see text.

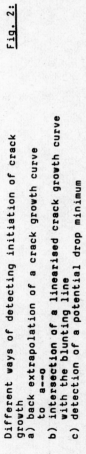

Clip gage displacement

Fig. 1: Different ways of detecting initiation of crack growth
a) back extrapolation of a crack growth curve to a→o
b) intersection of a linearised crack growth curve with the blunting line
c) detection of a potential drop minimum

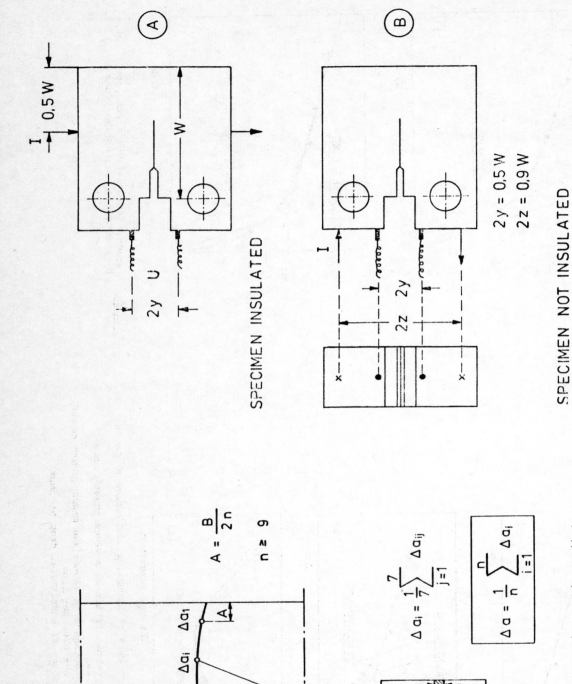

Fig. 4: Contact geometries used for the DCPD method

$2y = 0.5\,W$

$2z = 0.9\,W$

SPECIMEN INSULATED

SPECIMEN NOT INSULATED

$A = \dfrac{B}{2\,n}$

$n \approx 9$

$\Delta a_i = \dfrac{1}{7} \sum_{j=1}^{7} \Delta a_{ij}$

$\Delta a = \dfrac{1}{n} \sum_{i=1}^{n} \Delta a_i$

SEM SCREEN

Fig. 3: Procedure of measuring crack growth in the SEM.

- 44 -

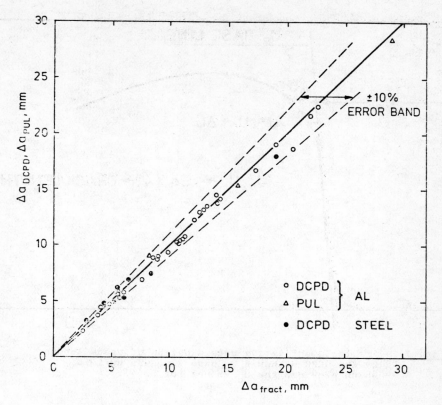

Fig. 5: Final crack extension values meausred on the fracture surface compared with the values predicted by the DCPD and the PUL techniques.

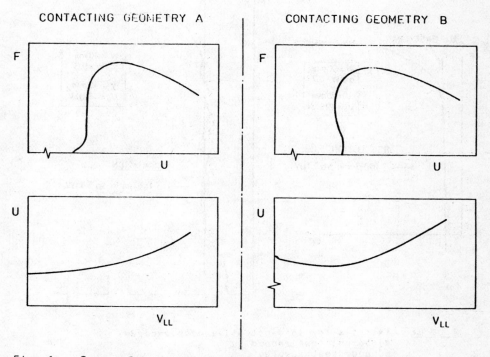

CONTACTING GEOMETRY A   CONTACTING GEOMETRY B

Fig. 6: Types of test records obtained with the DCPD test.

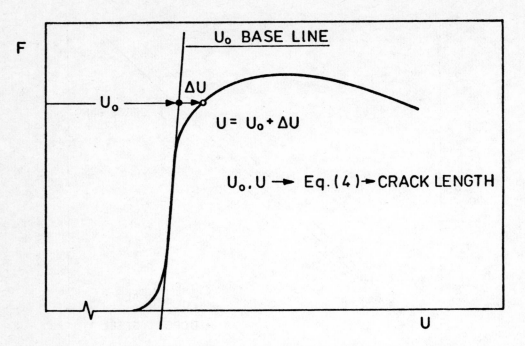

Fig. 7: For the quantitative determination of crack length by the DCPD method a U_o base line was defined which provided the U_o values for Eq(4).

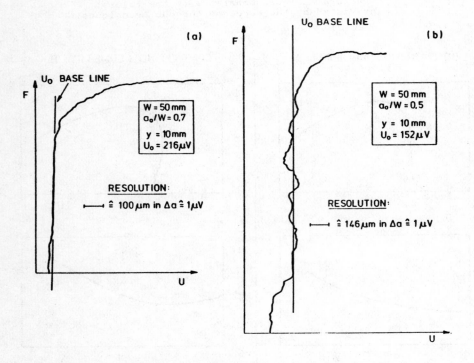

Fig. 8: Examples for load-potential drop records:
a) normal test record
b) poor test record.

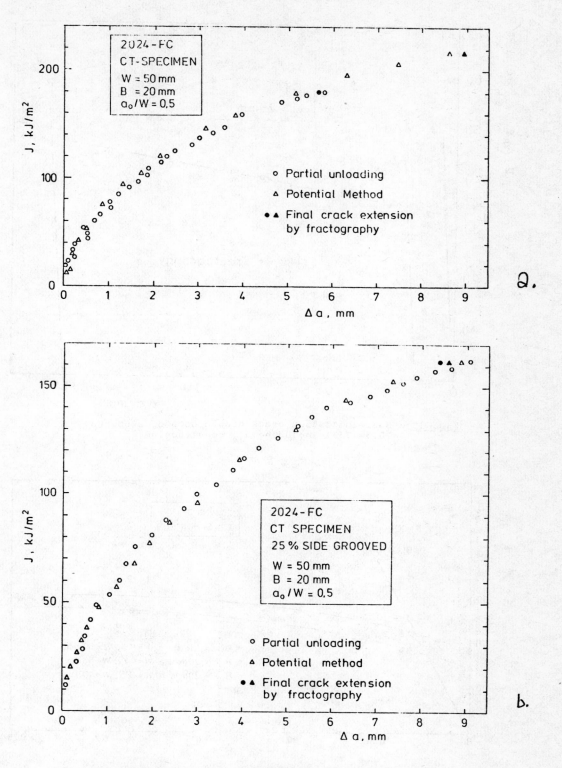

Fig. 9: R-curves obtained on two identical specimens using the DCPD and the partial unloading method respectively
a) smooth specimens
b) side-grooved specimens.

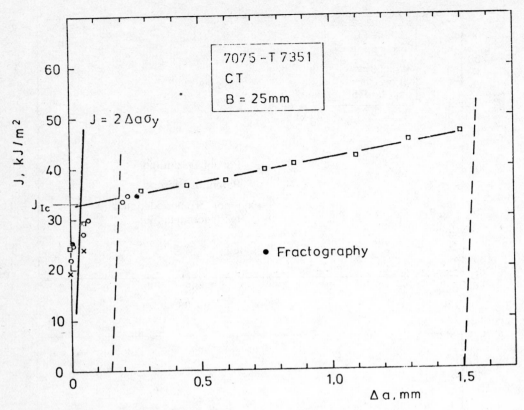

Fig. 10: Near-initiation crack growth data of aluminum
7075-T7351 including J_{Ic} construction.

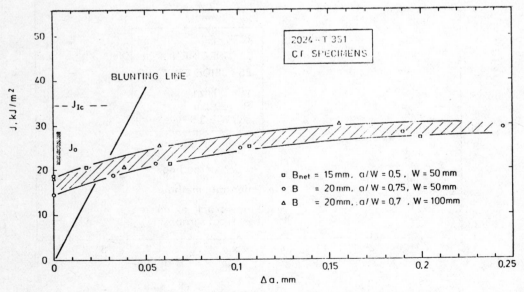

Fig. 11: Near-initiation crack growth data of aluminium
2024-T351. J_o was obtained by fractography of
10mm thick specimens. This value can be considered as representative of the 20mm thick specimens
since (as it will be shown in a later report) the
origins of the R-curves are independent of thickness range considered here.

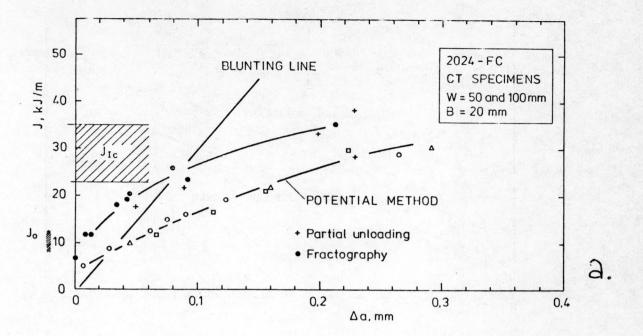

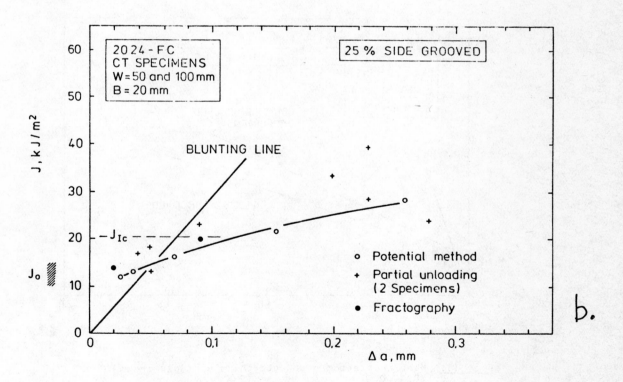

Fig. 12: Near initiation crack growth data of aluminium 2024-FC.
a) smooth specimens,
b) side-grooved specimens,

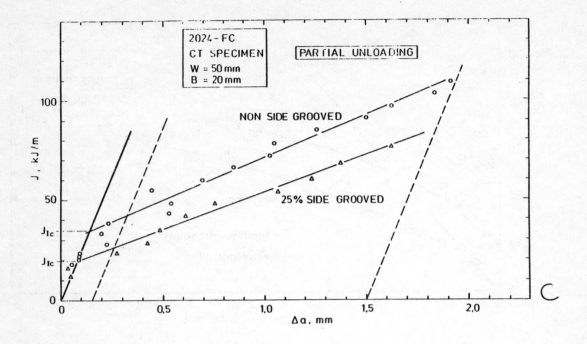

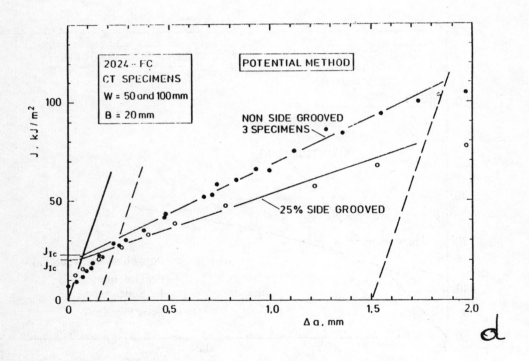

Fig. 12: Near initiation crack growth data of aluminium
2024-FC.
c) J_{Ic} construction based on R-curves obtained
on a side-grooved and a smooth specimen
using the partial unloading technique,

d) as Fig. 12c, using the potential method.

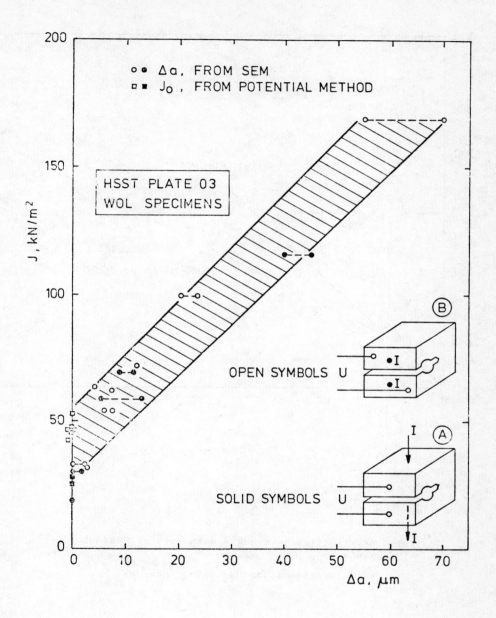

Fig. 13: Near initiation crack growth data of HSST steel plate 03
a) WOL-X specimens for the determination of Jo,

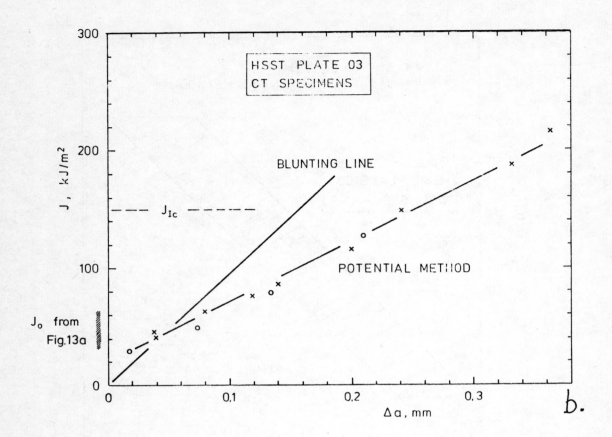

Fig. 13: Near initiation crack growth data of HSST steel plate 03

b) CT specimens for J_{Ic} determination.

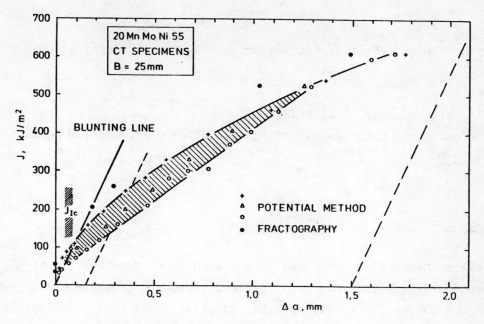

Fig. 14: Near initiation crack growth data of steel
20MnMoNi55.

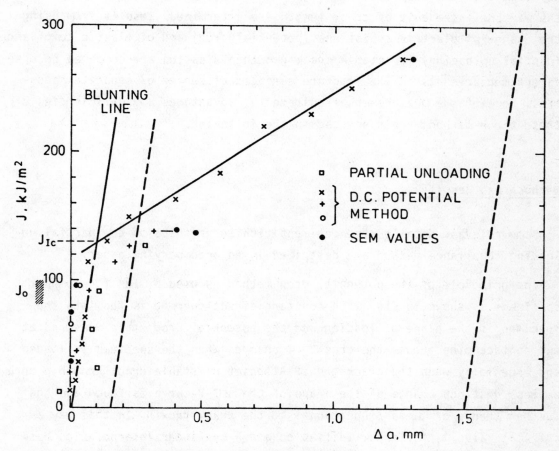

Fig. 15: Near initiation crack growth data of steel A542
including J_{Ic} construction.

Comparison of Different Methods for J_R-Curve Determination

T. Hollstein, B. Voss, J.G. Blauel
Fraunhofer-Institut für Werkstoffmechanik, D-7800 Freiburg, W-Germany

Introduction

Besides the multiple specimen interrupted loading method [1] single specimen procedures are more and more frequently used to generate crack growth
resistance curves for elastic plastic material characterization and failure
assessment. Whereas the crack driving parameter J is determined in the same
way from the load displacement diagram for all these methods they differ in
the way the increments of crack length are determined. Results from using
the change of electric resistance (potential drop) and of elastic compliance
(partial unloading) for stable crack growth evaluation are compared here to
direct measurements on the fracture surfaces of series of separate specimens. Some of the measurement requirements, advantages and deficiencies of
these three methods are discussed in the following.

Methods and Results

Some relevant details of experimentation and results of the partial unloading compliance method are described in an accompanying paper.

The principle of the potential drop method as used by IWM for compact
specimens is shown in Fig. 1. A constant direct current is fed into the
specimen in the plane of loading and the potential drop $\Delta\varphi$ is measured at
two contact pins across the crack. $\Delta\varphi$ changes when the specimen is loaded
and especially when the crack grows. At onset of stable crack growth a more
or less distinct change of the shape of the $\Delta\varphi$-V-curve is found and the
further change of $\Delta\varphi$ is proportional to the crack growth. In this way a
critical value J_i for crack initiation and - by linear interpolation between the values at initiation φ_i, J_i and at termination of the experi-

ment φ_{max}, J_{max} [2] - a complete J_R-curve may be evaluated. Figures 2 and 3 show a force/potential-displacement diagram and a continuous $J(\Delta a)$-curve evaluated from it for a structural steel A 542 Cl.3 tested at room temperature. The interrupted loading results from specimens of the same kind confirm this curve. The J_{IC}-value defined according to ASTM E 813-81 [1] by an extrapolation to $\Delta a = 0$ of the regression line through four data points between $0,15 \leqslant \Delta a \leqslant 1,5$ mm is about 12 % higher than the J_i-value calculated for the displacement at "i" in Fig. 2 which is well confirmed by the first interrupted loading point (Fig. 3).

In the test of Fig. 4 [3] initiation is uncertain between $0,5 \leqslant V_i \leqslant 0,7$ mm delivering a potential drop critical value $43 \leqslant J_i \leqslant 81$ kJ/m^2 but the ASTM procedure yields with only small scatter $143 \leqslant J_{IC} \leqslant 150$ kJ/m^2.

Therefore evaluating J_{IC} from potential drop measurements does not require an exact determination of the point of initiation, but to evaluate J_i testing of an additional specimen up to very little crack growth is recommended.

Figure 5 [4] shows an evaluation of the $J(\Delta a)$-curves from the potential-drop and partial-unloading raw data of Fig. 4. Both methods deliver similar curves which differ in slope only for $\Delta a \geqslant 1,5$ mm as a consequence of pronounced crack front tunneling in the non side grooved specimen (causing a stronger underestimation of crack growth by the compliance measurement). For this comparison a critical opening displacement of $V_i = 0,55$ mm for first real crack growth after crack tip blunting was read from the compliance measurements (6th unloading) and was then used to define the initiation value of the PD-curve, the second PD-calibration point was adjusted to the 13th unloading point at $\Delta a = 0,9$ mm assuming negligible effect of tunneling for small amounts of crack growth.

A comparison of results from the three test procedures - partial unloading compliance method, DC-potential drop technique, multiple specimen interrupted loading procedure - is shown in Figs. 5 and 6 for a pressure vessel steel 20 Mn Mo Ni 5 5 for test temperatures of 150°C and 300°C. As far as investigated the crack resistance curves are independent of test procedure, specimen size (B = 25 mm to 100 mm) and relative crack length $0,50 \leqslant a/W \leqslant 0,71$.

Conclusions

Different procedures can be used to determine material parameters J_i, J_{Ic} and J-crack growth resistance curves for the characterization of crack initiation and stable crack growth. Single specimen procedures are superior to multiple specimen techniques because of the reduced expenditure and improved information.

Equivalent results can be derived from potential drop (DCPD) and partial unloading (SSPUC) tests. And both procedures can be used at temperatures down to $-196^{o}C$ and up to $300^{o}C$ (or even more).

For each single specimen a complete $J_R(\Delta a)$-curve can be derived the quality of which (at least for SSPUC) may be assessed by comparing the predicted initial and final crack lengths with those measured on the fracture surface. Testing of several specimens of the same kind in principle delivers equivalent J-R-curves. Therefore additional information is gained about the material scatter. J_{Ic} may be determined as a technical initiation parameter following the ASTM extrapolation procedure. An undoubted determination of the real onset of crack growth (J_i) is not possible by SSPUC.

The potential drop method allows for the evaluation of a continuous $J(\Delta a)$-curve if initiation is given. There remains some uncertainty with this point of initiation because the physical sources of the change of potential under loading are not well understood. Therefore J_i should be confirmed by testing one additional specimen with very little crack growth. The J_{Ic}-extrapolation is less sensitive to the uncertainty in J_i. The results of both methods, SSPUC and DCPD, are in good agreement with the multiple specimen interrupted loading procedure.

Literature

[1] ASTM E 813-81, Annual Book of ASTM Standards, Part 10, Philadelphia (1981)

[2] G. Kloss: Proceedings of the 13. Sitzung des Arbeitskreises Bruchvorgänge, Hannover 1981, DVM Berlin (1982)

[3] B. Voss, J.G. Blauel: Proceedings of the 4th ECF-Conference, Leoben 1982, Vol. 1, edit. K.L. Maurer, F.E. Matzer, EMAS, UK (1982)

[4] B. Voss, Proceedings of the 14. Sitzung des Arbeitskreises Bruchvorgänge, Mülheim 1982, DVM Berlin (1982)

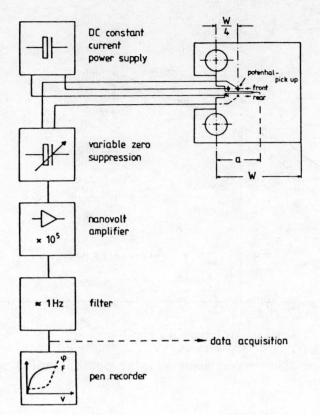

DC constant
current
power supply

$\frac{W}{4}$

potential-
pick up
front
rear

a

W

variable zero
suppression

nanovolt
amplifier
$\times 10^5$

Fig. 1: Principle of
direct current potential
drop method (DCPD)

$\approx$ 1Hz   filter

data acquisition

φ
F

pen recorder

V

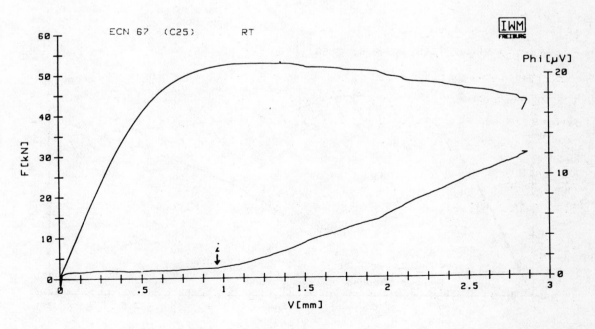

ECN 67   (C25)   RT

IWM
FREIBURG

Fig. 2: Force (F) / DC potential drop ($\Delta\varphi$) vs. displacement (V) change in
slope at initiation of stable crack growth

- 57 -

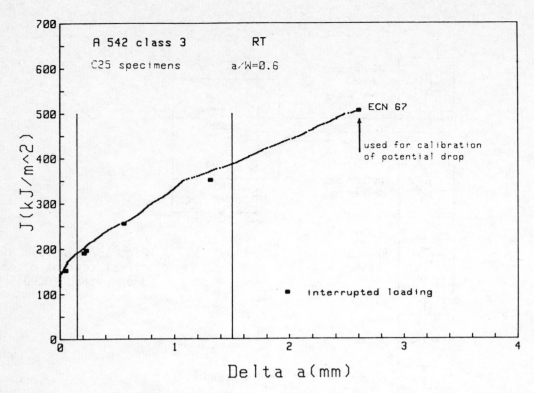

Fig. 3: J_R-curve from DCPD measurement - interrupted loading points for comparison

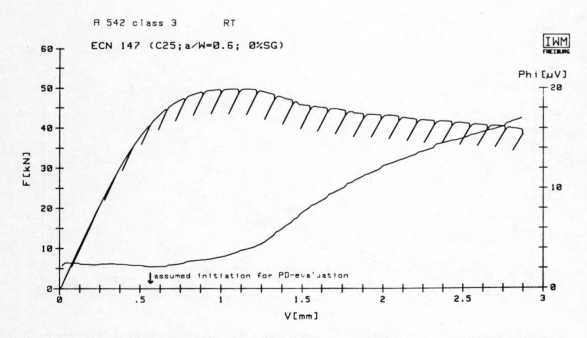

Fig. 4: F/$\Delta\varphi$ vs. V-diagram of simultaneous SSPUC- and DCPD measurement

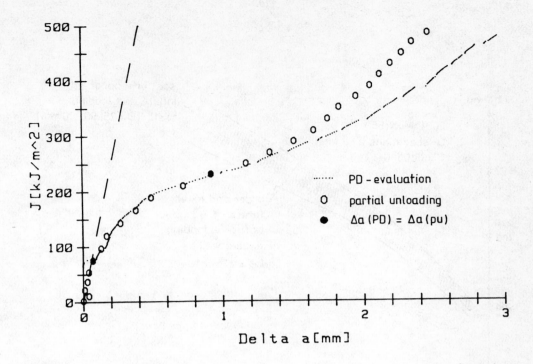

Fig. 5: J-R-curves evaluated from SSPUC- and DCPD-(Δa adjusted to two partial unloading points) measurements in Fig. 4

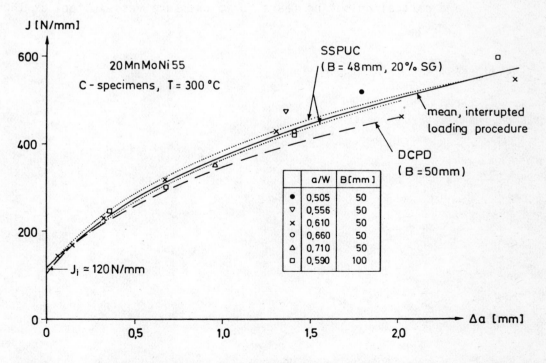

Fig. 6: J-R-curve evaluation from interrupted loading tests, SSPUC and DCPD for a pressure vessel steel at 300°C

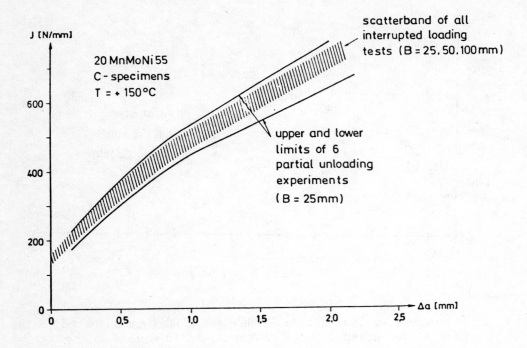

Fig. 7: Comparison of scatterbands for J-R-curves from interrupted loading
and partial unloading tests for a pressure vessel steel at 150°C

SIMULTANEOUS MEASUREMENT OF CRACK EXTENSION BY UNLOADING COMPLIANCE AND POTENTIAL DROP METHODS

Wallin, K., Saario, T., Auerkari, P. and Törrönen, K.
Technical Research Centre of Finland (VTT)
Metals Laboratory
SF-02150 ESPOO 15, Finland

Both the unloading compliance and potential drop methods have been simultaneously applied to crack length measurement in J-R curve determination. The materials used were bainitic pressure vessel steel A533B Cl. 1 and two Q & T steel OX522 and OX602 welds. Specimen geometries were 1TCT and 3PB, respectively.

The computer-interactive system used for fracture toughness measurements consists of a table computer, a 250 kN servo-hydraulic testing machine, a data aquisition/control unit, a digital volt meter and several auxiliary devices for real time data aquisition as well as for data storage. Unloading compliance method is used to indirectly monitor the crack extension during the experiment. Compact tension (CT), round compact tension (RCT), or three point bend (3PB) specimen geometries can be used. The system as a whole meets the requirements of ASTM standard E813-81 for determining elastic-plastic fracture toughness J_{IC}.

The potential drop device developed at VTT is of AC-type with variable phase-locked frequency (10 Hz - 5 kHz) and feed current (1 - 10 A). A low current (1 A) was used in the testing, and potential changes in the range of 1 nV were repeatably detectable. Due to the design the potential drop device is insensitive to external disturbancies.

The two methods have been found to give consistent results in the ductile tearing regime.

The results for a A533B 1T-CT specimen are given in Fig. 1. It is seen that both methods react similarly to the crack tip blunting by sensing the same "effective" crack growth. It is also abvious that the minimum in the potential does not correspond to the beginning of the actual crack growth, which will begin considerably later. Reasons for this behaviour are discussed.

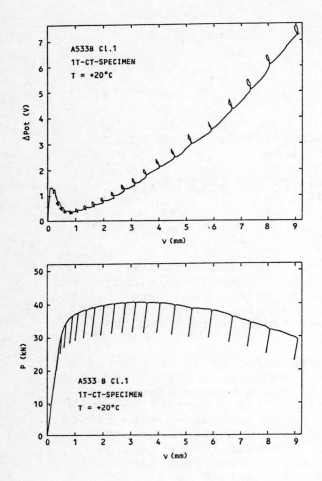

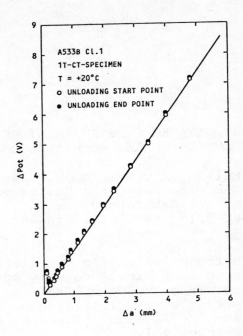

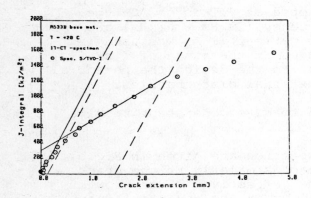

Fig. 1.
Comparison of potential drop and unloading compliance methods in determining ductile crack extension.

ELASTIC-PLASTIC FRACTURE STUDIES USING THE DC-EP METHOD

by

Gery M. Wilkowski

For CSNI Workshop on Test Methods for Ductile Fracture,
Paris, France, December 1-3, 1982

Introduction

During the last five years, the direct current (dc) electric potential (EP) method has been used at Battelle in various fracture mechanics research studies. Although some of them have involved subcritical crack growth, i.e., fatigue, corrosion fatigue, stress-corrosion cracking and creep crack growth, many of the applications have involved elastic-plastic fracture evaluations in laboratory specimens as well as in pipe and pressure vessel experiments. The cross-sectional areas of specimens tested have varied from as small as 51 mm^2 up to 33,550 mm^2.

Determining Crack Initiation

Differentiating crack blunting from crack initiation with the EP method requires recording the electric potential versus either the crack mouth-opening (CMO) or the load-line displacement. A typical dc-EP versus crack-mouth-opening record is shown in Figure 1. Here, there are four distinct stages to the test record. The first is during linear elastic loads where the CMO increases slightly, but there is no change in the dc-EP.

The second stage corresponds to plastic stress at the crack tip and crack blunting. For direct current applications there is a linear change between the CMO and the dc-EP during the crack blunting. We have termed this linear region the EP-blunting line. The first deviation from the EP-blunting line corresponds to ductile crack initiation. This has been documented by comparing CMO versus dc-EP results to multiple specimen testing[1-3] and unloading compliance data[3].

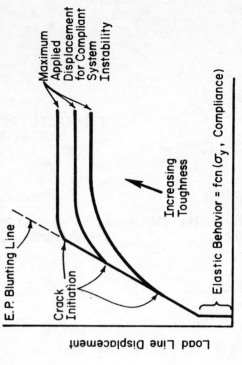

(a) Generalized Crack-Mouth-Opening Versus dc-EP Record

(b) Generalized Load-Line Displacement versus dc-EP Record

FIGURE 2. EFFECT OF TOUGHNESS ON GENERALIZED ELECTRIC POTENTIAL TEST RECORDS

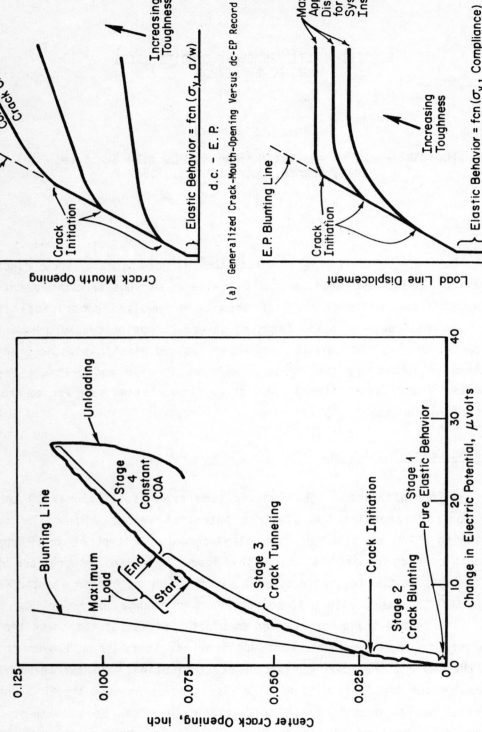

FIGURE 1. CENTER-CRACK-OPENING DISPLACEMENT VERSUS CHANGE IN ELECTRIC-POTENTIAL FOR CIRCUMFERENTIAL THROUGH-WALL CRACKED PIPE EXPERIMENT

The third stage is a nonlinear region which frequently corresponds to crack tunnelling in through-wall cracked specimens.

The fourth and final stage corresponds to crack growth occurring with a constant crack-opening angle. In small-scale specimens this generally occurs after maximum load has been attained.

Alternatively, one may wish to record the load-line displacement rather than CMO versus the dc-EP. For compact specimens such data are identical. For other types of specimens (i.e., center-cracked plates, SEN, three-point-bend specimens, etc.) reasonable results can be obtained as long as homogeneous specimens are being tested. For example, cracks in specimens containing a higher strength weld may yield questionable results since the load-line displacement may reflect plasticity remote from the crack. In such cases the CMO displacement should be used to detect crack initiation.

Figure 2 shows trends that could occur as toughness increases for CMO versus dc-EP (Figure 2a) and load-line displacement versus dc-EP records (Figure 2b). In the CMO versus dc-EP record, as toughness increases then crack initiation occurs further up the EP-blunting line. The COA line also increases slope since generally as the initiation toughness increases the crack growth resistance also increases. In the load-line displacement versus dc-EP generalized record, as the toughness increases, initiation occurs at higher displacements. For displacement-controlled tearing of compliant systems, a maximum applied displacement occurs which coincides with the onset of instability. This is a convenient way to record data for testing instability tests.

Elastic-Plastic Crack Growth Monitoring

The dc-EP can also be used to determine the crack growth during ductile tearing. This involves using the dc-EP signal at the start of ductile tearing and the theoretical or experimental calibration, rather than the dc-EP before crack blunting. Figure 3 shows a comparison of dc-EP predicted crack lengths to experimentally determined crack lengths from nine-point-average crack lengths measured from fracture surfaces. The only difference is the EP from the crack blunting. This is consistant with observations of others[4].

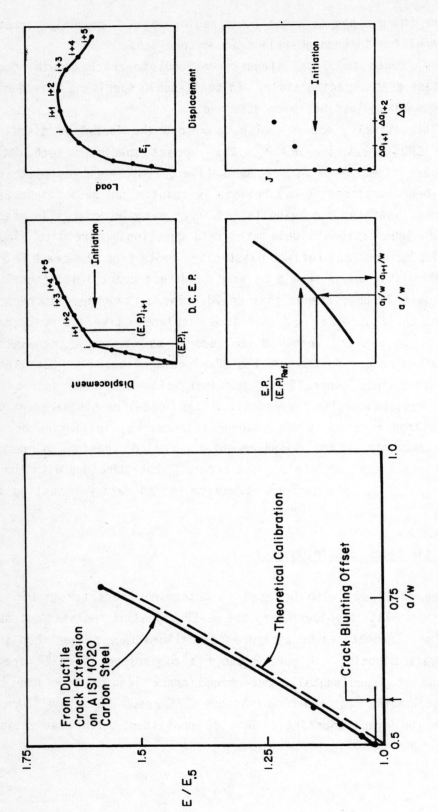

FIGURE 4. SCHEMATIC OF CALCULATION PROCEDURE

FIGURE 3. EXAMPLE OF EFFECT OF PLASTICITY ON DC EP CRACK GROWTH CALIBRATION CURVE

Single Specimen J_R-Curve Evaluations

Since the dc-EP can be used for determining both crack initiation and stable ductile crack growth, it is therefore only natural to apply it to single specimen J-resistance curve evaluation. Recently the author[2] showed how the dc-EP could be easily adapted to a computerized data acquisition system for J-resistance testing. Figure 4 shows a schematic of the calculational procedure used in the computerized data acquisition system. Unlike computerized unloading compliance systems[5], the dc-EP system need not control the testing machine, but only acquire the data. The computer hardware and software is therefore much simpler and hence cheaper for the dc-EP system than for the unloading compliance systems. Furthermore, the dc-EP can also detect crack tunneling, be used with dynamic testing rates, high and low temperatures, and environments where the unloading compliance method has difficulties or cannot be used. Figures 5 and 6 show several examples of single specimen J_R-curves data obtained by a computerized data acquisition system.

Surface Crack Applications

One of the main incentives for developing the dc-EP method at Battelle has been for studying elastic-plastic surface crack behavior. A past paper by the author[6] describes applications of the dc-EP to studies on surface cracked pipes and pressure vessels. These results are briefly described below.

One of the first concerns with application of the dc-EP to surface flaw monitoring, was whether the dc-EP was an average signal of the entire surface crack geometry, or if it represented the crack depth and behavior at the location of the potential probes. One critical test to evaluate this concern involved tight semi-circular machined surface flaws of different aspect ratios and depths in 24-inch-diameter pipes. Figure 7 shows a typical EP scan across a surface flaw. These EP data were then compared to the actual

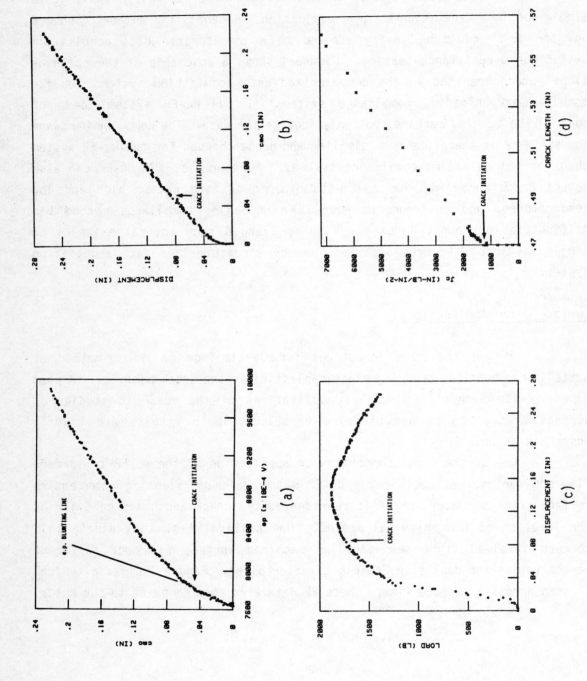

FIGURE 5. SAMPLE OF DATA ACQUIRED AND CALCULATED J_{Ic} FOR C-Mn-Mo STEEL

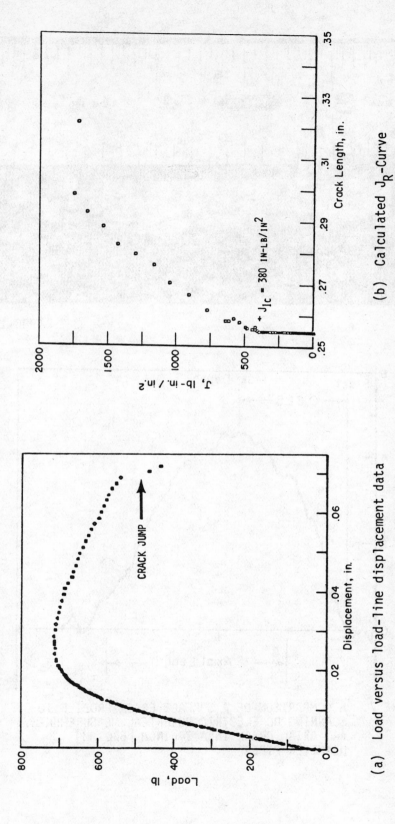

(a) Load versus load-line displacement data

(b) Calculated J_R-Curve

FIGURE 6. EXPERIMENTAL DATA AND J_R-CURVE FOR ZIRCONIUM--2-1/2 PERCENT NIOBIUM MINIATURE CT SPECIMEN AT 149 C (300 F)

4980-1

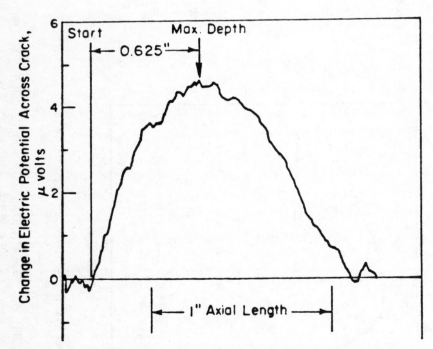

FIGURE 7. A COMPARISON OF A SURFACE CRACK PROFILE TO
SCANNING DC ELECTRIC POTENTIAL MEASUREMENTS.
(AN AXIAL CRACK IN A 24-INCH [610-MM]
DIAMETER PIPE)

crack depths along the semi-circular flaws. As can be seen in Figure 8, the data from all the surface flaws falls on one unique calibration curve. Hence, the EP signal reflects the crack depth at its actual location rather than averaging the signal over a large area. This was true for the flaws with the aspect ratios examined, however, one could imagine that for large aspect ratio flaws that the dc-EP may give an average of the crack depth.

A number of surface crack pipe experiments were subsequently conducted. Some of these were on Type 304 stainless steel, while others were on carbon steel pipes. Figures 9 and 10 show some typical crack-mouth-opening versus dc-EP records from circumferentially surface cracked pipe experiments. As with laboratory test specimens, the deviation from the EP-blunting line determines the point of crack initiation. For surface cracks, the point of initiation is more distinct than in laboratory specimen or through-wall cracked pipe experiments. As in laboratory test specimens, there are also four stages in the test record, with the final stage representing crack growth occuring with a constant crack-opening angle.

Summary

The dc-EP method has been found to be a versatile experimental tool making possible elastic-plastic fracture studies that are either not possible or intolerably expensive with other methods. It has been found to be a simple method for determining the onset of ductile crack growth, detecting tunneling, monitoring ductile crack extensions, applicable to dynamic loading, all of which make it an ideal tool for single specimen J_R-curve testing. For surface flaw testing it is an essential tool, making such studies practical and providing the data necessary for analysis which in the past has been too difficult to obtain.

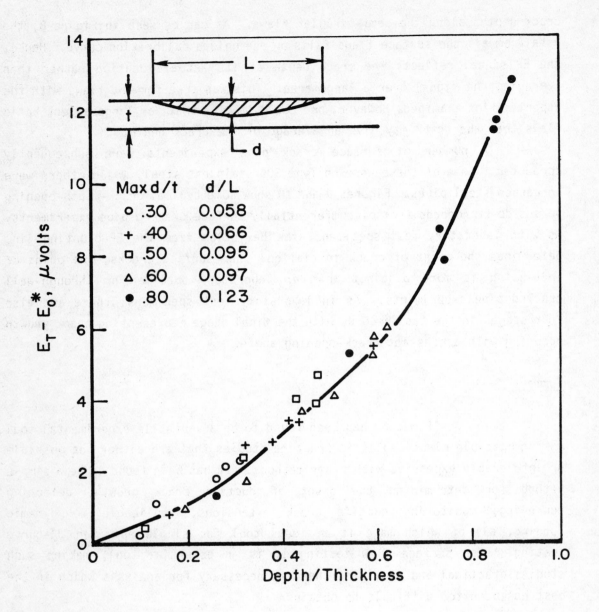

FIGURE 8.   ELECTRIC POTENTIAL CALIBRATION FOR VARIABLE DEPTH AXIAL
SURFACE CRACKS IN 24-INCH (610-MM) DIAMETER PIPE FROM
SCANNING METHOD

---

* E_o = Potential across base metal without flaw.

   E_T = Total electric potential across the crack.

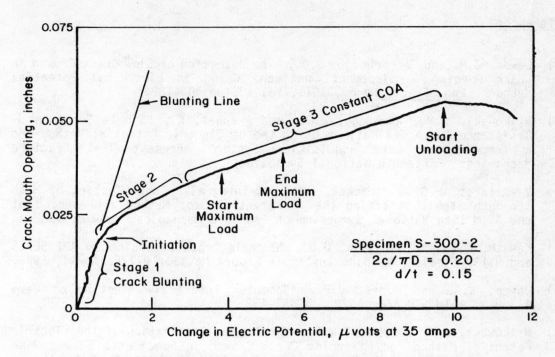

FIGURE 9. ELECTRIC POTENTIAL VERSUS CENTER-CRACK-MOUTH OPENING RECORD FROM
CIRCUMFERENTIALLY SURFACE FLAWED CARBON STEEL PIPE EXPERIMENT

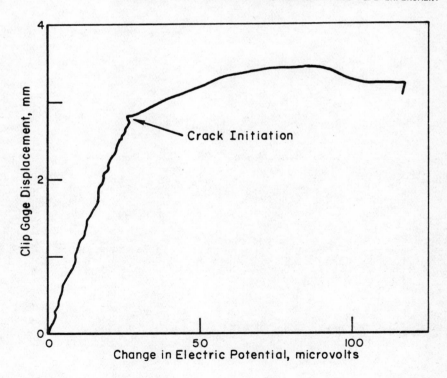

FIGURE 10. CRACK-MOUTH-OPENING DISPLACEMENT VERSUS
ELECTRIC POTENTIAL ACROSS THE CENTER OF
AN INTERNAL VARIABLE DEPTH SURFACE FLAW
ON 4-INCH-DIAMETER SCHEDULE 80 TYPE 304
STAINLESS STEEL PIPE

References

(1) Lowes, J.M. and Fearnehough, G.D., "The Detection of Slow Crack Growth in Crack Opening Displacement Specimens Using an Electrical Potential Method", Eng. Fract. Mech., 1971, Vol. 3, pp 103-108.

(2) Wilkowski, G.M., Wambaugh, J.O., and Prabhat, K., "Single Specimen J-Resistance Curve Evaluations Using the dc Electric Potential Method and a Computerized Data Acquisition System", Presented at Fracture Mechanics: Fifteenth National Symposium.

(3) Vassilaros, M.G. and Hacket, E.M., "J-Integral R-Curve Testing of High Strength Steels Utilizing the dc Potential Drop Method", Presented at the ASTM 15th National Symposium on Fracture Mechanics, July 8, 1982.

(4) Kamath, M.S. and Harrison, J.D., "Ductile Crack Extension in API 5LX65 and HY130 Steels", Welding Institute Report No. 36/1977/E, April, 1977.

(5) Joyce, J.A. and Gudas, J.P., "Computer Interactive Testing of Navy Alloys", ASTM STP 668, 1979, pp 451-468.

(6) Wilkowski, G.M. and Maxey, W.A., "Review and Application of the Electric Potential Method for Measuring Crack Growth in Specimens, Flawed Pipe and Pressure Vessels", Presented at the ASTM 14 National Symposium on Fracture Mechanics, Los Angeles, California, June 30-July 2, 1981.

Application of A.C. Potential Drop Method
to the Detection of Initiation in Static and Dynamic Testing.

P. De Roo and B. Marandet

IRSID, 185 Rue Président Roosevelt
78105 Saint Germain en Laye CEDEX, France

Outline

A number of experimental methods have been proposed for the detection of crack growth initiation during a fracture mechanics test. Among them, the A.C. potential drop (ACPD) technique has been developped for several years at IRSID [1]. The main reasons for having chosen the A.C. rather than the D.C. are :

- to avoid the thermal EMF voltages due to the connections between the wires and the specimen ;

- to avoid the voltage drift in the D.C. operational amplifiers ;

- to take advantage of the skin effect around the crack tip ;

- to increase the specimen impedance which is very low in D.C.

This ACPD method is used in both static and dynamic tests with similar electronic devices.

Description of the ACPD technique

A schematic description of the ACPD device is given on Figure 1 for the static loading conditions i.e. with the current frequency equal to 50 Hz and the intensity equal to 20 A.

The method is a differential one in order to cancel the output voltage changes due to specimen resistivity variations with temperature. Moreover, a zero measurement technique is usually more sensitive than an absolute measurement one. The tested specimen and an identical reference specimen are series wound and electrically insulated from the machine.

The two output voltages in phase opposition are separately amplified (about x 10.000) and then summed. Finally, the D.C. component of the voltage is recorded along with the load and the displacement signals. Before testing, the potential signal is set to zero. As the specimen is loaded, its impedance is varying and therefore the system is out of balance. A typical load and potential versus displacement diagram is shown on Figure 2 ; it can be divided in three zones :

- Zone I : the increase of impedance is due to the separation of the fatigue crack surfaces ;

- Zone II : the decrease of impedance is associated with the inverse magnetostrictive effect ;

- Zone III : the increase of impedance is due to crack growth.

Two examples of such records are given on Figures 3 and 4 in the case of a ductile initiation prior to a cleavage fracture and for a fully ductile behaviour. The minimum between the zones II and III results from a competition between the two phenomena and is considered to be associated with the initiation of stable crack growth. Fractographic examination of specimens after interrupted loading shows that the minimum of potential does correspond to the first stages of ductile tearing. It has been found [2] that, at low current frequency such as 50 Hz, there is a close agreement between J_i calculated at the ACPD minimum and J_{Ic} derived from the J_R - curve procedure (ASTM E 813). This is illustrated on Figure 5 for a 9 % Ni steel at - 163°C. Nevertheless, the engineering value J_{Ic} is usually larger than the J value at the "true" initiation and consequently J_i from ACPD is usually smaller than J_{Ic} .

The agreement between the ACPD minimum and the initiation point becomes poor with specimens larger than 3T CT specimens. The possibility of reducing the delay between the two events in thick specimens by increasing the current intensity must be investigated.

Use of the ACPD technique in dynamic testing

The same ACPD method is used for fracture mechanics testing under rapid loading conditions [3]. The test is carried out on 1T CT specimens at a K rate equal to 2.10^4 MPa$\sqrt{m}$/s. Such a test lasts about 20 ms and therefore the current frequency has been increased. After an investigation was performed with frequency ranging from 500 Hz to 50 kHz, a 10 kHz and 2A A.C. was chosen in order to get the best sensitivity.

The tests records are similar to those obtained in static conditions, as shown on Figure 6.

The good agreement between the minimum of potential and the initiation of stable crack growth can be checked by fractographic examination of specimens tested in the upper knee of the transition curve where cleavage fracture may be preceeded by a ductile initiation and a very small crack extension.

This technique is a very useful tool that allows for the evaluation of the loading rate effect on the transition curves, as illustrated on figure 7.

R-curve determination

The voltage increase ΔV_p (figure 2) beyond the minimum is proportional to the mean crack extension $\Delta \bar{a}$. The ACPD technique can thus be employed to determine R-curves from single specimens in both static and dynamic loading conditions. The ΔV_p versus $\Delta \bar{a}$ relationship depends on the current parameters and on the material. A calibration curve is therefore necessary to construct an R-curve and it can be approximated by a straight line given by the initiation point and the final loading point [2], as done with the DCPD technique. Figure 8 shows the very good agreement between the partial unloading results and the R-curve derived from the A.C. potential increase through a linear calibration.

REFERENCES

(1) B. MARANDET, G. SANZ - Experimental verification of the J_{Ic} and equivalent energy methods for the evaluation of the fracture toughness of steels. ASTM STP 631, (1977), p. 462-476.

(2) N. OKUMURA et al. - Application of the A.C. potential drop technique to the determination of R-curves of tough ferritic steels. Engineering Fracture Mechanics, Vol. 14, (1981), p. 617-625.

(3) B. MARANDET, G. PHELIPPEAU, G. SANZ - Experimental determination of dynamic fracture toughness by J-integral method. Advances in Fracture Research, ICF 5, (1981) p. 375-383.

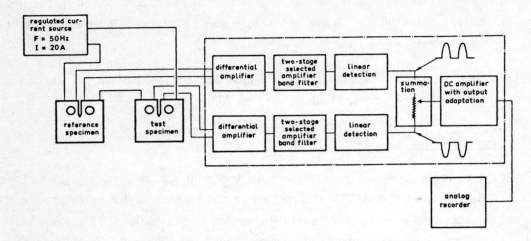

Figure 1 - Schematic description of the ACPD device.

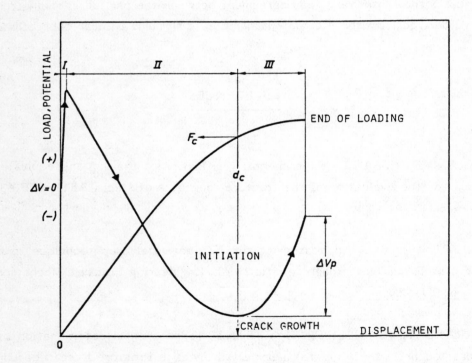

Figure 2 - Typical load and potential drop versus displacement diagram.

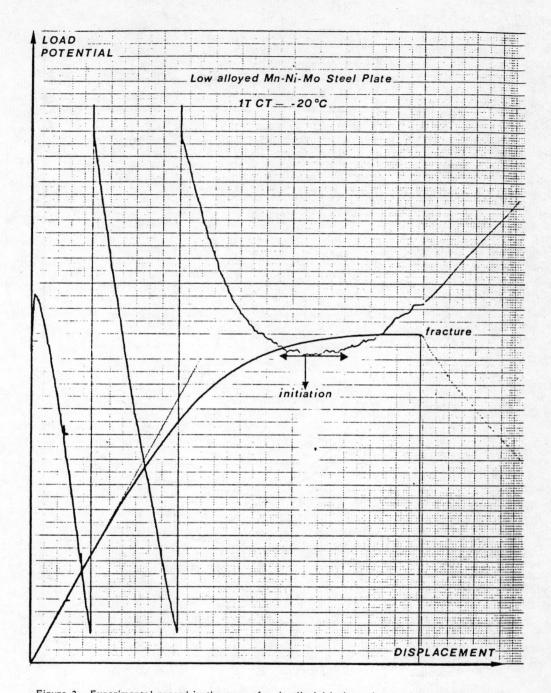

Figure 3 - Experimental record in the case of a ductile initiation prior to cleavage fracture.

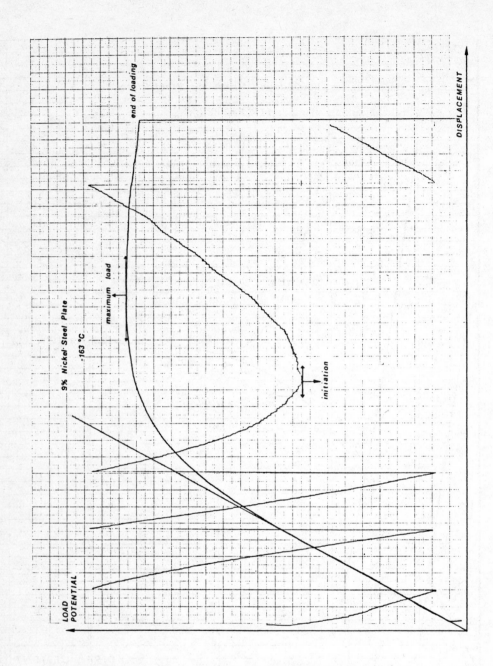

Figure 4 - Experimental record in the case of a fully ductile behaviour.

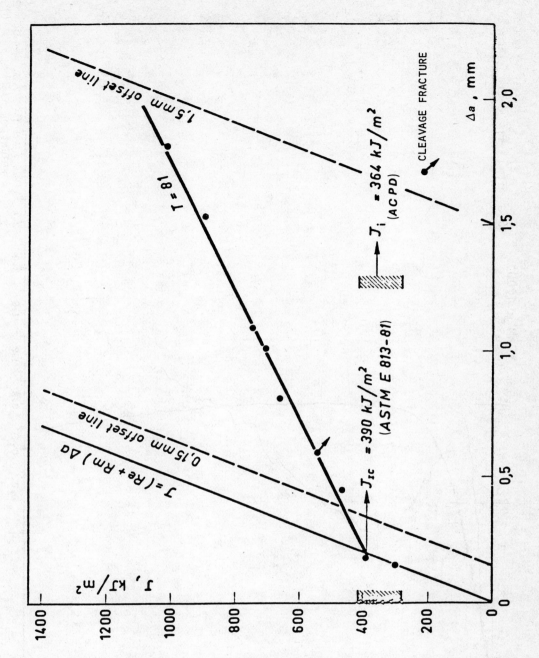

Figure 5 - J_R-curve showing a good agreement between J_{Ic} (ASTM E 813) and J_i determined by the ACPD technique.

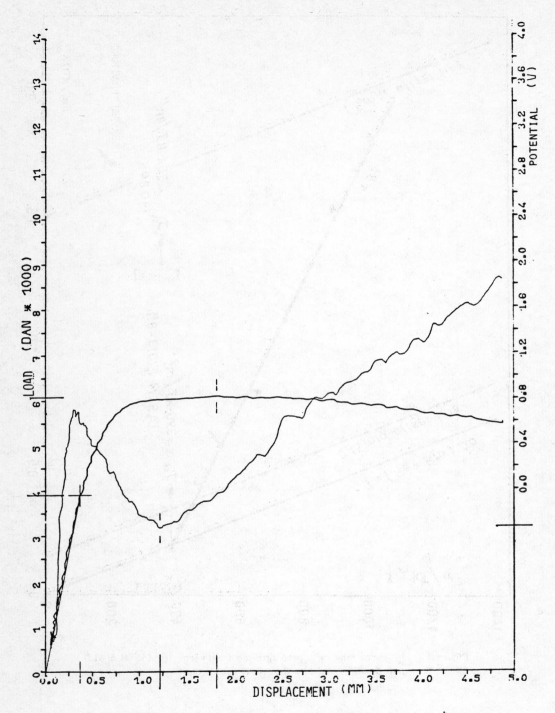

Figure 6 – Example of a record obtained under rapid loading conditions ($\dot{K} \simeq 2.10^4$ MPa $\sqrt{m}$/s) with the modified ACPD technique (F = 10 kHz, I = 2 A).

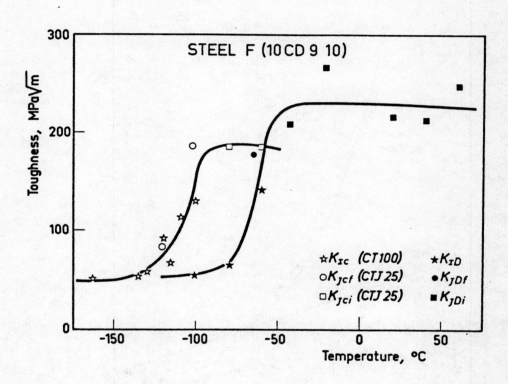

Figure 7 - Influence of the loading rate on the fracture toughness
transition curve of a 2.25 % Cr - 1 % Mo pressure vessel steel.

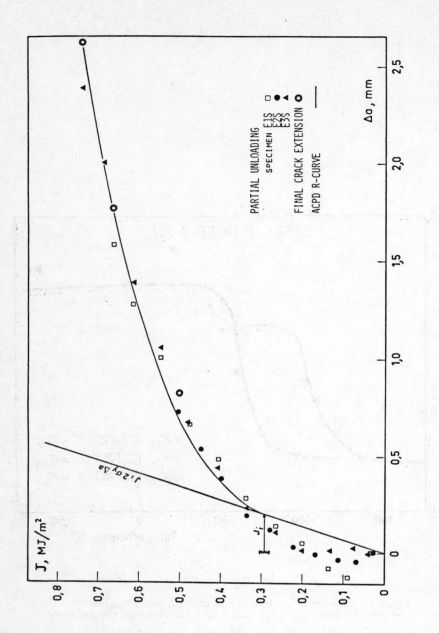

Figure 8 - Comparison between J_R - curves obtained with the partial unloading method and the ACPD technique.
(Ni-Cr-Mo steel plate at room temperature, 20 % side grooved specimen).

POSSIBILITY OF USING SMALL CT SPECIMENS
FOR THE DETERMINATION OF J RESISTANCE CURVE
OF PRESSURE VESSEL STEELS

P. SOULAT - C.E.A. SACLAY

INTRODUCTION

The tests have been conducted with two steels of different toughnesses.

The toughness at the upper shelf of Charpy V notch transition curve was :

- 175 Joules for steel A
- 100 Joules for steel B

Tests have been made in irradiated and unirrradiated condition, at different temperatures. The purpose of this paper is to compare tests conducted with 1 TCT (25 mm) and 0.5 TCT (12.5 mm) specimens.

TEST METHOD

We have employed partial unloading compliance technique.

Specimens was 20 % side grooved after a fatigue precracking to a/w = 0.6.

The computation was made in accordance with ASTM recommendations, without any correction for rotation.

A difference was still adopted, by taking in account a limit of 0.1 b for the stable crack growth (b is the ligament size) instead of 1.5 mm in ASTM standard E. 813.

GENERAL DISCUSSION

We have got some discrepancy with unloading results in the area of the blunting line and for three tests we had to suppress some of them.

For crack extension we have generally good results and it is possible to compare tests with 1 TCT and 0.5 TCT specimens.

The examination of the curves shows generally a good agreement between the tests with the two size of specimens but with a relatively great scatter, essentialy in the J_{1c} results.

If we consider the average results, we have approximately the same influence of the temperature with the two specimen size.

For example if we consider the steel B in irradiated condition, for an increase of temperature between 80 °C to 290 °C, J_{1c} value decreases of 28 % for 1 TCT and of 23 % for 0.5 TCT specimens.

The slope value dJ/da decreases of 10 % for 1 TCT and 15 % for 0.5 TCT specimens.

CONCLUSIONS

The scatter of results requires some suplementary testing but we did not find out any significant size effect beetween, 1 TCT and 0.5 TCT side groove specimens.

The use of 0.5 TCT specimens seems to be technically possible for determinating of elastoplastic toughness of pressure vessel steels in surveillance program.

Temperature of test , °C	Treatment	Size of specimen	J_{1c} KJ/m²	dJ/da MPa

STEEL A

Temperature of test , °C	Treatment	Size of specimen	J_{1c} KJ/m²	dJ/da MPa
70	Irradiated	1	284	208
		0,5	306	333
290	Irradiated	1	252	118
		0,5	183	133

STEEL B

Temperature of test , °C	Treatment	Size of specimen	J_{1c} KJ/m²	dJ/da MPa
25	Unirradiated	1	141	121
		0,5	192	83
80	Irradiated	1	132	47
		0,5	149	80
290	Unirradiated	1	127	61
		0,5	138	54
290	Irradiated	1	95	42
		0,5	114	68

TABLE 1 _ AVERAGE OF RESULTS

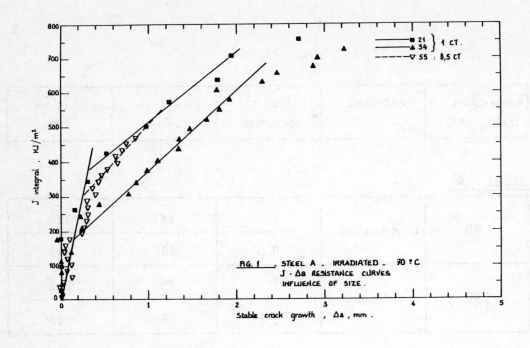

FIG. 1 . STEEL A . IRRADIATED . 70 °C
J - Δa RESISTANCE CURVES
INFLUENCE OF SIZE.

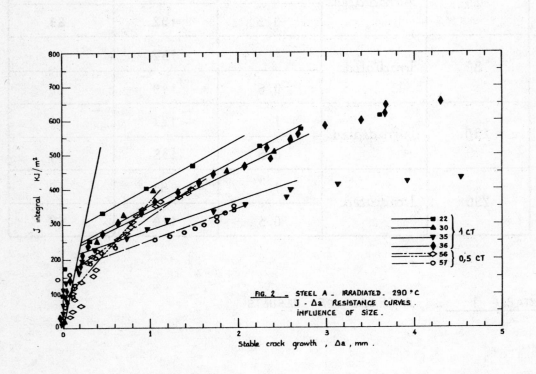

FIG. 2 . STEEL A . IRRADIATED . 290 °C
J - Δa RESISTANCE CURVES .
INFLUENCE OF SIZE.

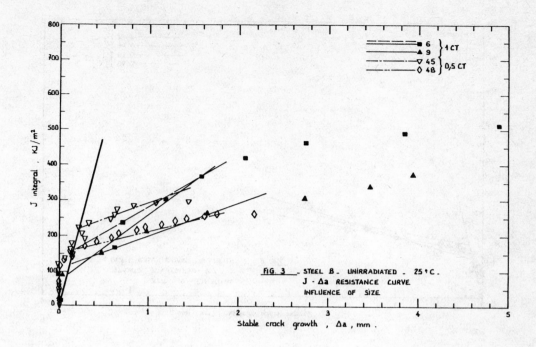

FIG. 3 _ STEEL B _ UNIRRADIATED _ 25°C _
J - Δa RESISTANCE CURVE
INFLUENCE OF SIZE

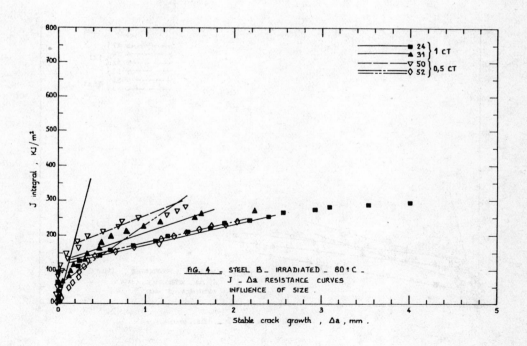

FIG. 4 _ STEEL B _ IRRADIATED _ 80°C _
J - Δa RESISTANCE CURVES
INFLUENCE OF SIZE

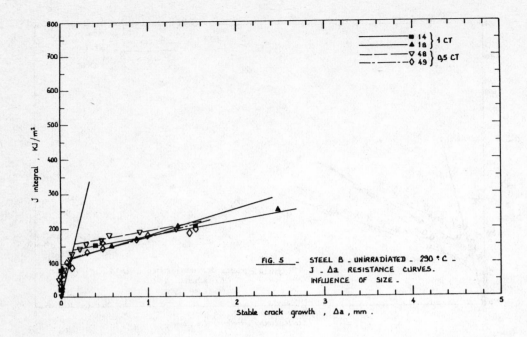

FIG. 5 _ STEEL B _ UNIRRADIATED _ 290 °C _
J _ Δa RESISTANCE CURVES.
INFLUENCE OF SIZE.

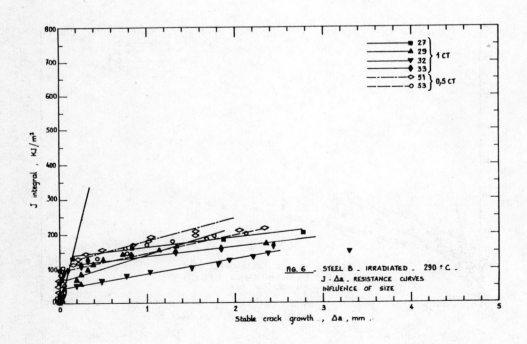

FIG. 6 _ STEEL B _ IRRADIATED _ 290 °C _
J _ Δa _ RESISTANCE CURVES
INFLUENCE OF SIZE

RESISTANCE CURVE MEASUREMENT TECHNIQUES FOR LOW STRENGTH STEELS AND WELDMENTS

by

S. J. Garwood
The Welding Institute
Abington
Cambridge
CB1 6AL.
United Kingdom

By far the majority of resistance curve determinations performed at The Welding Institute employ the multiple specimen method. No testing standard exists for multiple specimen R-curve procedures except as a means for determining J_{Ic} or δ_i. Such a standard is long overdue!

The procedures adopted by The Welding Institute for maximum load and multiple specimen R-curve evaluations (see Fig. 1) are documented in reference 1. The three point bend geometries recommended are defined, and various experimental techniques for testing both parent materials and weldments are outlined in this reference.

A number of detailed experimental points are discussed below:

1. Of particular importance in weldment testing is the local compression treatment[2] used to obtain valid fatigue crack fronts to ASTM E813 and BS5762 standards (see Fig. 2). Fatigue crack validity can lose significance after substantial tearing however (see Fig. 3).

2. Displacement control is normally adopted for R-curve determinations. In cases where maximum load toughness is being evaluated, and there is concern over the possibility of cleavage interventions in ferritic steels, load control may be stipulated by some organisations test procedures.

    Use of load control can have the effect of causing a higher maximum load crack tip opening displacement (CTOD) to be evaluated[3] (see Fig. 4).

    The Welding Institute now recommends displacement control for all conventional CTOD testing[4]. For time dependency studies, the use of clip gauge control is particularly useful[5].

3. Current theoretical statements on J validity restrict the extent of crack growth for J resistance curve determinations. Although no such restriction exists for CTOD the emphasis on the determination of δ_i has the same effect. In many cases where full section thickness is adopted 'non-valid' determinations are the most relevant. Also if notch orientation or temperature effects are being examined in terms of R-curves, large effects can be demonstrated (see Fig. 5) for extensive crack extension (Δa) which tend to be hidden for $\Delta a < 2mm$ by material scatter.

4. The Welding Institute recommends the use of CTOD referred to the original crack tip position (see Fig. 6). CTOD can then be determined from a clip gauge using BS5762 procedures and the initial crack length a_0. In certain cases this value of CTOD will differ to that obtained by infiltration

methods due to the application of an assumed rotational factor of 0.4 in the standard. For most applications this slight inaccuracy is not significant and this deficiency is outweighed by the advantage of having a simple, standardised formula.

When testing to BS5762 specifications a J equivalent to the CTOD formula can be derived from the formula given in Fig. 7. Alternatively, the ASTM formula can be employed if comparator bar or double clip gauge methods are used (Fig. 8).

5. In specialised testing, such as wide plates, the cost of extensive multiple specimen testing would be punitive. However, R-curve methods can be used by marking the crack front by periodic excursions to zero load (Fig. 9a). CTOD determinations, particularly for surface notched specimens, can be difficult. In these wide plate tests photographs (Fig. 9b) cross referenced to displacements made via transducers or clip gauges are invaluable for the derivation of calibration curves relating mouth opening to CTOD.

6. With the advent of computers, instant data analysis is possible provided accurate compliance calibrations are available.

   For three point bend specimens the formula of reference 6 is being evaluated with a correction factor based on a finite element study[7] to allow for knife edge height as shown in Fig. 10.

   This formula is not accurate for short cracks however (see Fig. 11) and an alternative is being evaluated.

7. At present crack measurement procedures in BS5762 and ASTM E813 are different. The effect of excluding the stretch zone and eliminating the awkward blunting line procedure is well known (Fig. 12). However, in the case of non-side grooved specimens where crack tunnelling occurs, the crack growth excluding stretch zone using the BS5762 procedure can exceed that including stretch zone obtained via the ASTM E813 method (Fig. 13).

A number of single specimen R-curve methods have been evaluated at The Welding Institute for both three point bend and CT geometries. Unloading compliance, AC and DC PD, and Back Face Strain methods (Fig. 14) have all demonstrated limited success for certain applications.

Where many identical specimens on well characterised materials are being tested, as in surveillance work for example, this type of procedure is very valuable however, it is difficult to recommend any single specimen method for general laboratory R-curve testing. Particularly as a multple specimen method invariably has to be adopted to provide relevant calibration curves.

Of the various techniques the potential drop method has a particular disadvantage in that the equipment has not been standardised to date.

One problem commonly encountered with unloading compliance using the three point bend geometry is the tendency towards initially negative crack growth predictions (Fig. 15). Improved accuracy can be obtained with unloading compliance methods by using the physical crack measurements to define the initial and final crack lengths and adjusting the compliance formula to interpolate between these points.

References

1. GARWOOD, S. J. and WILLOUGHBY, A. A. 'Fracture toughness measurements on materials exhibiting stable ductile crack extension'. Int. Conf. on "Fracture Toughness Testing - Methods, Interpretation and Application", London, June 1982.

2. DAWES, M. G. 'Contemporary measurements of weld metal fracture toughness'. Welding Journal, December 1976.

3. GARWOOD, S. J., WILLOUGHBY, A. A. and RIETJENS, P. 'The application of CTOD methods for safety assessments in ductile pipeline steels'. Proc. of Conf. on Fitness for Purpose Validation of Welded Constructions, London, November 1981.

4. GARWOOD, S. J. et al. 'The use (and abuse) of CTOD'. Metal Construction, Vol. 14, No. 5, May 1982.

5. GARWOOD, S. J. 'Time dependent ductile crack extension of reactor pressure vessel steels under contained yielding conditions'. NII Specialist Seminars held at EPRI, July 1982 and MPA, October 1982.

6. ALBRECHT, P. et al. 'Tentative test procedure for determining the plane strain J_I R-curve'. ASTM Jrnl of Testing and Evaluation.

7. WILLOUGHBY, A. A. and GARWOOD, S. J. 'An unloading compliance method of deriving single specimen R-curves in 3 point bending'. ASTM Conference, November 1981.

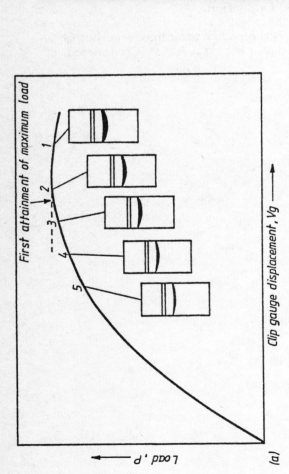

(a)

First attainment of maximum load

Load, P

1
2
3
4
5

Clip gauge displacement, Vg

(b)

CTOD, δ₀

δᵢ (initiation CTOD)

δₘ (maximum load) CTOD

1
2
3
4
5

Slow crack growth Δₐ (excluding stretch zone width)

Fig. 1. Determination of the initiation of tearing value of CTOD following the procedure of Appendix A, BS5762.

a) load clip gauge displacement record
b) CTOD v. slow crack growth (R-curve).

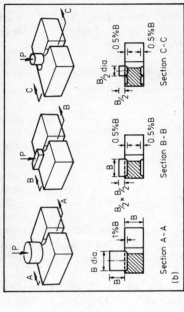

Fatigue crack growth prior to local compression

After local compression

10mm

7G03

40650

(a)

P

P

P

A

B

C

A

B

C

B dia.

B¼

1%B

B

Section A-A

B/2

B₂

B/2

B

0.5%B

0.5%B

Section B-B

B/2 dia.

B/2

0.5%B

0.5%B

Section C-C

(b)

Local compression treatment is frequently a necessary preliminary to CTOD testing of weldments: a) Example of fatigue crack front bowing before and after local compression in a 38mm thick MIG weld (as-welded condition); b) Alternative methods of local compression using single and double platens.

Fig. 2.

- 94 -

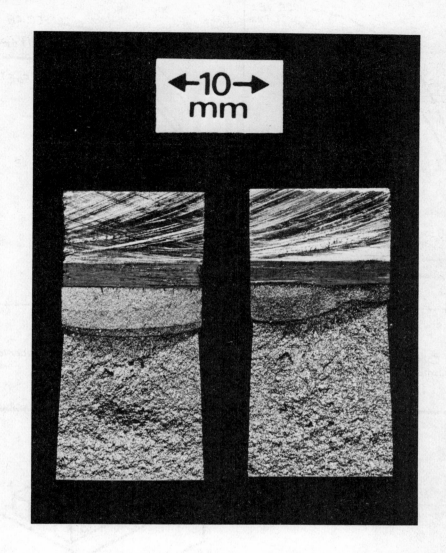

Fig. 3.

Comparison of the development of stable tearing
on specimens with:

a) a valid shape
b) an invalid fatigue crack shape
to BS5762 requirements

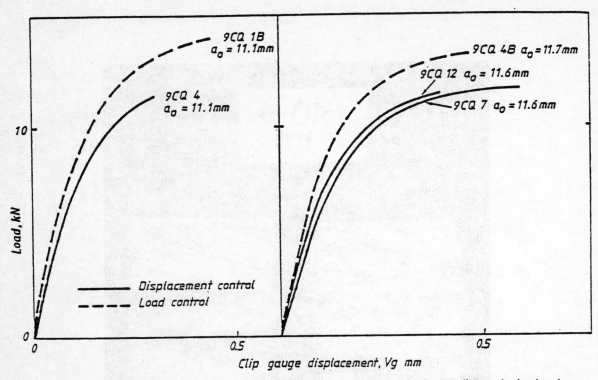

Fig.4   Comparison of load versus clip gauge displacement traces in three point bending under load and displacement control.

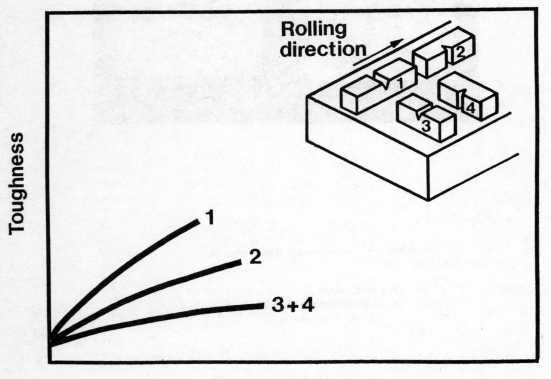

Fig. 5.

Fig. 6. *Definition of crack opening angle in terms of COD.*

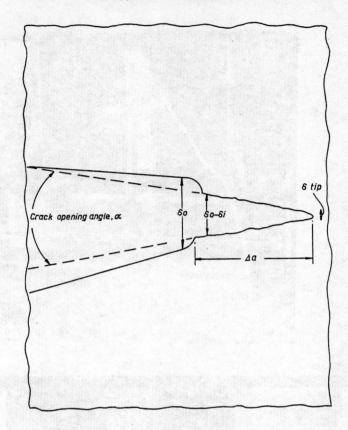

Fig. 7.

THE ESTIMATION OF J VALUES

ASTM E.813-81 describes the evaluation of J at initiation of stable tearing J_{Ic}. As yet no standard employing J equivalent to BS5762 exists.

For the three point bend geometry

since

$$J = \frac{\eta_e U_e}{B(W-a_0)} + \frac{\eta_p U_p}{B(W-a_0)}$$

where U_e and U_p are the elastic and plastic components of the energy absorbed by the specimen

η_e and n_p are geometric factors and B is the specimen thickness.

$$J = \frac{2 U_T}{B(W-a_0)}$$

U_T is determined by measuring the total area under the load-load point displacement record. In terms of the area under the load-plastic clip gauge displacement record U_v assuming a rotational factor of 0.4, a J equivalent to the CTOD forumla is: given by:

$$J = \frac{K^2(1-\nu^2)}{E} + \frac{2 U_v S/4}{(0.4W+0.6a_0+z)B(W-a_0)}$$

where S is the testpiece span.

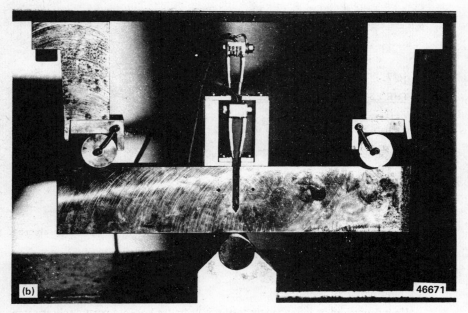

Fig. 8. *Load point displacement measurement methods:*

a) comparison bar technique
b) double clip gauge arrangement.

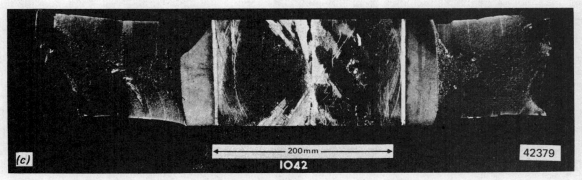

Wide plate fracture surfaces—A533B steel: (a) Surface notch; (b) semielliptical; (c) centre cracked

Fig. 9a.

Wide plate no. 2. Loading no. 9.

Wide plate no. 2. Loading no. 10.

Fig. 9b.

- 100 -

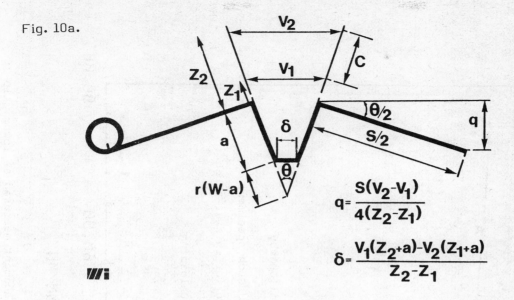

Fig. 10a.

$$q = \frac{S(V_2 - V_1)}{4(Z_2 - Z_1)}$$

$$\delta = \frac{V_1(Z_2+a) - V_2(Z_1+a)}{Z_2 - Z_1}$$

Fig. 10b.

CRACK LENGTH ESTIMATION

The consistency and accuracy of initial crack length estimations is currently under evaluation at The Welding Institute. At present reasonable estimates have been obtained using the formula for three point bend, i.e.:

$$a_o/W = 0.998265 - 3.81662U - 1.80596U^2$$
$$+32.3104U^3 - 44.1566U^4$$
$$-52.6788U^5 \hspace{5cm} [1\ ]$$

where U is a function of the mouth opening displacement of the specimen.

A finite element study has enabled the influence of the knife edge height (z) to be allowed for. Based on the results of this study the crack mouth opening (V_o) can be estimated from the clip gauge displacement (V_g) using:

$$V_o = \frac{V_g}{(1+1.7z/W)}$$

Thus U in equation 1 is given by:

$$U = \frac{1}{\left(\frac{4W\ E\ B}{S(1+1.7z/W)} \times \frac{V_g}{P}\right)^{\frac{1}{2}} + 1}$$

where P is the load at the clip gauge displacement V_g.

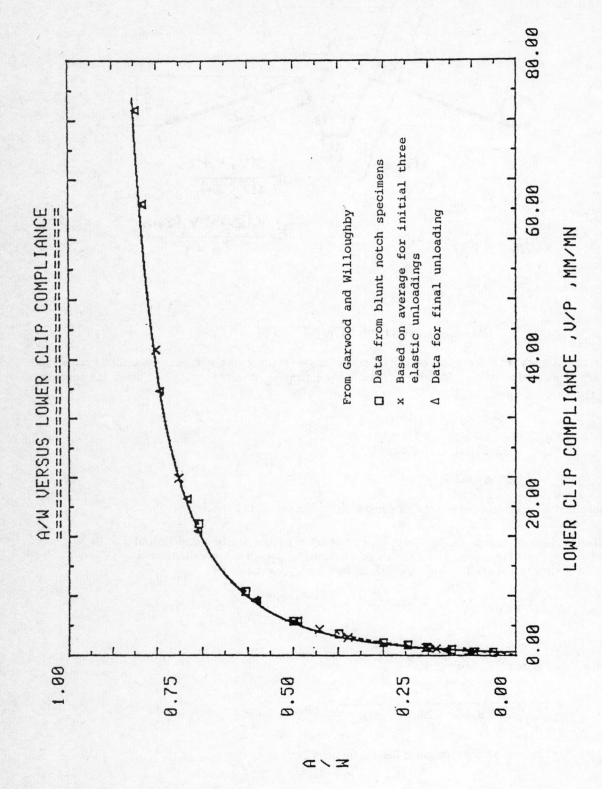

Fig. 11a.   Compliance versus crack length for displacement measured at lower clip gauge.

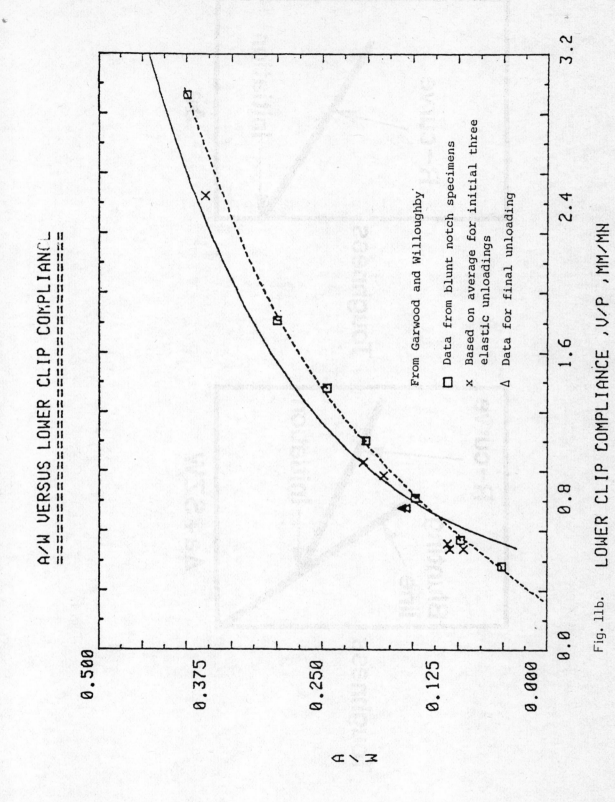

A/W VERSUS LOWER CLIP COMPLIANCE
=================================

From Garwood and Willoughby

□ Data from blunt notch specimens

× Based on average for initial three elastic unloadings

△ Data for final unloading

Fig. 11b.   LOWER CLIP COMPLIANCE ,∪/P ,MM/MN

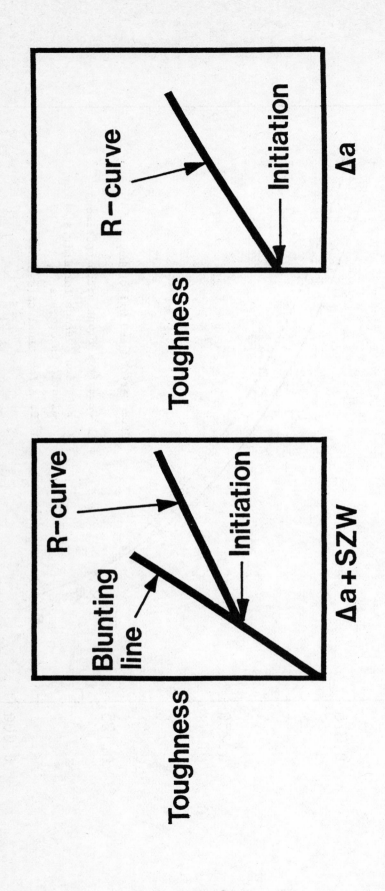

Fig. 12.

Fig. 13. FRACTURE FACE DETAILS

Job No:

Specimen No:

Investigator: S. J Garwood

Date: 17/11/82

B =   mm (gross)

B = 13·045   mm (nett)

W = 26·000   mm

machine notch = 8·56 mm

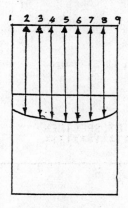

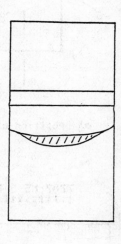

	Fatigue Crack Length, mm, (a)	stretch zone	Δa				
1	12·215	—	—				
2	12·735	0·055	0·130	a^0 = 12·896			
3	13·000	0·110	0·755	$a_0^{(9k)}$ = 12·808			
4	13·015	0·155	1·110	Δa = 0·821			
5	13·080	0·110	1·210	$\Delta a^{(9k)}$ = 0·817			
6	12·995	0·110	1·430				
7	12·845	0·160	0·955				
8	12·600	0·100	0·160				
9	12·175	—	—				
ASTM Ave.	12·808	0·099	0·718				
BSI Ave.	12·896	0·114	0·821				

Fig. 14.

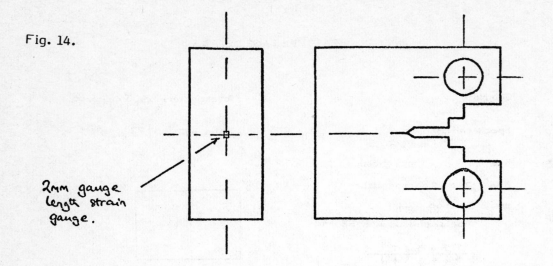

2mm gauge
length strain
gauge.

a) Position of back face strain (BFS) gauge

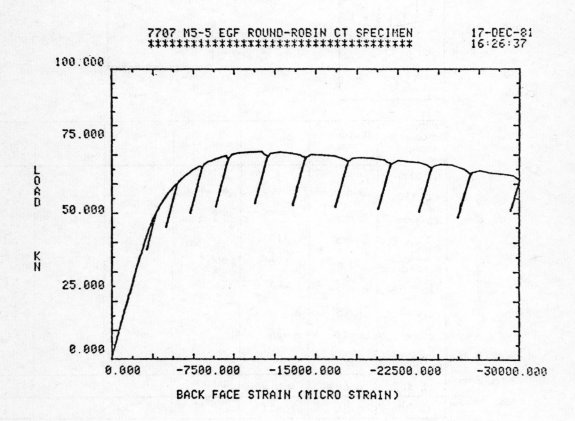

7707 M5-5 EGF ROUND-ROBIN CT SPECIMEN   17-DEC-81
***   16:26:37

b) Load versus back face strain (BFS) trace for specimen No M5-5.

- 106 -

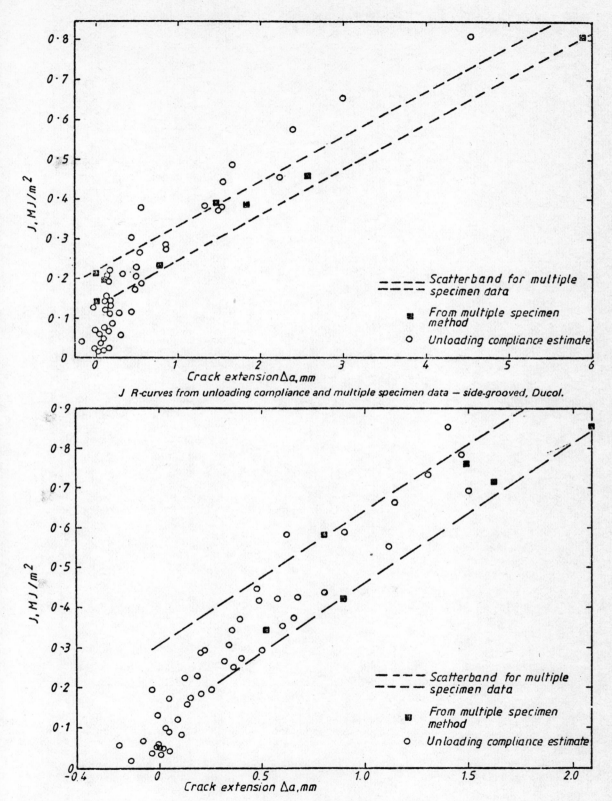

J R-curves from unloading compliance and multiple specimen data — side-grooved, Ducol.

Fig. 15.   J R-curves — plane-sided Ducol.

Session II

J-R CURVE TEST TECHNIQUES
(cont'd)

Chairman - Président

K.H. SCHWALBE
(Federal Republic of Germany)

Séance II

METHODES D'ESSAI POUR DETERMINER LES COURBES J-R
(suite)

J-R CURVES ON AUSTENITIC STAINLESS STEELS — COMPARISON OF UNLOADING COMPLIANCE AND INTERRUPTED TEST METHODS — EFFECT OF SPECIMEN SIZE — NOTCH EFFECT —

P. BALLADON

CREUSOT-LOIRE , Centre de Recherches d'Unieux
B.P. 34 , 42701 FIRMINY CEDEX , FRANCE

SYNOPSIS

J — R Curves have been determined at room temperature and at 350°C for a 304 L Stainless Steel Plate. The aim of the study was to compare results obtained by two experimental methods, unloading compliance and interrupted loading methods, and two different specimen sizes, 2T CT and 1T CT specimens with or without side-grooves.

Main results

Interrupted loading method

Using optical measurement of crack extension by heat-tinting lead to take a theoritical blunting line of equation $J = 4\sigma_f.\Delta a$ instead of $J = 2\sigma_f.\Delta a$ to determine J_{1c}. But a measurement of Δa including the stretched zone determined by micrographie and microfratographie examinations lead to take the equation $J = 2\sigma_f.\Delta a$ for the theoritical blunting line. However, J_{1c} values obtained by the two methods are the same.

2 T CT and 1 T CT give the same values of J_{1c}, about 780KJ/m2 at room temperature. Side-grooving gives a more linear crack front and lowers $\frac{dJ}{da}$, from 380 MPa for plain specimens to 190 MPa for 20% side-grooved specimens at room temperature. The same values of $\frac{dJ}{da}$ are obtained with 2T CT and 1T CT specimens.

Unloading compliance method

J-R curves obtained by unloading compliance allow to determine J_{1c} values at the point where the curves leave an apparent blunting line. These values are the same that those obtained by the interrupted loading method. However, it appears difficult to obtain the real J-R curves because of the discrepancy between crack extensions measured by compliance and real crack extensions measured optically. This discrepancy is due to the large opening of the specimens and it is necessary to correct experimental compliance measurements to take account of the rotation of the specimens. The use of rotation corrections is discussed further by B. HOUSSIN. Some experiments show however that CLARKE'S correction is rather good for tests performed at + 350°C.

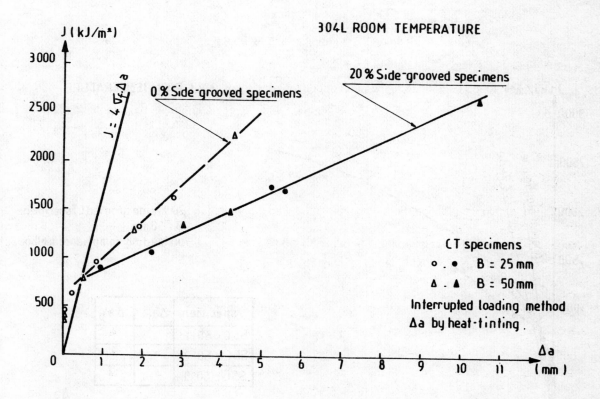

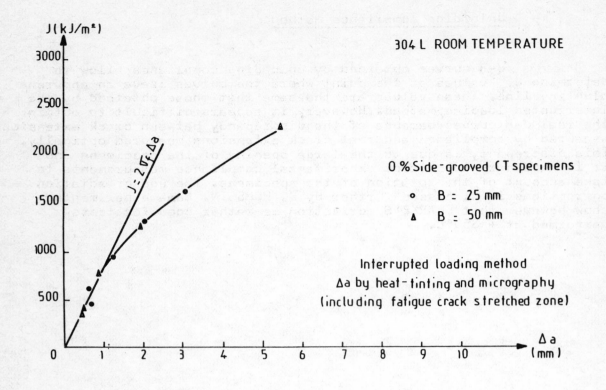

Interrupted loading method
Δa by heat-tinting and micrography
(including fatigue crack stretched zone)

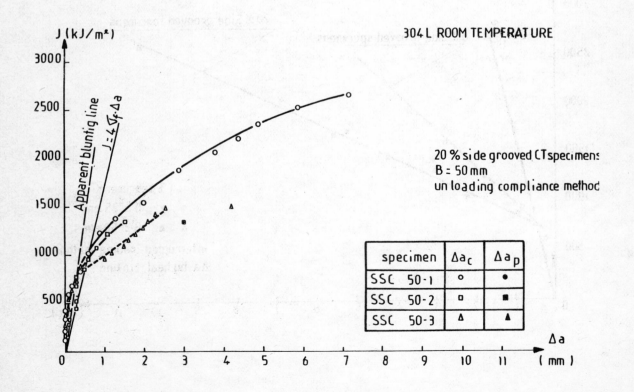

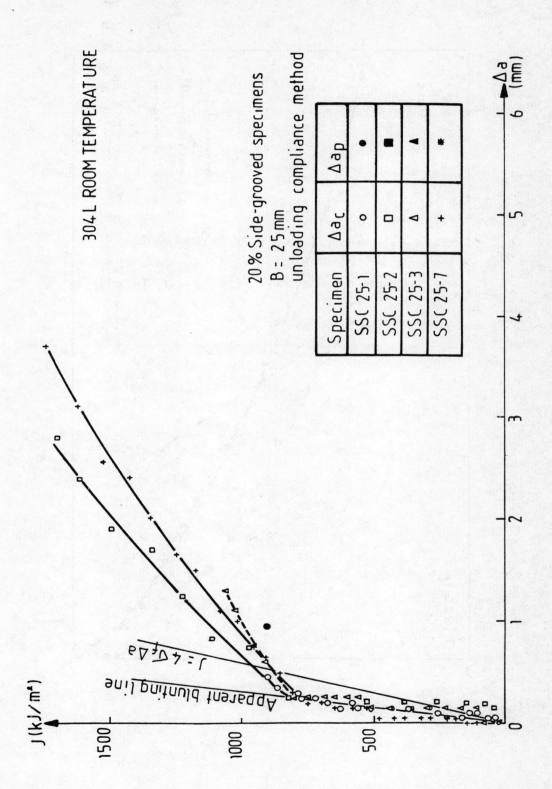

304 L ROOM TEMPERATURE

20% Side-grooved specimens
B = 25 mm
unloading compliance method

Specimen	Δa_c	Δa_p
SSC 25-1	o	●
SSC 25-2	□	■
SSC 25-3	△	◀
SSC 25-7	+	✳

$J = 4\overline{\sigma_f} \Delta a$

Apparent blunting line

$J (kJ/m^2)$

1500

1000

500

0

0   1   2   3   4   5   6 Δa (mm)

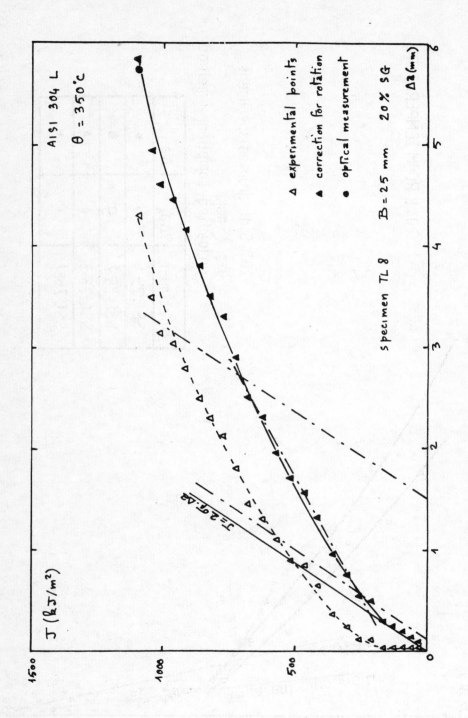

Assessment of Crack Extension by Different Methods.

G.Prantl,   Laboratory for Stress Analysis and Material
             Behaviour, Gebr.SULZER AG.Winterthur,Switzerland.

Abstract:

Some problems in static and dynamic fracture testing are
addressed.
The first point is concerned with tunneling stable cracks
in Compact Tension Specimens. A compliance approximation
is proposed which may be used to calculate average or
maximum crack extensions from unloading phases.
The second part gives an example of the d.c.potential
drop technique and the third part deals with the problem
of load oscillations in testing precracked Charpy type
specimens to determine the dynamic fracture toughness.

1. CT - Specimen with a Curved Crack

A simple analytical model for the specimen can be used to
derive a good approximation for the elastic compliance and
the stress intensity calibration factor.
The method is explained in Fig.1 . Fig.2 shows results for
two types of stable cracks which may be described by a
quadratic parabola or a sine curve.
For comparison,the result of the analytical model for a
straight crack and the two dimensional compliance solution
are given.
In Fig.3, the stress intensity calibration factors, as
derived from the results shown in Fig.2, are shown.
Both C* and Y are smaller for a tunneling crack than for a
straight one having the average length of the former.
Fig.4 contains R-curves, evaluated with the commonly used
2-D compliance solution and the analytical model on the
assumption of a sine - shaped crack.
It is interesting to note the influence of the compliance
approximation on the tearing modulus.
Since J is an average value for the resistance to crack
growth which varies along the crack front, it is probably
reasonable to plot it versus the average crack extension.
Accounting for the curvature of the crack front yields a
smaller T- modulus.
In a specimen with $B = \infty$, the crack would grow with a
straight front by an extent at least equal to (or perhaps
even larger than) the maximum extension found in the center
part of the finite specimen.
In such a specimen, J would be constant along the crack front,
its value being larger than the average J in a specimen with
finite dimensions.
A real plane strain tearing modulus would thus result from
a plot of the described quantities and differ from the one
of the finite specimen.
Consideration of the curvature of the crack is herein
demonstrated to lower the T-modulus significantly. It is

suggested that the plane strain tearing modulus is larger
than the one given by the average J versus the maximum crack
depth but smaller than the one which results from the use of
a 2-D compliance solution.

Specimens with side grooves approach plane strain conditions
and therefore yield flatter R-curves.

Since the degree of plane strain in a side grooved specimen
is not known, it gives a very special R-curve which can not
be applied to plane specimens or structures without side
grooves.

2. D.C.- Potential Drop Technique.

Fig.5 shows the testing arrangement. In Fig.6, the change
in potential is plotted versus the load line displacement
for two specimens with sharp notches. Note the effect of
the unloadings on the potential. Fig.7 shows more such
curves and indicates the effect of an electrically insulated
clip gauge. The blunting phase is rather well reproducible,
as is the slope of the curves in the crack growth phase.
In Fig.8, J_{Ic} as indicated by the knees in the potential
curves is compared to J_{Ic} determined by the unloading
compliance technique according to E 813.

3. The Role of Crack Length in Instrumented Impact Testing.

Fig.9 shows that the load oscillations strongly depend
on the crack or notch depth.

In a series of tests, the notch depth was varied from 1
to 5 mm and the average load as well as the amplitude
of the first load oscillation measured. With increasing
notch depth the amplitude increases,whereas the average
load decreases. In order to meet the conditions given in
the Swiss Recommendation for K_{Id} - Testing, ASK-AN-425,
the notch or crack depth should not be more than about
2,5 mm. On the other hand, the lower limit of about 1,8 mm,
which is given by the condition that yielding must be

confined to the notched cross section, has to be observed. The measured phenomenon can be explained by a superposition of the inertial load oscillations and the rise in average load which is determined by the compliance of the specimen, which in turn depends on the crack length.

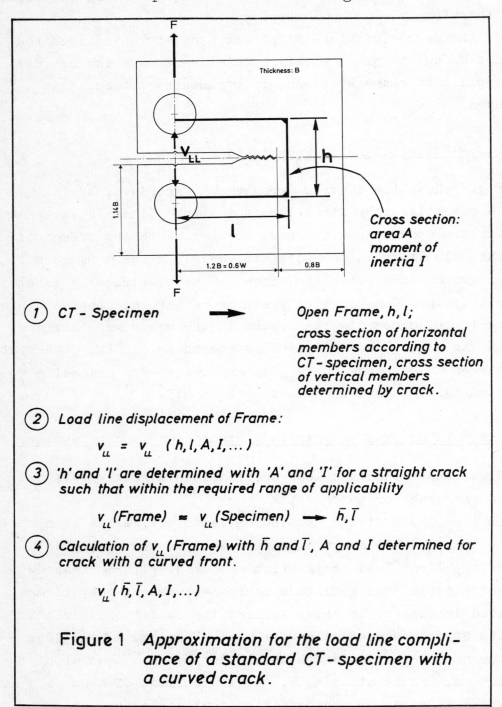

① CT - Specimen ⟶ Open Frame, h, l;

cross section of horizontal members according to CT - specimen, cross section of vertical members determined by crack.

② Load line displacement of Frame:

$$v_{LL} = v_{LL} \; (h, l, A, I, \dots)$$

③ 'h' and 'l' are determined with 'A' and 'I' for a straight crack such that within the required range of applicability

$$v_{LL} (Frame) \approx v_{LL} (Specimen) \longrightarrow \bar{h}, \bar{l}$$

④ Calculation of v_{LL} (Frame) with $\bar{h}$ and $\bar{l}$, A and I determined for crack with a curved front.

$$v_{LL} \, (\bar{h}, \bar{l}, A, I, \dots)$$

Figure 1 *Approximation for the load line compli-ance of a standard CT-specimen with a curved crack.*

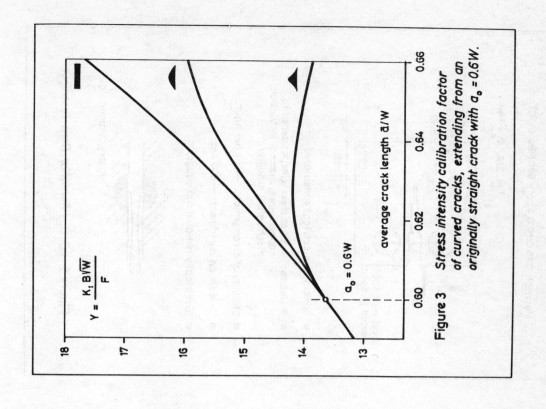

Figure 3   Stress intensity calibration factor of curved cracks, extending from an originally straight crack with $a_o = 0.6W$.

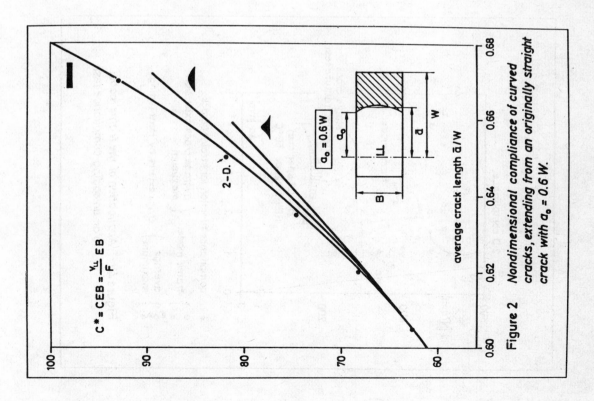

Figure 2   Nondimensional compliance of curved cracks, extending from an originally straight crack with $a_o = 0.6W$.

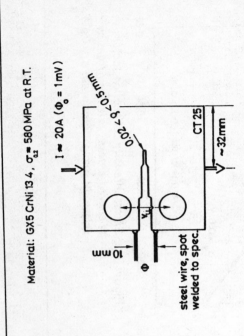

Material: GX5 CrNi 13 4, $\sigma_{0.2} \approx$ 580 MPa at R.T.

$I \approx$ 20A ($\Phi_0 = 1$ mV)

$0.02 < \delta < 0.5$ mm

CT 25

~32mm

10 mm

Φ

steel wire, spot welded to spec.

- Specimen electrically insulated.
- Φ–leads: ≈ 0.5m steel wire, soldered to copper wires, thermal insulation, mechanical stability.
- Stability of arrangement: ~1μV per hour.
- Time for an experiment: ~500 sec.
- Electrically isolated clip gauge:

Epoxy,Ceramic Mat.

Figure 5 *D.C. – Potential Drop Technique*

J [N/mm]

600

400

200

0

2 D compliance

frame model

stable crack after the fracture test

2T – CT specimen RPV steel at +50°C

$J = 2\sigma_f \cdot \Delta a$

0   1   2   3   4   5   Δa [mm]

•······ compliance function for straight crack.

$\begin{cases} \bullet \\ \circ \\ \triangle \end{cases}$ frame model $\begin{cases} \text{average crack extension} \\ \text{maximum} \quad '' \quad '' \end{cases}$

$\left.\begin{matrix} \circ \\ \triangle \end{matrix}\right\}$ $\begin{matrix} \text{average} \\ \text{maximum} \end{matrix}$ crack extens. by 'Heat Tinting'

Figure 4 *Application of the frame model to an unloading compliance test.*

- 120 -

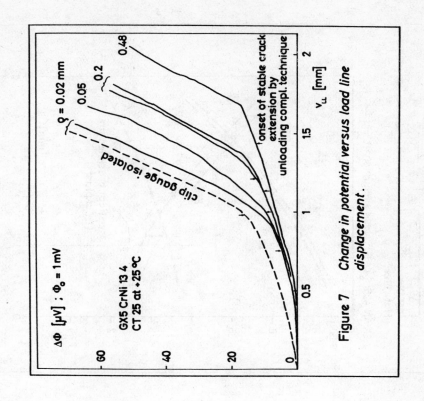

Figure 7   Change in potential versus load line displacement.

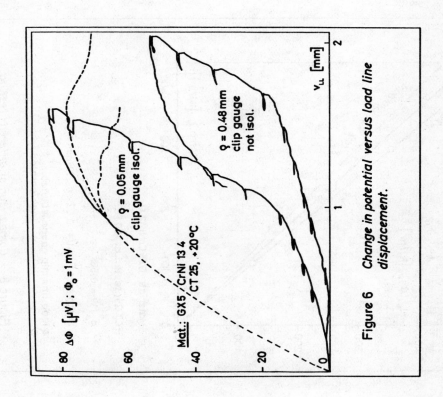

Figure 6   Change in potential versus load line displacement.

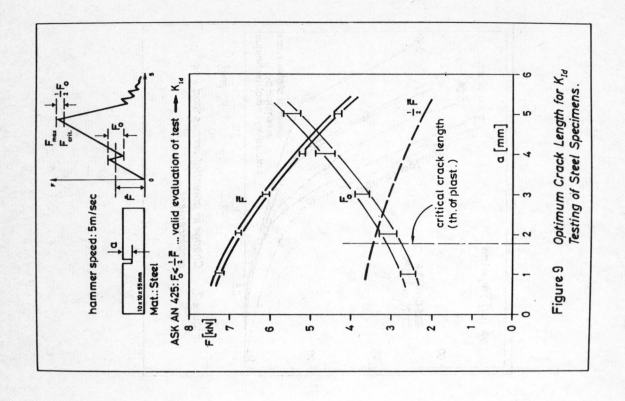

ASK AN 425: $F_o < \frac{1}{2} \bar{F}$ ... valid evaluation of test $\longrightarrow K_{Id}$

hammer speed: 5 m/sec

Mat.: Steel

Figure 9   *Optimum Crack Length for K_{Id}*
            Testing of Steel Specimens.

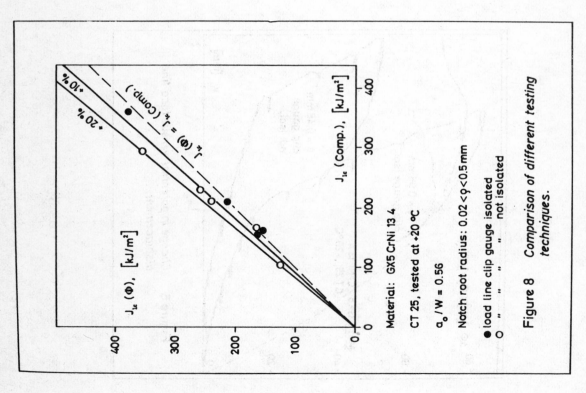

Material: GX5 CrNi 13 4

CT 25, tested at +20 °C

$a_o / W = 0.56$

Notch root radius: $0.02 < \varrho < 0.5$ mm

● load line clip gauge isolated
○ " " " not isolated

Figure 8   *Comparison of different testing*
            techniques.

INTERRUPTED TEST METHOD :
THE EFFECT OF SPECIMEN GEOMETRY

G. ROUSSELIER

Electricité de France, Département Etude des Matériaux,
Les Renardières, 77250 Moret-sur-Loing.

1 - INTRODUCTION

The J-integral concept has two applications : fracture toughness testing of materials and mechanical analysis of structures. In this short paper, the test methods are not reviewed : only the interrupted test method is used and the ASTM standard test procedures are not considered. The objective is to point out that due to the large effect of specimen geometry, the results of material testing should not be used for structural analysis by means of the J-integral concept without an extreme care.

2 - EXPERIMENTAL RESULTS

The J-Δa points obtained on a single forging of A 508 Cl 3 steel with the interrupted test method with heat tinting are shown in figure 1. All the specimens are compact tension specimens with machined knives on the load-line, taken at the quarter-thickness of a 250 mm-thick forging. The homogeneity of the steel at the quarter-thickness was checked with tension, hardness and Charpy-V tests. All the tests were performed in EdF test facilities (1).

Actually the scatter of figure 1 is not due to the material. It results almost totally from the effect of specimen geometry : size, thickness and side-grooves. A still larger scatter could be expected if more different geometries were considered, like center-cracked panels, etc.

The detailed results are shown in figure 2. Four geometries of 1T-CT specimens were tested : plain standard specimens, side-grooved specimens with 25 % and 50 % reduction of thickness, and specimens with double thickness (B = 50 mm). The higher the stress triaxiality, the lower the J-resistance curve : the effect of specimen thickness is important, but the lower results are obtained with the side-grooved specimens. If the specimen size is increased, see the side-grooved 3T-CT specimens in figure 2, the slope of the J-resistance curve is not modified, but the level of the curve is higher.

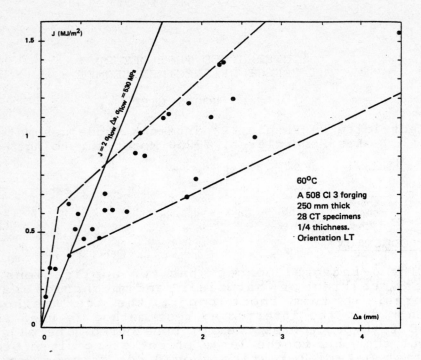

Figure 1 — J—Δa points obtained with the interrupted test method on 28 CT specimens taken from a single forging of A 508 Cl 3 steel.

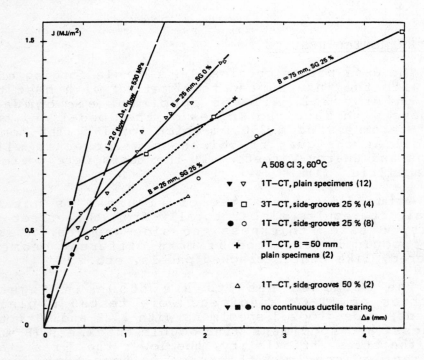

Figure 2 — J—resistance curves obtained with the interrupted test method with different geometries of CT specimens taken from a simple forging of A 508 Cl 3 steel. The J—Δa points are identical to those of figure 1.

This latter effect was also achieved in a systematic test program performed in IRSID laboratories (2). The effect of specimen size on the J-resistance curves of a low alloyed Ni-Cr-Mo steel* is shown in figure 3. The effect of geometry was assigned to the uncracked ligament length, irrespective of thickness. It was also demonstrated that the ASTM E 813-81 size requirement does not ensure a geometry-independent J-resistance curve.

3 - DISCUSSION

The effect of specimen geometry on the J-resistance curve is a complicated problem. Depending on the material, even within the same grade, it may be observed, or not. It may be related to the thickness (see figure 1) or not (2). In the present state of knowledge, there is no size requirement that would ensure the geometry independence in any case, including the structures with cracks.

4 - CONCLUSION

The characterization of ductile tearing of structural materials by means of resistance curves is now a basic method in material testing, and the development of improved test procedures like compliance unloading or potential drop is of great use. But for material comparative programs, only identical specimens should be tested.

For structural analysis, the J-integral concept may be useful if the conservatism of the method and the safety of the structure are both demonstrated. But the limitations and problems associated with this method make it necessary to devote an increasing effort to the development of alternative methods of structural analysis.

REFERENCES

(1) A. MENARD, G. ROUSSELIER, Etude de l'acier faiblement allié au manganèse-nickel-molybdène pour cuves de réacteurs à eau ordinaire sous pression approvisionné sous la forme d'une plaque forgée de 250 mm d'épaisseur auprès de la Société CREUSOT-LOIRE, EdF-Département Etude des Matériaux, internal report n° D 464 MAT/T 43, February 1981.

(2) B. MARANDET, G. PHELIPPEAU, P. DE ROO, G. ROUSSELIER, Effect of specimen dimensions on J_{IC} at initiation of crack growth the 15th Symposium on Fracture Mechanics, ASTM, July 1982, College Park, Maryland, U.S.A.

* The specimens were taken from a 180 mm-thick rolled plate for PWR reactor coolant pump fly-wheels. The chemical composition is C : 0.167 %, Ni : 3.23 %, Cr : 1.71 %, Mo : 0.55 %. σ_{YS} = 680 MPa, σ_{uts} = 800 MPa.

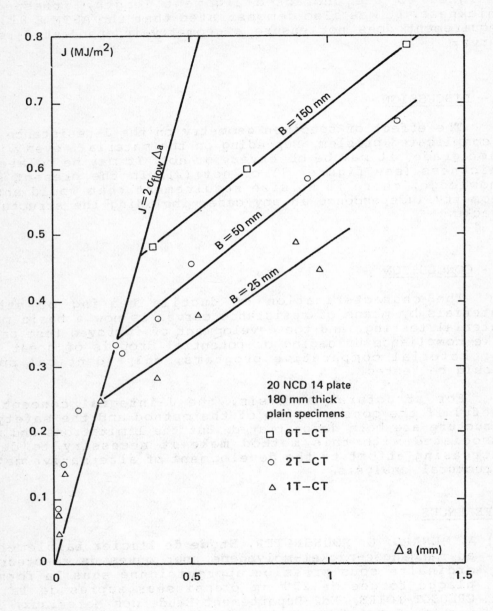

Figure 3 — J—resistance curves obtained with the interrupted test method on a low alloyed Ni - Cr - Mo steel plate.

Application of dc-potential method to prediction of crack initiation*

C. Berger, F. Vahle

Kraftwerk Union AG, Mülheim/Ruhr, Materials Engineering and Research

Abstract

Two direct current potential drop methods were used for determining the beginning of stable crack growth and crack growth resistance curves. One of these methods measured exactly the beginning of stable crack growth. The comparison of crack growth resistance curves evaluated with pd-method, that means single specimen technique and with multiple specimen technique, showed a good correlation for larger specimens.

Main Results

For determination of the beginning of stable drack growth and crack growth resistance curves (R-curves) two circuits of dc-potential methods were tested (fig. 1 and 2).

The beginning of the stable crack growth with method A (fig. 3) is at a distinct changing of the slope of the potential drop and with method B (fig. 4) in the minimum of the potential drop versus displacement curve. In some cases it is very difficult to find this distinct point with method A, especially at small specimens (for instance CT1-specimen), because of the large influence of plastic deformation on the potential drop. With decreasing amount of plastic deformation before and during stable crack growth it is easier to interpret the potential drop record as one can see at fig. 5 and 6 for CT4- and WOL-X150-specimens.

Not only with method A but also with method B there is a linear correlation between the potential drop Δ U and the crack growth Δ a (fig. 7 and 8). In these figures each point means one specimen, using the optical measurement of crack extension by heat tinting method. Different test series and materials show the same behaviour. Although it is easier to determine the minimum point of potential drop the method B (fig. 8) does not find exactly the beginning of stable crack growth. This increases with increasing specimen sizes. However, method A (fig. 7) determines this point very exactly also for larger specimens. Thus method A is used for determining of crack growth resistance curves (J-R-curves).

Fig. 9 to 11 show such J-R-curves for CT1-, CT2- and CT4-specimens evaluated with different test methods. For CT1- and CT2-specimens (fig. 9 and 10) the R-curves after pd-method A (single specimen technique) lead to other results than R-curves from multiple specimen technique if one determines the J_{Ic}-values per ASTM-E 813. But there is an excellent agreement for larger specimens as one can see in Fig. 11 for CT4-specimen. The reason for the different results at smaller specimens is not clear yet (perhaps material scatter). The application of the pd-method is very helpful for estimation the beginning of stable crack growth, i.e. J_{Ic}-value in large specimens. Fig. 12 shows as an example the R-curve of a WOL-X 150-specimen tested at $+80^{o}$C.

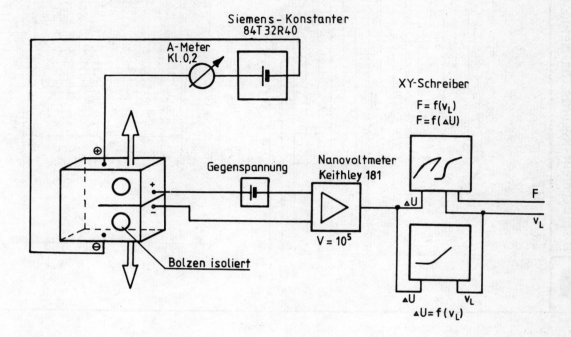

Siemens - Konstanter
84T 32R40

A-Meter
Kl.0,2

XY-Schreiber
$F = f(v_L)$
$F = f(\Delta U)$

Gegenspannung

Nanovoltmeter
Keithley 181

$V = 10^5$

ΔU

F

v_L

$\oplus$

$\ominus$

Bolzen isoliert

ΔU

v_L

$\Delta U = f(v_L)$

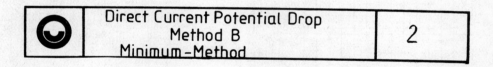

Direct Current Potential Drop
Method B
Minimum-Method

2

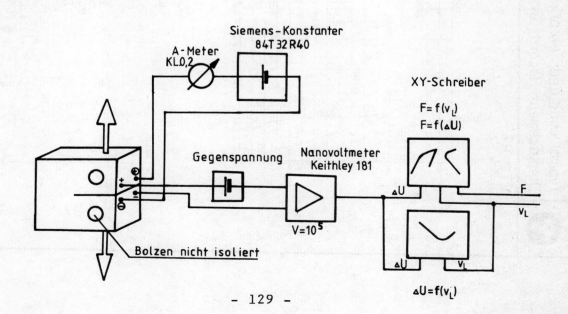

Siemens - Konstanter
84T 32R40

A-Meter
Kl.0,2

XY-Schreiber
$F = f(v_L)$
$F = f(\Delta U)$

Gegenspannung

Nanovoltmeter
Keithley 181

$V = 10^5$

ΔU

F

v_L

Bolzen nicht isoliert

ΔU

v_L

$\Delta U = f(v_L)$

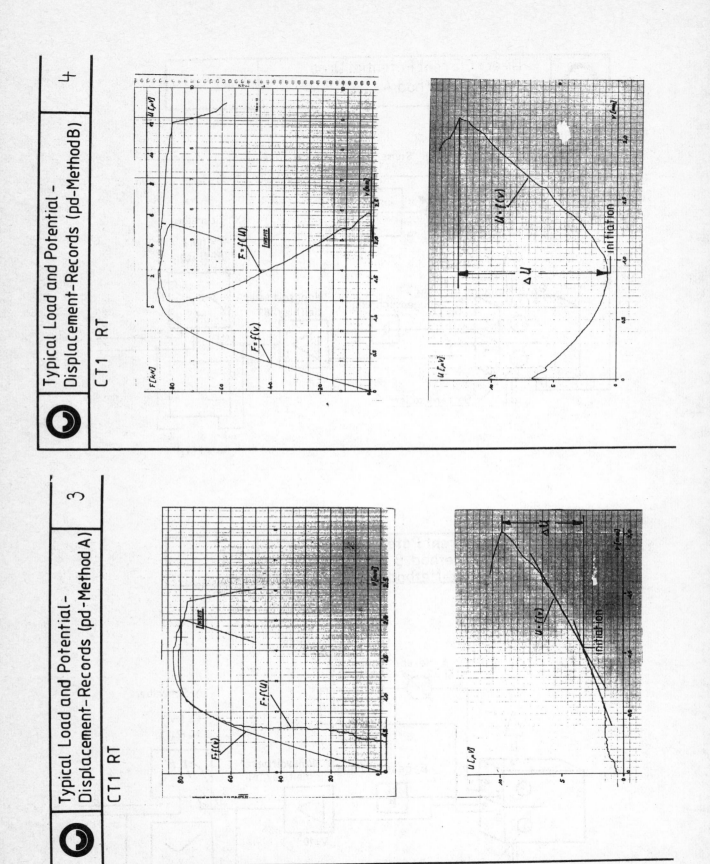

Typical Load and Potential – Displacement-Records (pd-Method A)

CT1 RT

Typical Load and Potential – Displacement-Records (pd-Method B)

CT1 RT

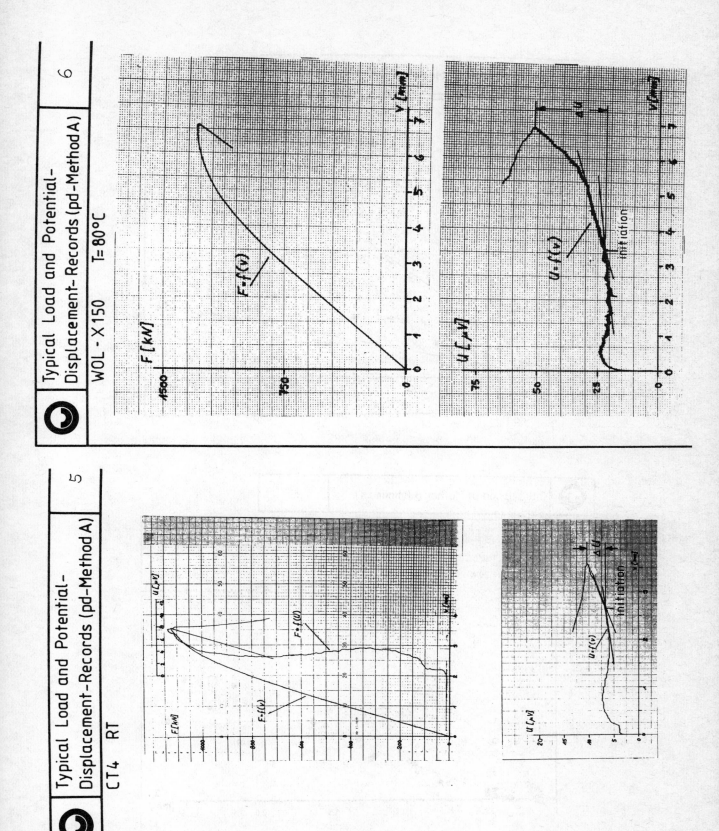

6 Typical Load and Potential-Displacement-Records (pd-Method A)

WOL - X150   T = 80°C

5 Typical Load and Potential-Displacement-Records (pd-Method A)

CT4   RT

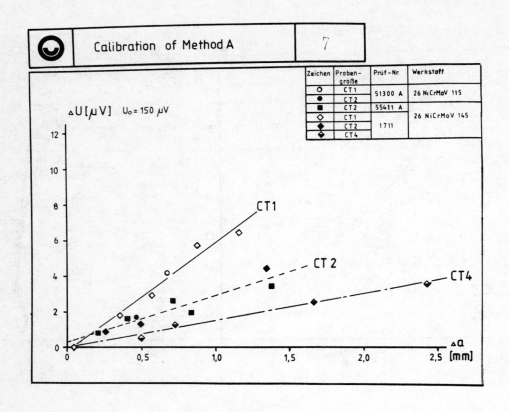

Calibration of Method A

Zeichen	Proben-größe	Prüf-Nr.	Werkstoff
○	CT1	51300 A	26 NiCrMoV 115
●	CT 2		
■	CT 2	55411 A	
◇	CT 1		26 NiCrMoV 145
◆	CT 2	1711	
◈	CT 4		

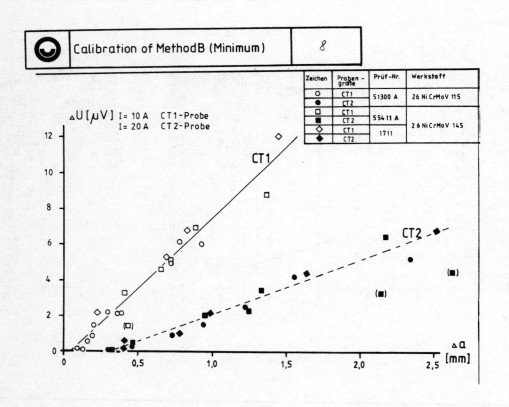

Calibration of Method B (Minimum)

Zeichen	Proben-größe	Prüf-Nr.	Werkstoff
○	CT1	51300 A	26 Ni CrMoV 115
●	CT 2		
□	CT 1	55411 A	
■	CT 2		2 6 NiCrMoV 145
◇	CT 1	1711	
◆	CT 2		

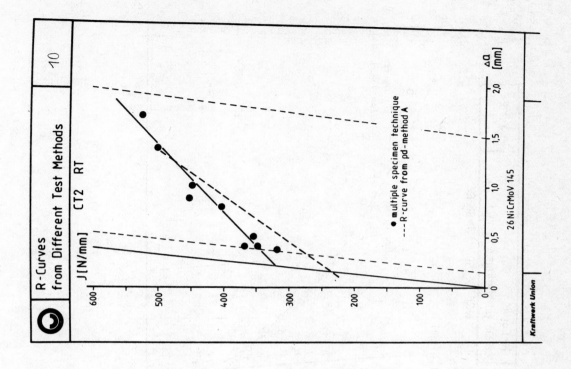

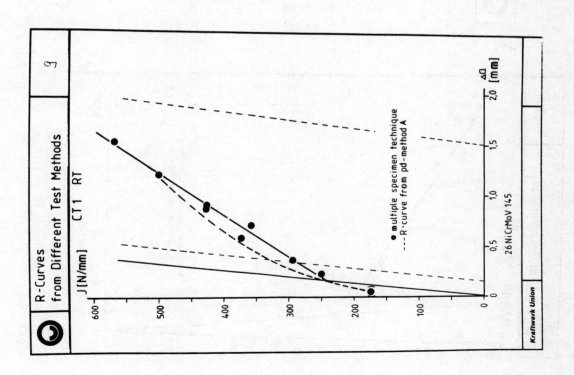

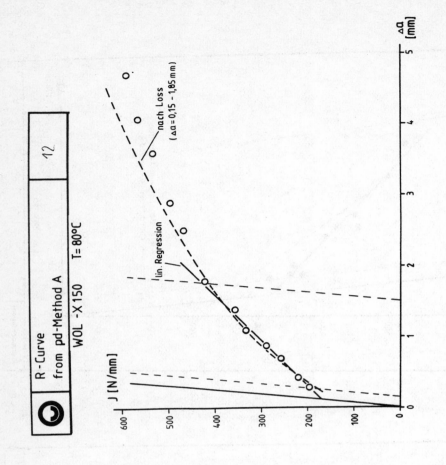

R-Curve
from pd-Method A   12

WOL-X150   T= 80°C

lin. Regression

nach Loss
(Δa = 0,15 - 1,85 mm)

J [N/mm]

Δa [mm]

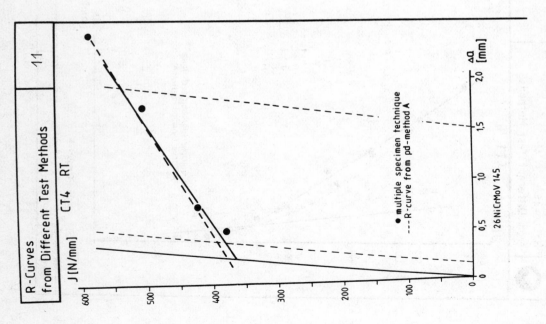

R-Curves
from Different Test Methods   11

CT4   RT

J [N/mm]

● multiple specimen technique
--- R-curve from pd-method A

26 NiCrMoV 14-5

Δa [mm]

KWU - experience with J_{Ic}-testing by partial unloading compliance method and multiple specimen instrumentation.

E.N.Klausnitzer, A.Gerscha, H.D.Aßmann
Kraftwerk Union AG, Erlangen
Federal Republic of Germany

It is KWU-practice, to instrument CT-specimens for ductile fracture tests with the following measuring methods

> Potential drop
>
> Load line displacement
>
> Front face displacement
>
> Crack-tip-opening by TV-camera

The methods and results are shown in Fig. 1 - 6 and explained as follows:

Fig. 1 shows the experimental set up for J_{Ic}-testing by the partial unloading compliance method with additional instrumentation for measurement of potential drop and crack-tip-opening-displacement by means of TV-camera.

Fig. 2 shows positioning of power supply leads and voltage probes for measurement of potential drop.

Fig. 3 shows a modified CT-specimen providing locations for measurement of load line displacement and crack-mouth-opening-displacement in parallel.

Fig. 4 shows the results of a J_{Ic}-test by the partial unloading compliance method with additional potential drop- and crack-tip-opening-displacement measurement. The slopes of the unloading steps are plotted on a second x-y-recorder and processed by a data acquisition and analysing unit.

Fig. 5 shows the correlation of J vs. COD of selected examples. Thus a J-determination might be unnecessary, e.g. with wide-plate tests.

Fig. 6 shows the correlation of COD measured on one surface of the specimen vs. COD calculated with the Wells formula (constant rotationfactor of 0,55).

Fig. 7 shows a simple production technique for manufacturing specimen fixtures, which is paticularly useful in case of large specimen dimensions. Thereby the demand for high quality steels in reduced as well.

Conclusion

The described KWU-practice has been applied successfully to CT-specimens of 25, 50 and 100 mm thickness. Tests with specimens of 150 and 200 mm thickness are in preparation.
The experiments have shown, that the determination of J_{Ic} is possible either with the partial unloading technique or the potential drop method. The scatter is negligible. It is also possible to determine J-R-curve via the COD-R-curve.

The KWU-practice of multiple specimen instrumentation can be recommended even in such cases, where only one or a few specimens are available. The determination of the material properties by different methods ensure the results to be redundant.
The additional costs for multiple specimen instrumentation are in the range of 10 %, if the measuring equipment is available at all.

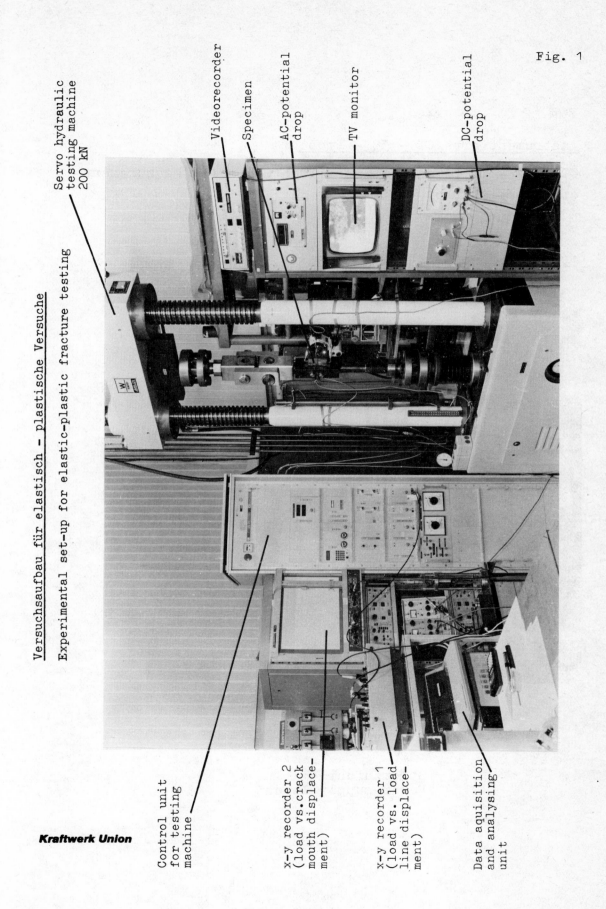

Versuchsaufbau für elastisch – plastische Versuche

Experimental set-up for elastic–plastic fracture testing

Fig. 1

Servo hydraulic testing machine 200 kN

Videorecorder

Specimen

AC-potential drop

TV monitor

DC-potential drop

Control unit for testing machine

x–y recorder 2 (load vs. crack mouth displacement)

x–y recorder 1 (load vs. load line displacement)

Data aquisition and analysing unit

Kraftwerk Union

Fig. 2

<u>Instrumentierte CT - 50 - Probe</u>

CT - 50 - Specimen with Instrumentation for Measurement
of Potential Drop

Potentialabgriff
Voltage probes

Stromzufuhr
Power supply leads

Fig. 3   CT-Specimen for Double Clip Gage Instrumentation

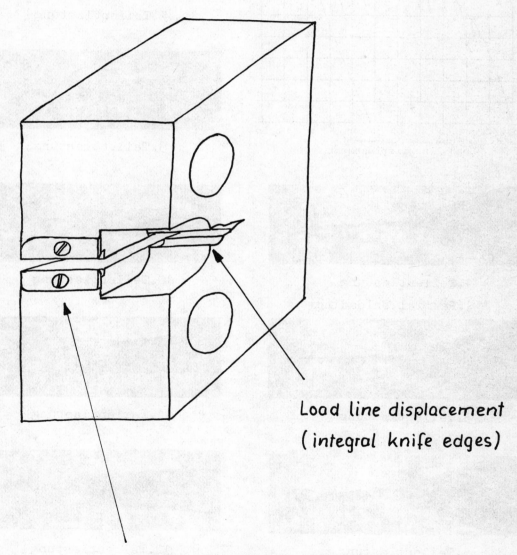

Load line displacement
(integral knife edges)

Notch mouth displacement
(screwed on knife edges)

Fig. 4 J_{Ic} - Testing by Partial Unloading Compliance Method with Additional Potential Drop - and COD - Measurement

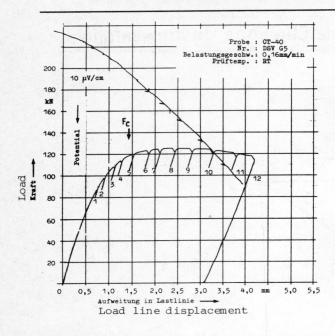

Probe : CT-40
Nr. : DSV G5
Belastungsgeschw.: 0,16mm/min
Prüftemp. : RT

10 μV/cm

Potential

F_C

Load
Kraft

Aufweitung in Lastlinie
Load line displacement

Teilentlastungsversuch mit Potentialsonde und COD-Messung mittels Fernsehkamera

7.Teilentlastung

9.Teilentlastung

1.Teilentlastung
1.Partial Unloading

10.Teilentlastung

3.Teilentlastung

11.Teilentlastung

5.Teilentlastung

12.Teilentlastung

1mm

Fig. 5

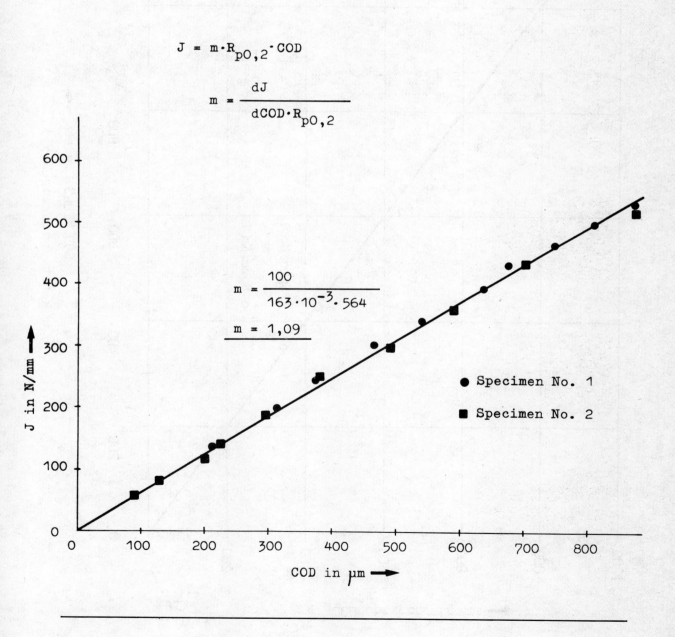

Zusammenhang J-COD für CT-40 Proben aus X 10 CrNiNb 18 9

Correlation of J vs. COD for CT-40 Specimens of X10CrNiNb 18 9

$$J = m \cdot R_{p0,2} \cdot COD$$

$$m = \frac{dJ}{dCOD \cdot R_{p0,2}}$$

$$m = \frac{100}{163 \cdot 10^{-3} \cdot 564}$$

$$m = 1,09$$

J in N/mm

● Specimen No. 1

■ Specimen No. 2

COD in µm

Fig. 6   Vergleich von gemessenem und berechnetem COD
          Comparison of measured and calculated COD

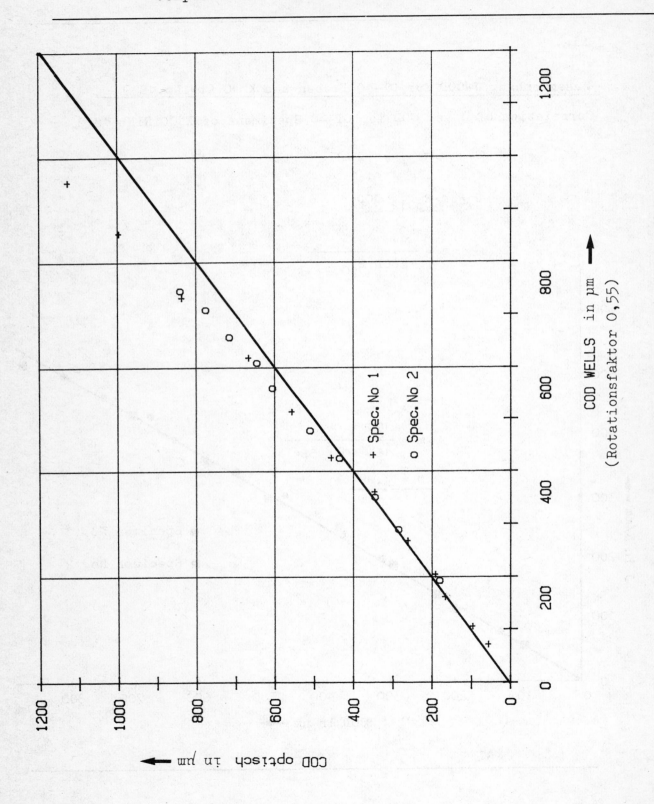

Fig. 7   Specimen Fixtures

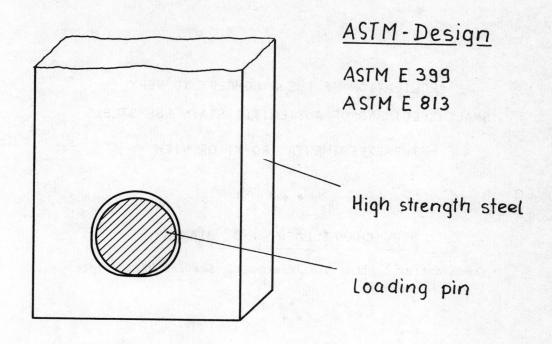

ASTM-Design

ASTM E 399
ASTM E 813

High strength steel

Loading pin

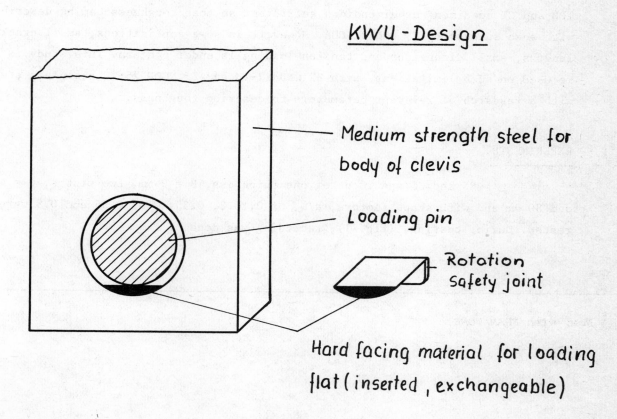

KWU-Design

Medium strength steel for body of clevis

Loading pin

Rotation safety joint

Hard facing material for loading flat (inserted, exchangeable)

APPLICATION OF THE J CONCEPT TO VERY
SMALL SPECIMENS OF AUSTENITIC STAINLESS STEELS
- THE EXPERIMENTAL POINT OF VIEW -

J.P. CHAVAILLARD[*] , D. MIANNAY

Commissariat à l'Energie Atomique - Service Métallurgie

Usually, toughness in the elasto-plastic range is determined with TPB and CT specimens of dimensions sufficient so that toughness can be described with some success by J, COD or COA. However, in some applications, small crack lengths, small dimensions and tension loading is under concern. This study is focused on such limitations, with Z2 CND 17-13 and Z1 NCDU 25-20 stainless steels, with a research of relevant parameters to describe toughness.

I - EXPERIMENTS.

DEN and CC specimens of one thickness, B = 9 mm, two widths, w = 40 and 80 mm and with shape factors, a/w, of 0, 0.2, 0.35, 0.5, 0.65 and 0.8 were tested. During testing, (fig. 1), recording was done of :

----

[*] *Now, with FRAMATOME.*

- load.
- elongation for an external gauge length of 1.5 w through an inductive transducer.
- elongation for a gauge length of 40 mm externally for DENS and of 13.5 mm axially for CCP with a double cantilever clip gage.
- notch root contraction with a special clip gage. This record was not used later on.

Moreover, many photographs were taken on a face on which a grid pattern formed by adjacent 2 mm diameter circles was deposited (fig. 2). Theses photos were intended to determine crack shape evolution (blunting ...), crack growth and the deformation of the specimen everywhere.

Moreover to determine crack growth resistance curves, the unloading method and the heat tinting technic were applied.

II - RESULTS.

Examples hereafter presented are on one or other steel and the 40 mm width except when mentionned.

II.1. The mechanical behavior.

II.1.1. *Elastic compliance.*

The elastic behavior of the materials is in a fairly good agreement with the litterature. However, the small variation of the compliance versus the shape parameter does not allow the determination of the crack growth in spite of the good records (fig. 3). This small variation is in contrast with variation when bending occurs.

II.1.2. *The overall behavior.*

Figures 4 and 5 show that normalizing load through the net section stress and displacement by the ligament gives a big discrepancy wich may be advocated for the unadequacy of the RICE estimation for tension, as described below. This fact may be due partially to the plasticity development, because the effect is more for the DENS than for the CCP, the plasticity development in the ligament being more lasting in the CCP.

II.2. The relevant toughness parameter.

The preceding behavior must be corrected for the crack growth occuring. However correction is very small.

II.2.1. *The net-section stress citerion*.

When analyzing net section stress versus current shape parameter, we observe that for the smaller width (fig. 6), plasticity localisation is more and more on the ligament with the shape factor for the DENS and that the behavior may be described roughly by a net section stress criterion in plane stress. However for the bigger width, growth occurs well below the net section stress criterion. So such a criterion may be suspicious ever for the smaller width.

II.2.2. *The J criterion*.

II.2.2.1. *The J evaluation*.

Five values for J are used

- the J calculated from the pseudo-potential as proposed by BEGLEY and LANDES (fig. 7),
- the J estimation by SUMPTER and TURNER,
- the J estimation by RICE, PARIS and MERKLE,
- the J calculated by integration along different paths from the deformation of the grid pattern and total plastic flow theory,
- the J estimation by SHIH and HUTCHINSON with estimated plastic flow parameters deduced from the tensile test.

Figure 8 shows such a compilation. The result is that for deep cracks, some agreement is observed but for small cracks, the discrepancy observed depends on geometry and crack length.

II.2.2.2. *J as describing the HRR field*.

Results of deformation around the crack tip field as illustrated on figure 9 shows that if HRR field occurs, its extent is very small.

II.2.3. *The crack opening displacement*.

Crack opening was determined conventionally by the intercept of the 90° angle localized at the crack tip.

II.2.4. *The J - COD relationship*.

The relationship obtained is shown in figure 10.

II.3. Fracture toughness.

The crack growth resistance curve obtained for the Z2 CND 17-13 steel is shown in figure 11 with data from litterature. Due to the good agreement exhibited, it is concluded that toughness can be determined with small specimens under tension, in spite of the theoretical limitation.

To determine crack growth initiation, two methods are suggested :

- data plotting of true crack growth without blunting,
- for the CCP, plotting of COD versus displacement across the small gauge length (fig.12) : two linear relationships are obtained, the intercept of which is occuring at initiation.

III - PERSPECTIVE.

Such a study is planned with bending.

FIGURE 1 - Photograph of the experimental set-up for tension testing.

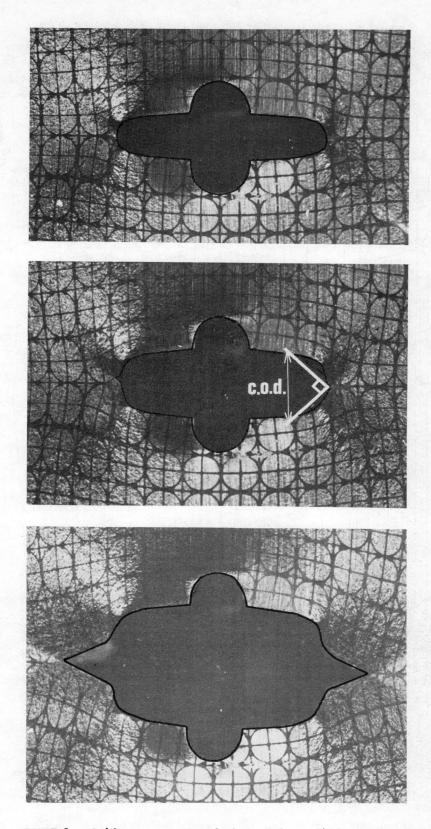

FIGURE 2 - Grid pattern around the notch at different loads.

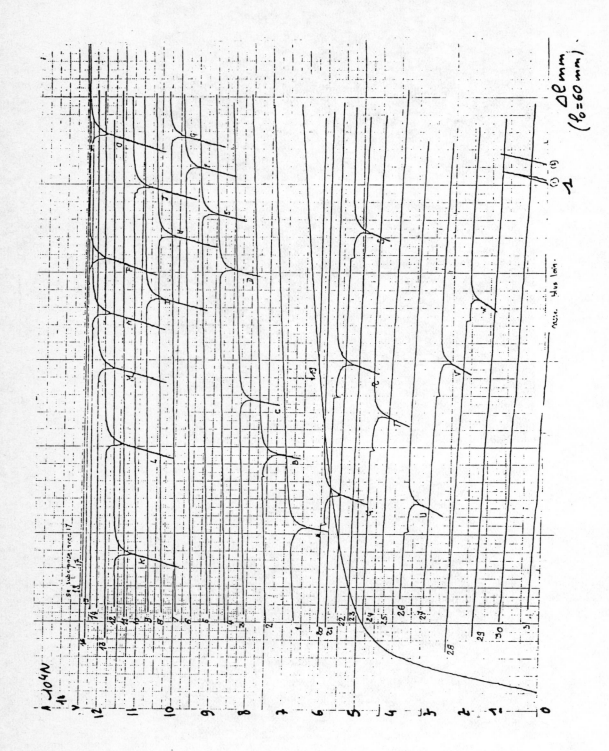

FIGURE 3 — Load - Load point displacement curves, under controlled displacement, with partial unloading.

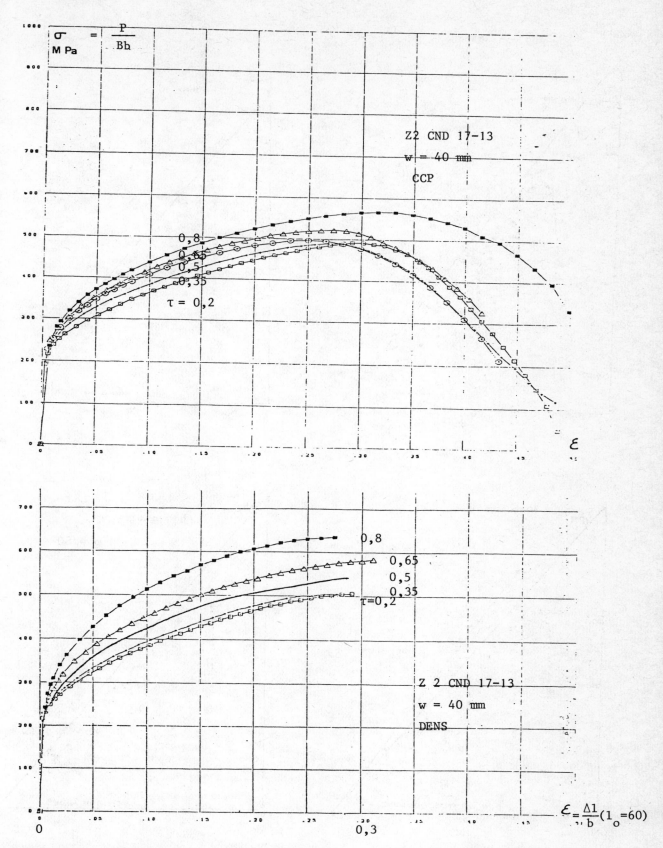

FIGURES 4 and 5 - Normalised load - Normalised load point displacements
curves for different crack lengths.

- 151 -

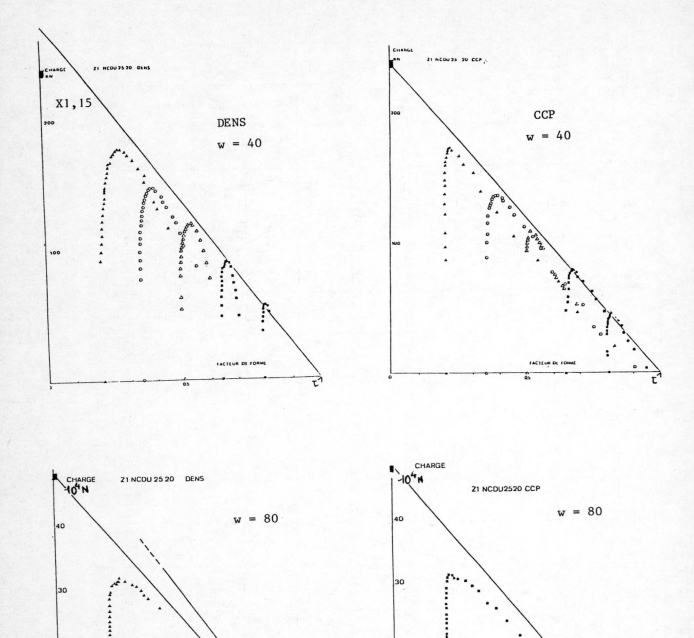

<u>FIGURE 6</u> - Relationship between load and crack length (ultimate load for
a/w = 0 ; current load for a/w ≠ 0)

- 152 -

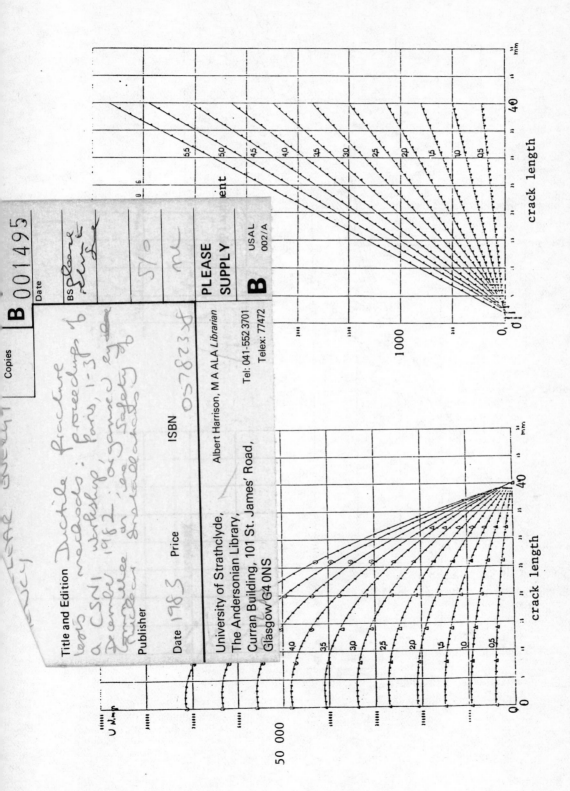

FIGURE 7 - Establishment of the J – U relationship by the compliance method.

- 153 -

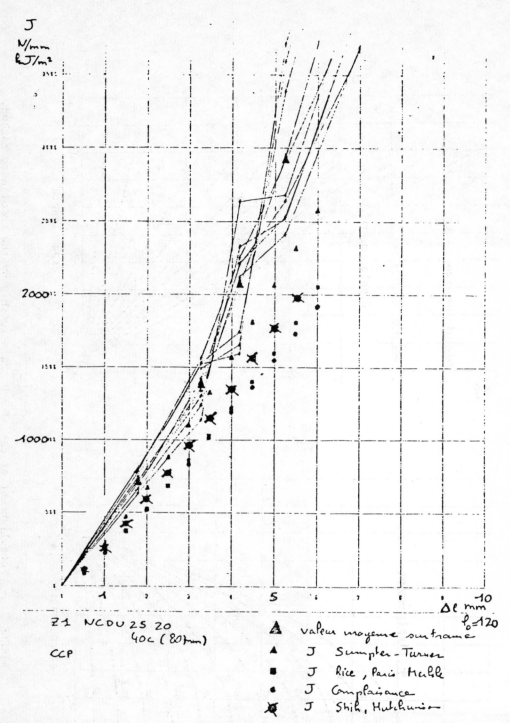

FIGURE 8 - Comparison of J estimations.

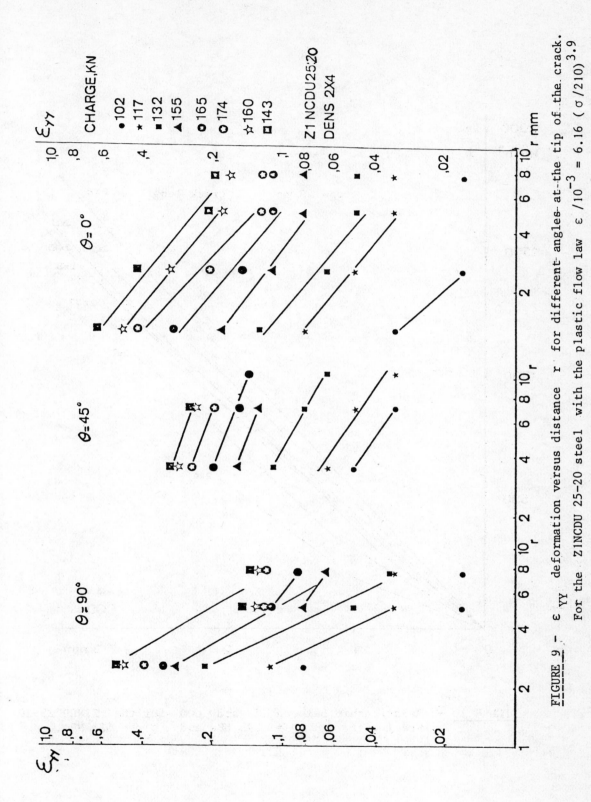

FIGURE 9 - ε_{YY} deformation versus distance r for different angles at the tip of the crack. For the Z1NCDU 25-20 steel with the plastic flow law $\varepsilon/10^{-3} = 6.16 \, (\sigma/210)^{3.9}$ with σ in MPa.

CHARGE, KN

- ● 102
- ★ 117
- ■ 132
- ▲ 155
- ◉ 165
- ○ 174
- ☆ 160
- ▣ 143

Z1 NCDU 25-20

DENS 2X4

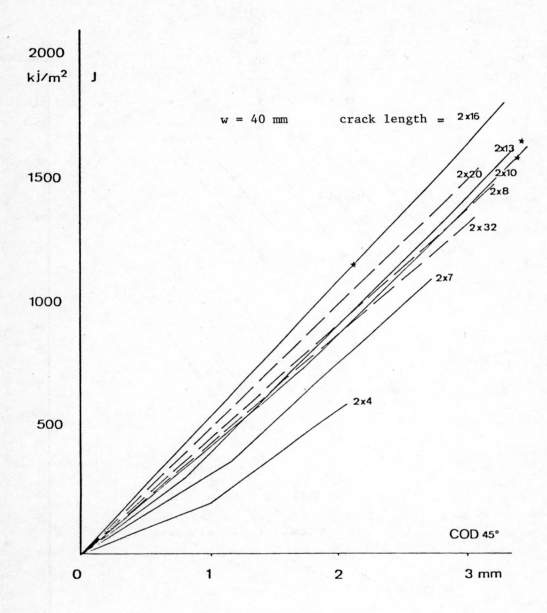

FIGURE 10 - Relationship between J and COD for the Z1NCDU25-20 Steel with $\sigma_{Y0,2}$ = 275 MPa and σ_u = 631 MPa.

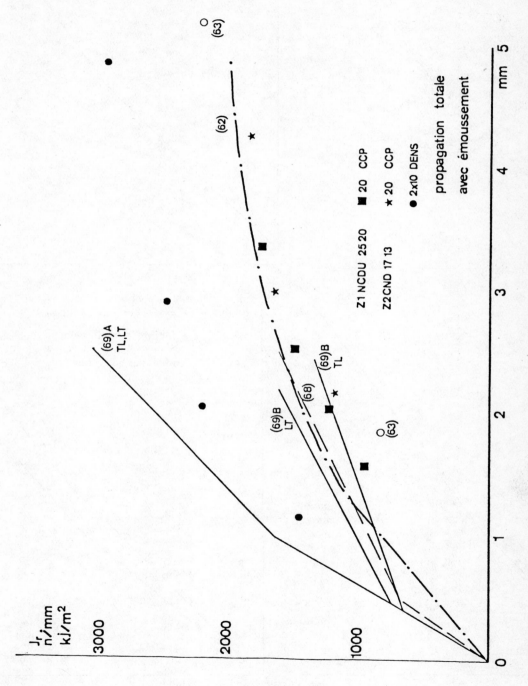

FIGURE 11 - Comparison of our results and the results from the litterature : (62) TANAKA, HARRISON ; (63) BAMFORD, BUSH ; (68) HODKINSON ; (69) BALLADON, HERITIER, RABBE.

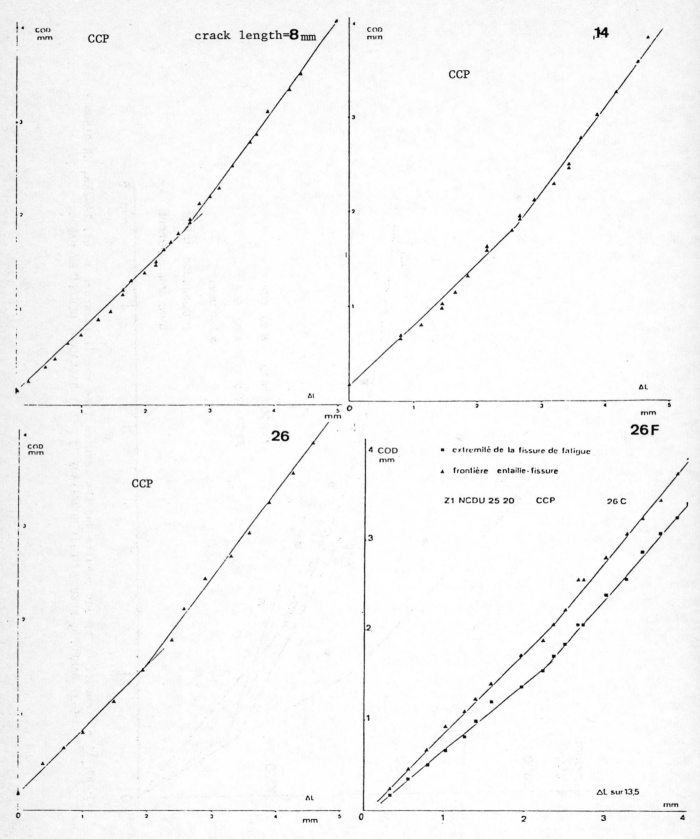

<u>FIGURE 12</u> - COD versus elongation of the 13.5 mm gauge length and determination of initiation - for the Z1NCDU 25-20 steel.

- 158 -

EXPERIMENTAL ASPECTS OF UNLOADING-COMPLIANCE J-TESTS

M.I. DE VRIES

Ductile crack growth resistance J_R-curves characterize stable crack growth in terms of the elastic-plastic parameter J and the corresponding crack extension Δa. The unloading compliance test technique provides a series of crack length data from a sequence of measurements of the specimen compliance during stable crack growth. With this method a "continuous" R-curve can be measured from one single test specimen.

Compact-tension (1T-CT) specimens of stainless steel Type 304 and A542 steel have been tested to evaluate the validity of the UC-technique for materials with different ductility. The stainless steel is a very ductile, high strain-hardening material whereas the A542 steel is a material with moderate ductility and limited strain-hardening, as can be seen from the tensile curves in Fig. 1.

The "blunting-line procedure" (1),

$$J = 2 \sigma_y \Delta a \quad \text{with} \quad \sigma_y = \frac{\sigma_{0.2} + \text{UTS}}{2},$$

uses the effective yield strength σ_y to estimate apparent crack growth due to crack tip blunting. However, comparison of the effective yield strength values from Table 1, in terms of tensile flow stresses, shows that both values represent quite different amounts of tensile deformation. The value of 610 MPa for A542 represents the tensile flow stress at about 3% strain, being 40% of the uniform elongation. For Type 304 the effective yield strength value of 400 MPa represents the flow stress at 9% strain (20% of the uniform elongation).

Typical characteristics of displacement controlled stable crack growth tests of the two materials are compared in Table 2. The inaccuracy of the Δa-data from the compliance measurements of the stainless steel specimens is due to the extended measuring-range of the displacement gages, the large rotation and the contraction at the crack tip. The scatter of the compliance data is mainly due to additional movements of the displacement gages caused by the rotation of the stainless steel specimen. There are no such disadvantages for the tests of the A542 specimens. The UC-method, being strongly affected by the ductility of the material, is a suitable technique for the less ductile A542 material.

In Fig. 2a the load-displacement curves for A542 specimens with and without side-grooves are shown. In Fig. 2b the relative position of the curves, normalized with respect to the net thickness of the specimen, is reversed. The resistance curves are given in Fig. 3, showing non-linearity between the exclusion lines and a lower R-curve for the side-grooved specimen. Loss has based his analyses on the non-linearity, excluding crack tunneling (2). Non side-grooved specimens exhibit steeply rising R-curves and crack tunneling during the later stages of crack extension. Crack fronts of side-grooved specimens are relatively straight, Fig. 4.

The unloading compliance method predicts within few percent accuracy the real crack extension of side-grooved specimens. However, crack tunneling causes significant underestimation of the crack length for non side-grooved specimens. In Fig. 5 the Δa-values are plotted versus the load-line displacement. There is an excellent agreement between the predicted and the 7 mm real crack extension measured as an average of 25 points on the fracture surface of the side-grooved specimen (Fig. 5a). For the non side-grooved specimen an underestimation of about 40% of the real final crack extension of 2.6 mm is shown in Fig. 5b.

For ductile crack growth a stabilized condition at the crack front after a small amount of early crack extension is indicated by the linear part of the $\Delta a - V_{LL}$ curve. The initial non-linear part represents crack initiation and the development of a "steady state" shape of the crack front prior to "stable" crack growth. Local crack initiation and crack extension in the middle of the specimen cause differences between the first parts of the curves of side-grooved and non side-grooved specimens. Fig. 5 shows an initial non-linear stage of about 0.2 mm for the non side-grooved specimen and about 1.0 mm for the side-grooved specimen. The effect of inhomogeneous, local crack initiation on Δa as averaged over the specimen thickness is demonstrated in the schematic presentation in Fig. 6 for 0.2 mm and 1.0 mm crack extension.

To avoid significant underestimation of the crack length from crack tunneling effects, side-grooving is a necessary condition for measuring continuous R-curves with the unloading compliance technique. The crack growth phenomenology indicates that it might be useful to distinguish between an "initiation" stage and a "stable" (constant CTOA, ref. 3) crack growth stage. As a consequence critical J-values, which have physical meaning, should be measured accordingly from different parts of the R-curve. For instance the physical onset of "stable" crack growth should be determined from the linear part of the Δa-V_{LL} curve. In this way the "blunting-line procedure" can be avoided. A "technical" J-value for crack initiation, measured at a distinct Δa value, can be related (e.g. 10%) to the crack length at the physical onset of "stable" crack growth.

(1) Annual Book of ASTM Standards, Part 10, 1981, E183-81, Standard Test for J_{1C}, A Measure of Fracture Toughness.
(2) Loss, F.J., in Effects of Radiation on Materials, ASTM STP 725, 1979, pp. 278-294.
(3) De Koning, A.U., in Proceedings ICF4, Waterloo 1977, Vol. 3, pp 25-31.

Table 1.

Tensile properties of the two materials, Type 304 and A542.

		Type 304	A542
0.2-Yield stress	(MPa)	230	550
Ultimate Tensile Strength	(MPa)	590	670
Total Elongation	(%)	65	15
Uniform Elongation	(%)	50	8
Effective Yield Strength	(MPa)	400	610

Table 2.

Comparison of the UC-test characteristics of J-tests on ductile (Type 304) and less ductile (A542) material, non side-grooved specimens.

	Ductile Type 304	Less ductile A542
Displacement at max. load	7 mm	1 mm
Contraction	up to 20%	up to 2%
Rotation	much	less
Crack initiation	local	uniform
Crack front	straight	curved
Compliance accuracy	low	high
Scatter	much	less
UC-technique	not suitable	suitable

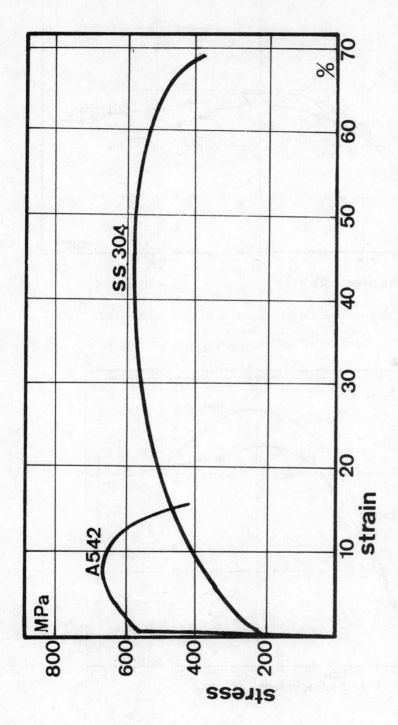

Fig. 1. Tensile curves of the two materials, Type 304 and A542.

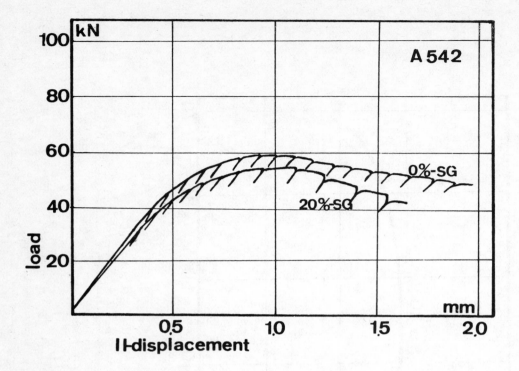

Fig. 2a. Load-displacement diagrams from UC-tests of 20% side-grooved
and non side-grooved specimens (A542).

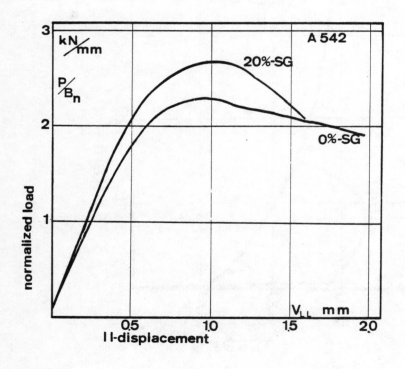

Fig. 2b. The load-displacement diagrams, normalized with respect to the
net thickness of the specimens.

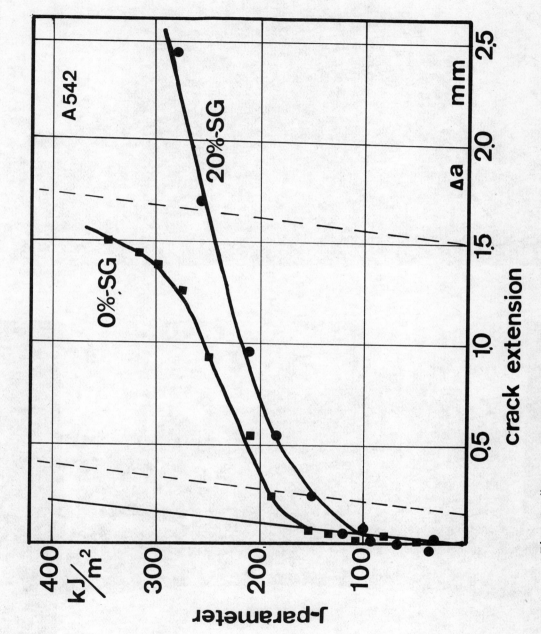

Fig. 3. "Continuous" R-curves from side-grooved and non side-grooved specimens (A542, 1T-CT).

0%-SG

20%-SG

Fig. 4. Fracture surfaces of stable crack growth tested 1T-CT specimens,
showing crack tunneling in the non side-grooved specimen.

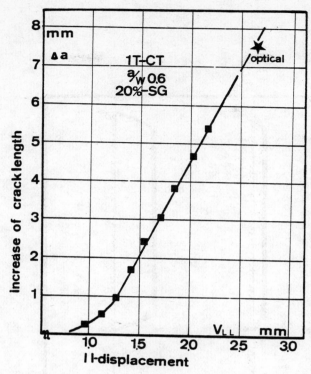

Fig. 5a. Curve of predicted crack extension Δa versus the load-line
displacement for 1T-CT specimen with 20% side-grooves, showing
good correspondence with the optical measured real crack extension.

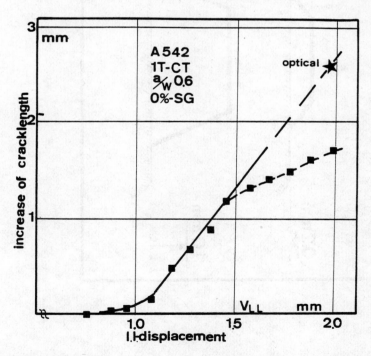

Fig. 5b. The Δa-V_{LL} curve for 1T-CT specimen of A542 without side-grooves,
showing underestimation of the final crack extension.

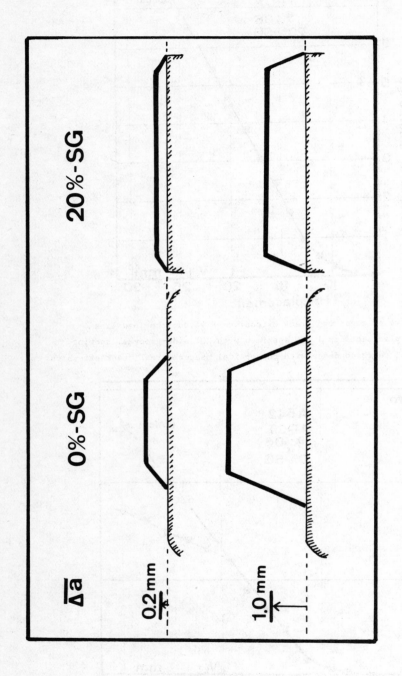

Fig. 6. Schematic presentation of the crack fronts of side-grooved and non side-grooved specimens at 0.2 mm and 1.0 mm crack extension.

SOME FINITE ELEMENT RESULTS OF CTS SPECIMEN

by

J. Prij
Stichting Energieonderzoek Centrum Nederland

1. INTRODUCTION

In fracture mechanics tests with a monotonic loading such as a J test
one often observes that the crackgrowth in the specimen is not uniform
along the crackfront. These observations can only be understood if the
three dimensional character of the specimen is taken into account.
In this paper some results are presented of three dimensional finite
element analyses of CTS viz. elastic-plastic analysis of a CTS with
$W = 40$ mm and $a_0 = 27.5$ mm and 28.6 mm and thickness $B = 25.4$ mm
and an elastic analysis of a CTS with a curved crackfront.
Experimental aspects and observations are described in a paper of
M.I. de Vries of this workshop [7]

2. NUMERICAL MODELS AND METHODS

A typical element lay-out of a symmetry-quarter of a CTS is shown in fig. 1.
In this model 20 node brick elements are used. The elements along the
crackfront are collapsed into wedges. All nodes at the cracktips have
different degrees of freedom being capable to describe crackblunting.
It has been shown that this modelling gives very good results for elastic-
plastic analyses [1]. The pinhole is not taken into account, which does
not lead to serious errors as is stated by several authors [2,6].
The elastic-plastic response is calculated with the MARC program [3] using:
small strain theory, Von Mises type of equivalent stress, an associated
flow rule, isotropic hardening and a tangent modulus solution technique
with a residual load correction. The elastic-plastic properties are given
in Table I. Energy releases are calculated by means of a virtual crack
extension. Local virtual crack extension δA_j of a portion of the crackfront
around nodal point j leads to an energy release δU_j. The local energy
release rate J_j is calculated on basis of the assumption that J is locally
constant:

$$J_j = 2 \frac{\delta U_j}{\delta A_j} \tag{1}$$

This value of J is supposed to be the local value of J in nodal point j.
The global value of J is calculated from the energy release δU due to
a virtual extension δa of the total crackfront, thus leading to:

$$J = 2 \frac{\delta U}{B \delta a} \qquad (2)$$

This global J value can also be calculated in accordance to the experimental practice where J is taken from the measured load-displacement curve.

$$J = \frac{A}{Bb} f\left(\frac{a_0}{W}\right) \qquad (3)$$

where: A   : the area under the load-displacement curve, here
              the calculated curve

       B   : width of the specimen

       b_a : remaining ligament

       $f\left(\frac{a_0}{W}\right)$: tabulated function from [4].

From the local values J_j a mean value $\overline{J}$ can be derived:

$$\overline{J} = \frac{\sum_j J_j \, \delta A_j}{\delta A} = \frac{\sum_j \delta U_j}{\delta A} \qquad (4)$$

TABLE I   Elastic-plastic properties

σ_{yield} = 534 MPa; E = 2.0376 10^5 MPa; ν = .3

Hardening modulus:

$T = \frac{\Delta\sigma}{\Delta\varepsilon^{pl}}$ [MPa] $\begin{cases} 5.1943 \ 10^4 & 0 < \varepsilon^{pl} \leq .11\% \\ 2.9985 \ 10^3 & \varepsilon^{pl} > .11\% \end{cases}$

3. RESULTS

3.1. Elastic-plastic analysis of CTS with a straight crackfront

In fig. 2 the calculated load versus load point displacement curves are given for a number of numerical models while an experimental curve is given for comparison.

It can be seen quite clear that the two-dimensional approximations plane stress and plane strain provide a lower and an upper bound for the 3D-solution. The 3D solution gives a good description of the real behaviour as long as the load level is low.

At higher loads the crack will grow and this crackgrowth is not taken into account in the analysis. Some results of the J evaluations are given in fig. 3. The dots represent J values according to eq. (3).

The solid curve of $\bar{J}$ according to eq. (4). From the numerical results it could be observed that $\delta U = \sum_j \delta U_j$ or using eqs. (2) and (4): $J = \bar{J}$.

The variation of J along the crackfront is given in fig. 4.
It can be seen that for increment 7 (max. load level where $P \simeq 70$ kN) the max. value in the middle is about 20% higher than the mean value. Combining the result of fig. 3 this means also that the maximum value along the crackfront is about 20% higher than the global J determined in accordance with the experimental practice.

3.2. Elastic analysis of CTS with a curved crackfront

A 3D elastic analysis has been performed of the geometry given in fig. 5. This geometry is modelled with 4 elements across half the thickness. These 4 elements have thickness ratio's 1:2:3:4.
Some energy release rate results are given in fig. 6.
The meaning of the symbols is:

$J(a)$: J from eq. (1) for the curved crackfront
$\bar{J}(a)$: $\bar{J}$ from eq. (4) for the curved crackfront
$\bar{J}(\bar{a})$: $\bar{J}$ from eq. (4) for a straight crackfront with a mean cracklength $\bar{a}$.

The results for the compliance $\dfrac{EBv_{LL}}{P}$ are given in the following table.

METHOD	straight crackfront a/W = .55	curved crackfront $\bar{a}$/W = .6015
3D elastic analysis	47.02	57.27
Standard form [5]	48.04	64.06

Based on the compliances it is common practice to estimate the crackgrowth. For this curved crack this procedure leads to the following result:

- 172 -

Compliance [5] $\Rightarrow$	a/W $\Rightarrow$	a [mm]
47.02	.5460	39.80
57.27	.5819	41.60

This means that the estimate for the crackgrowth Δa = 1.8 mm. Realising that the real Δa = 2.5 mm it must be concluded that the method of the partial unloading compliance gives an underestimation of about 40% in this case.

This is also confirmed by experiments [7].

In fig. 7 the stress distribution at the crackfront is given. As a measure of stress the equivalent stress of Von Mises is given.

It can be seen that the distribution of σ_{eq} along the crackfront is almost similar to the distribution of J. Based on this distribution one can imagine that a complex state of residual stresses due to unloading or plastic deformation will develop as a function of the hardening behaviour of the material.

4. SUMMARY AND DISCUSSION

Numerical results are presented of some three dimensional finite element analyses of compact tension testspecimen. A comparison has been made with elastic-plastic results and the experimentally determined loaddeflection behaviour. It can be concluded that the two-dimensional approximations, viz. plane stress and plane strain, give a lower and an upper bound for the real behaviour. The 3D-analysis gives good correspondence with the experiments as long as the crack does not grow. Energy release rates have been calculated using a virtual crack extension technique. Global J values derived from a virtual extension of the complete crackfront are compared with J values derived from the calculated load deflection curve.

It is demonstrated that the difference between both results is only marginal. On the other hand it is demonstrated that the global J value is the mean value of the local J values along the crackfront. Based on this observation and the calculated substantial variation of J along the crackfront it is concluded that the experimentally determined J value from the loaddeflection curve is also a global value and will be able to characterize a global behaviour and not a local initiation.

Based on the variation of the local J values it seems to be logical that the crack starts to grow in the middle of the specimen where the J value is maximal. After this local crackgrowth some local unloading and stress redistribution will occur leading to new crackfront which is stable as long as the load is not too high. This "final" shape does not depend solely on J or the variation of J but also on the stress redistribution and therefore on the hardening behaviour of the material. This is in accordance with the experimental practice where some materials show cracktunneling and others don't.

The influence of the strong curvature of a tunneled crack on the J and the compliance is studied further. As a first step a three-dimensional elastic analysis of a CTS having a curved crackfront due to cracktunneling during crackgrowth is presented.

Due to the fact that in the specimen crackgrowth always is preceded by plastic deformation and residual stresses this elastic results must be handled with care. Local J values have been calculated using the virtual crack extension technique. A comparison has been made between the mean value of the local J values, being the global J value for this curves crack, and the global J value of a crack with an average length and a straight crackfront.

It could be concluded that the differences are not substantial. Having calculated the local J values it is possible to give an estimation of the crackgrowth behaviour of the crack if one assumes that the local crackgrowth is an increasing function of the local J values.

As a result the crack grows at the location with the highest J. Based on these results it is not possible to say something with respect to the direction of crackgrowth, it might be possible to use a strain energy density concept for this purpose. Nevertheless as a result of the crack-growth assumption the crack grows to a shape with a less pronounced curvature. This is also observed in the laboratory if the specimen is unloaded and loaded again in a fatigue mode. If one, however, does not unload the specimen but continues the loading by means of a prescribed displacement the crackgrowth is different. For some materials the crack tunnels further while other materials show a crackgrowth leading to a less pronounced curvature. This is in correspondence with the earlier observation that some materials show cracktunneling and others don't.

In the experimental practice the unloading compliance is used to estimate the amount of crackgrowth. In this method the compliance of the specimen having a curved crackfront is compared with the tabulated compliances of specimen having straight crackfronts. The accuracy of this method is not very high if a pronounced crackfront curvature is present. Based on the calculated compliance of the tunneled crack it is shown that the real crackgrowth is 40% higher than the crackgrowth based on the difference in compliance.

5. REFERENCES

[1] Prij, J.
Two- and three-dimensional elasto-plastic finite element analyses of a CTS specimen. Paper presented at the IAEA Spec. Meeting on "Reliability Engineering and Life-time Assessment of Primary Components, 1-3 December 1980.

[2] Redmer, J., Dahl, W.
Two- and three-dimensional elasto-plastic FEM calculations of CT-specimen.
In: Numerical Methods in Fracture Mechanics. Ed. by D. Owen and A. Luxmore. Pineridge Press, Swansea, (1980).

[3] MARC General Purpose Finite Element Analysis Program.
Vol. I-V (1979).

[4] Clarke, G.A., Landes, J.D.
Evaluation of J for the compact specimen.
JTEVA, Vol. 7, No. 5, Sept. 1979, pp. 264-269.

[5] Saxena, A., Hudak, S.J., Jr.
Review and extension of compliance information for common crackgrowth specimen.
Int. J. of Fracture, Vol. 14, No. 5, October 1978.

[6] Riedel, H. et al.
Comparison of measured and calculated J-integral and COD-values.
Spec. Meeting on Elasto-plastic Fracture Mechanics, Daresbury, UK,
22-24 May, 1978.

[7] De Vries, M.I.
Experimental aspects of unloading compliance J-tests.
Paper presented at the CSNI workshop on test methods for ductile
fracture, Paris, 1-3 December 1982.

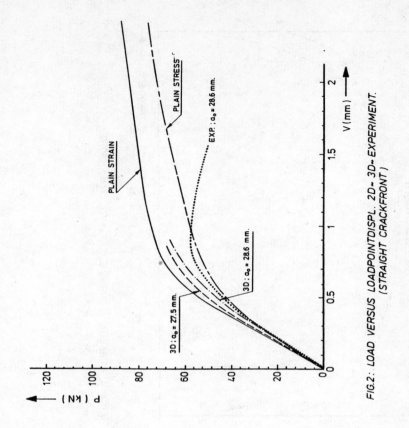

FIG.2: LOAD VERSUS LOADPOINTDISPL. 2D- 3D- EXPERIMENT.
(STRAIGHT CRACKFRONT)

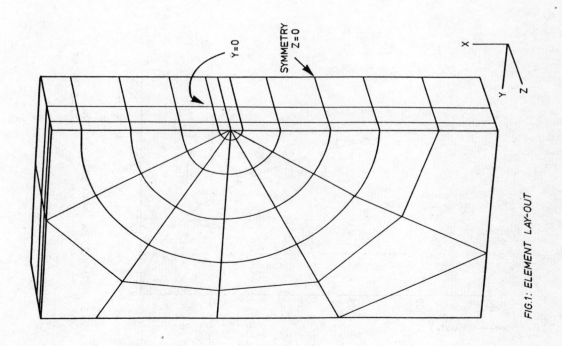

FIG.1: ELEMENT LAY-OUT

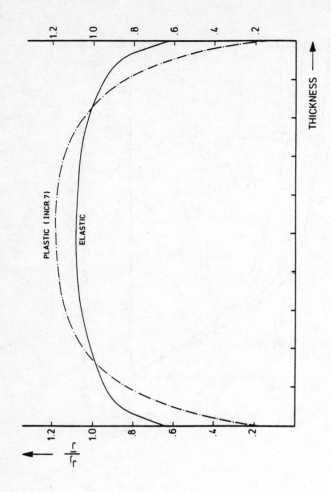

FIG.4: J VARIATION ALONG THE STRAIGHT CRACKFRONT.

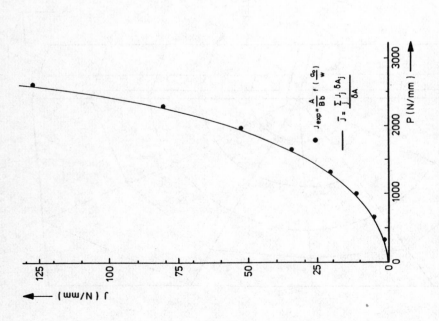

FIG.3: METHODS FOR J EVALUATION (STRAIGHT CRACKFRONT)

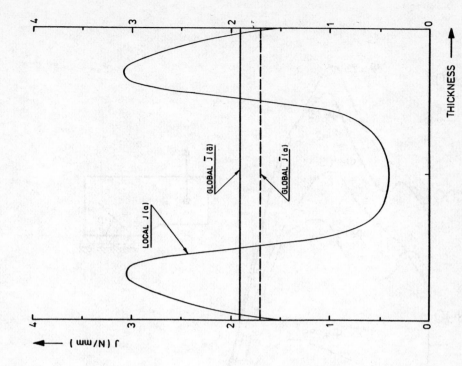

FIG.6: ELASTIC DISTRIBUTION OF $J(a)$, $\bar{J}(a)$, $\bar{J}(\bar{a})$.

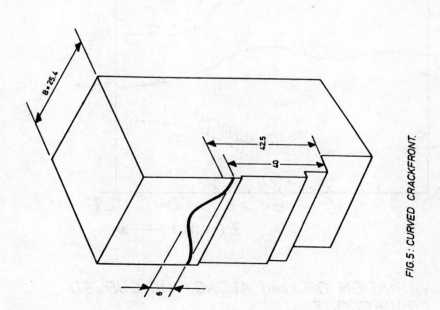

FIG.5: CURVED CRACKFRONT.

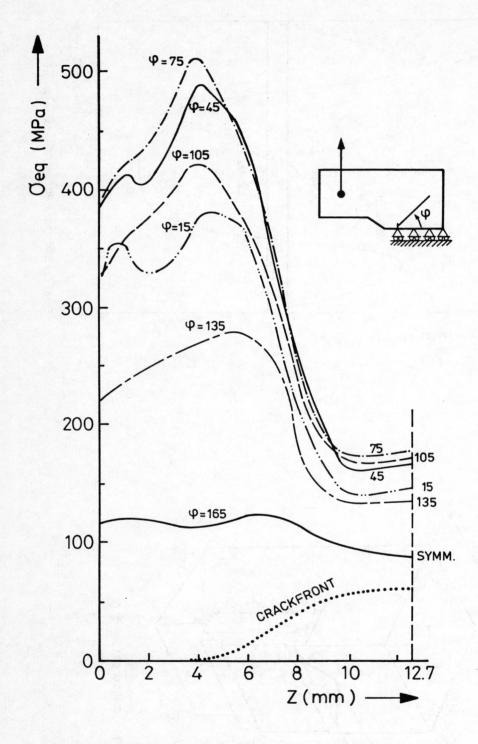

FIG.7: VARIATION OF σ_{eq} ALONG THE CURVED CRACKFRONT

- 180 -

EXPERIENCE WITH DUCTILE CRACK GROWTH
MEASUREMENTS APPLYING THE DC-PD TECH-
NIQUE TO COMPACT TENSION FRACTURE SPECIMENS

By

C.P. Debel and F. Adrian
Metallurgy Department
Risø National Laboratory
DK-4000 Roskilde, Denmark

EXPERIENCE WITH DUCTILE CRACK GROWTH MEASUREMENTS
APPLYING THE DC-PD TECHNIQUE TO CT SPECIMENS

1. INTRODUCTION

The objective of this paper is to present the experience obtained at Risø in applying the direct current potential difference (DC-PD) technique to the indication of ductile crack
initiation as well as to the quantitative measurement of the
ductile tearing itself. The technique is used in connection
with single specimen J_R-curve tests, employing compact tension
(CT) specimens which may or may not contain side-grooves. The
technique is based on the establishment of reference-curves,
relating the change in electrical potential to the increase
in ductile crack length in the CT-specimen.

2. EXPERIMENTAL

So far the technique has been applied to three steels: ASTM
A533B, BS1501-281A and BS4360-Gr50D. The chemical composition
and the tensile as well as CVN properties of these steels are
given in tables 1a and 1b. It is noted that the two former
steels are HSLA-steels, whereas the latter is a conventional
micro-alloyed and normalised structural steel.

All specimen were 25 mm thick (1TCT) with either no side-
grooves (0% SG) or 2.5 mm deep grooves on both sides (20% SG),
Figure 1a. The notch geometry always conforms to the ASTM configuration, [1], but nominal dimensions were changed slightly
in order to improve overall performance, [2]. The specimens
were tested in both displacement and load control, with a maximum displacement rate of about 13 mm/min., and all tests were
performed at ambient temperature.

The DC-PD circuit employed was very simple and similar to circuits often seen in the literature, [3]. No filter was used,
however, but the overall electrical noise level of the laboratory was carefully minimised and stabilised prior to the testing. The current leads (Cu) were soldered to the back face
of the specimens and the PD-wires (Fe) were spark-welded to

the lower step of the notch, against opposite faces, Figure
1a. Insulating members were introduced in the load chain, above
and below the clevises, and the clip-gauge employed incorporated
insulation in its spacer block, Figure 1b. A current level
of 50A was used in all tests.

3. ANALYSIS AND RESULTS

During each test, diagrams representing load P versus load
point displacement δ and potential difference PD versus δ were
recorded on an XYY-recorder, Figure 2, and some specimens were
tested to various smaller displacements in order to yield various values of PD and crack growth at the end of the test.
Previous to the test the PD circuit is stabilised (thermo-ef-
fects eliminated) by subjecting the specimen to the full current for several minutes, whereupon the potential difference
V_i is recorded.

During loading the PD initially increases rapidly with an amount
ΔV_i due to the opening of the fatigue crack, Figure 2, and
the sum of V_i and ΔV_i constitutes the potential difference
V_o, which is related to the fatigue crack length a_o. Subsequently crack tip blunting occurs, followed by crack growth, but
the PD-trace only displays an ever increasing curve with no
indication of the shift between these two processes. Consequently, the indication of crack initiation is based on a back extrapolation of end-of-test data relating Δa to the increase in
potential difference ΔV after the opening of the fatigue crack.

After a conventional heat-tinting and fracturing procedure,
the fatigue crack length a_o was measured employing an X-Y table
of high resolution, and a_o was computed as the mean of at least
7 measurements across the surface. The amount of ductile crack
growth Δa was established by measuring the total area of ductile tearing - excluding the region of crack tip blunting as
well as possible - on enlarged photographs of the fracture
surface using a semi-automatic picture analyser.

In Figures 3a and 3b values of V_o have been shown against

- 183 -

a_o for 0% SG and 20% SG specimens, respectively, and in Figures 4a and 4b the quantitative relationship between the normalised potential change $\Delta V/V_O$ and the ductile crack growth Δa is presented, again for 0% SG and 20% SG specimens, respectively. A straight line has been fitted to all of these four set of data, and the regression coefficients obtained are presented in Table 2. Furthermore, various limiting lines were introduced in Figures 3 and 4 in order to illustrate the degree of reproducebility obtained.

4. DISCUSSION AND CONCLUSION

From figures 3a and b it is seen that the reproducebility of the relation:

$$V_0(a_0) = A \cdot a_0 + B$$

is of the order of ±1 mm, and that a higher PD is measured on 20% SG specimens since the current density here is higher in the plane of fracture due to the side-grooves. It should be noted that these relations could be applied to on-line computer-systems, allowing an evaluation of the J_R-curve in the course of the test itself, since a_o can be obtained at the start of test. However, since absolut values are related, the figures are only applicable to specimens with an input/output configuration as shown in Figure 1a, and only for I = 50A.

The crack growth relations:

$$\frac{\Delta V}{V_0}(\Delta a) = A \cdot \Delta a + B$$

shown in Figures 4a and 4b are used to give both the onset of ductile crack extension, when:

$$\Delta V = B \cdot V_0$$

as well as a quantitative measure of the crack extension itself, by:

$$\Delta a = \frac{1}{A}(\Delta V/V_0 - B)$$

Overall reproducebility is of the order of ±0.2 mm and the relations are valid over a considerable amount of crack extension,
Figure 4. Since a relative measure of PD is considered the
latter relations probably are valid for any current level, but
a consistent input/output configuration is still important.

The above equation also is well-suited for on-line computer
application, as Δa can be calculated at, say, every 1/10 mm
of specimen displacement. If a simultaneous calculation of J
is performed at similar stages a full J_R-curve can be constructed. A typical example is shown in Figure 5. It appears
that the J_R-curve has a curved shape, and that no blunting line
is applied in determining the value of J at initiation, since
blunting was excluded in the analysis of the fracture surfaces.
The value of J_{IC} is, however, not readily obtainable from this
curve.

On the topic of the scatter in the data it should in conclusion
be recognized, that while some of the scatter seen in Figures
3 and 4 may be due to the equipment as well as the technique
used, most of the scatter probably orginates in the fact that
every data-point represents the result of the test of one specimen. Consequently, the performance of the equipment may easily
be superior to the results shown in Figures 3 and 4 - see for
example the PD-trace in Figure 2 - but the overall accuracy
or reproducebility may be governed more by the uniformity in
shape of the fatigue crack as well as the ductile tearing in
the different specimens than by the properties of the equipment.

5. REFERENCES

[1]   ASTM E 813-81: "Standard Test for J_{IC}, a Measure of Frac-
       ture Toughness", American Society for Testing and Ma-
       terials, 1981 Annual Book of ASTM Standards, Part 10,
       pp. 810-828.

[2]   Debel, C.P.: Report on the DC-PD technique, to be pub-
       lished as a Risø-M report.

[3]   Hollstein, T.; Voss, B.; Blauel, J.G.:
       "Comparison of Different Methods for J_R-Curve Determina-
       tion". Present proceedings.

Table 1a:

W(%) Steels	C	Mn	Si	Ni	Cr	Mo	V	Nb	Al	S	P
ASTM A533B Cl 1 (base plate)	0.20	1.35	0.23	0.61	0.13	0.50	/	/	.015	.005	.006
BS1501-281A	0.14	1.21	0.29	0.74	0.64	0.23	0.09	<.005	.026	<.005	.013
BS4360-Gr.50D	0.17	1.42	0.29	0.23	0.21	0.04	<.01	<.005	.027	<.005	.007

Table 1a: Chemical composition of the steels employed (base plates only).

Table 1b:

Properties Steels	σ_Y (MPa)	σ_{UTS} (MPa)	δ %	Area Reduc. %	CVN (Nm at 20°C)
ASTM A533B Cl 1 (base plate)	425	583	30	74	139 (at 3°C)
BS 1501-281A	451	601	31	71	210
BS4360-Gr.50D	354	550	35	71	275

Table 1b: Tensile and upper shelf CVN properties.

Relation	% SG	No. of datapoints	A	B	R^2
$V_O(a_O)$	0	41	41.89	-10.59	0.908
	20	68	40.82	+362.33	0.834
$\frac{\Delta V}{V_O}(\Delta a)$	0	31	0.041	0.0146	0.983
	20	63	0.029	0.0132	0.986

Table 2: Coefficients of the linear regression $y = Ax + B$ of the relations plotted in Figures 3 and 4.

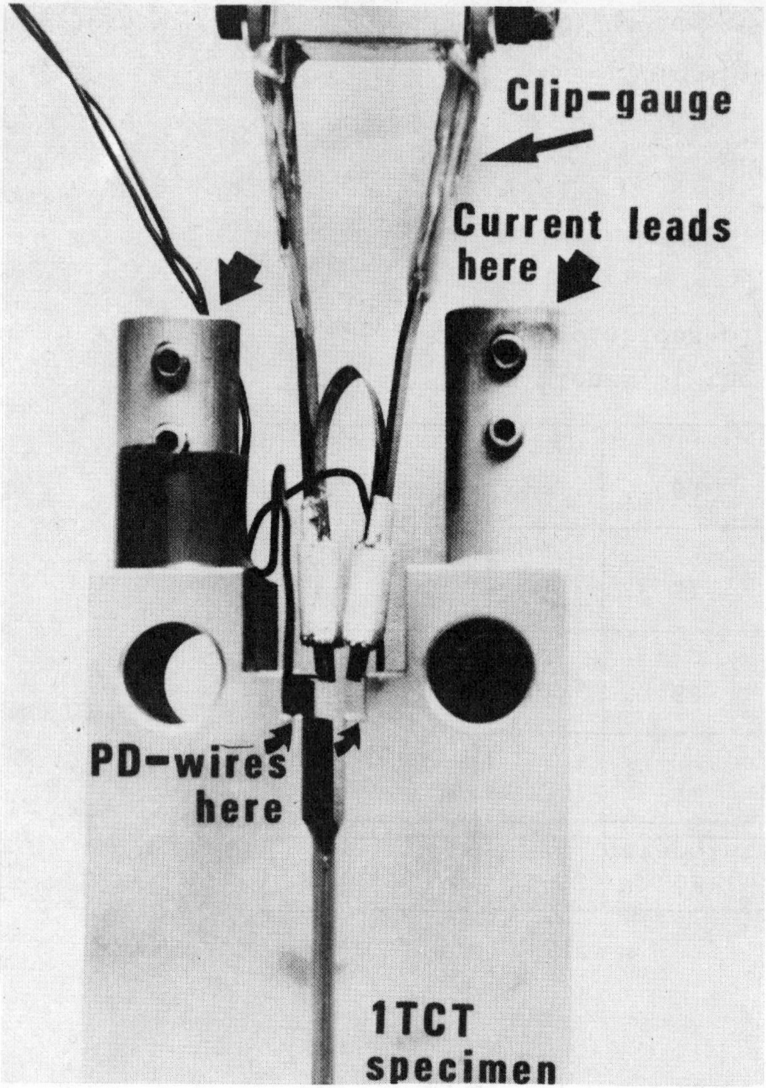

Figure 1b: Specimen mounted between clevises, with areas of insulation identified. No pin-insulation is used.

Figure 1a: Fully instrumented 1TCT specimen, with clip-gauge, current leads and PD-wires. 2o% side-grooves.

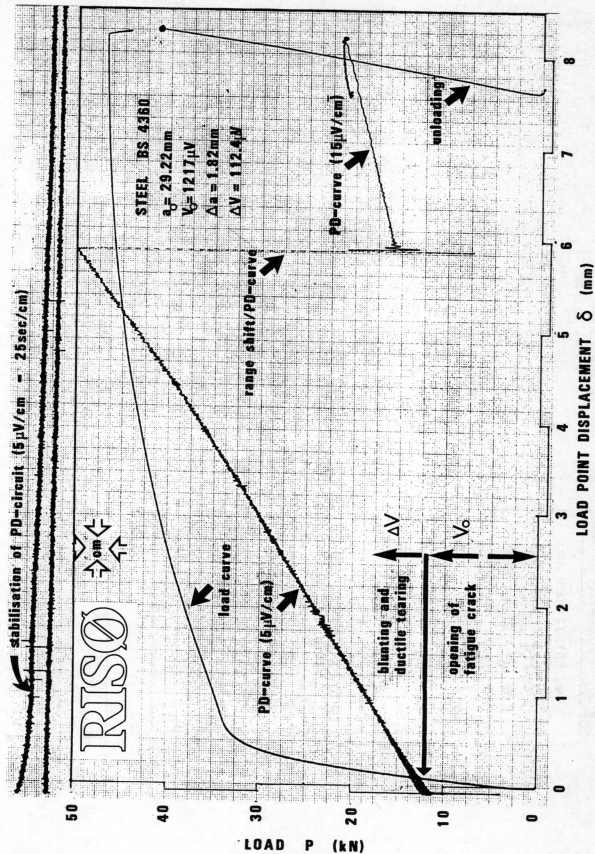

Figure 2: XYY-record of load P and pot.diff. PD versus load point displacement δ. (Spec: 2A52-3cht4, load control) Note stabilizing trace above.

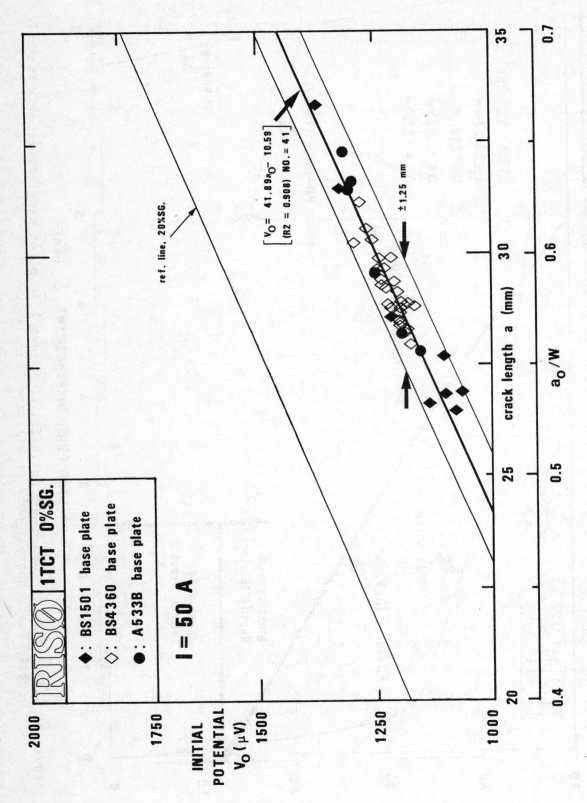

Figure 3a: Relationship between initial potential V_0 (including opening PD-change due to the opening of the fatigue crack) and the fatigue crack length a_0 for plane-faced 1TCT.

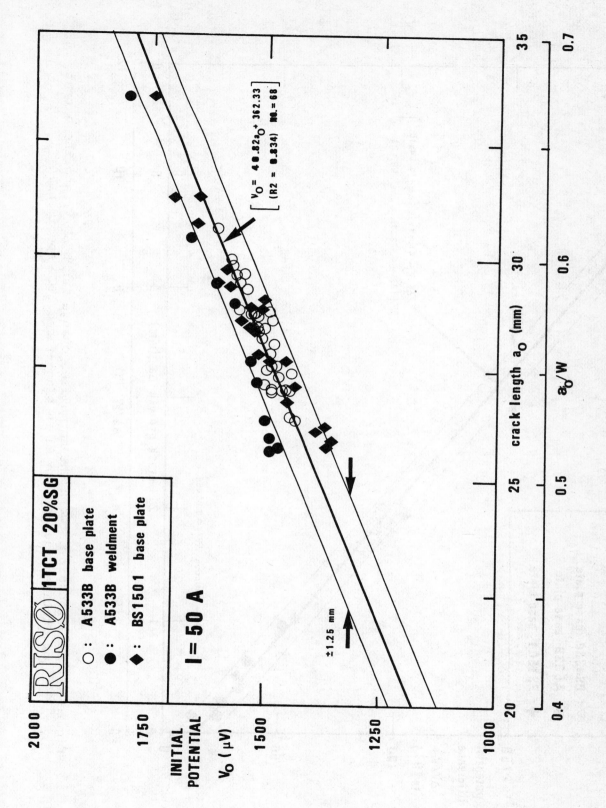

Figure 3b: as Figure 3a, but covering 20% side-grooved specimens.

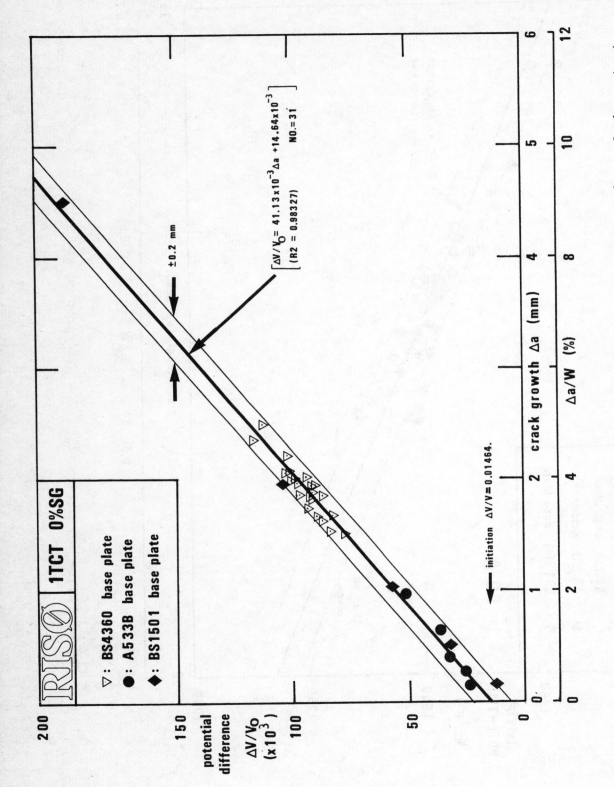

Figure 4a: Relationship between the normalised change in PD $\Delta V/V_0$ and the extent of ductile crack growth Δa in plane-faced 1TCT specimens.

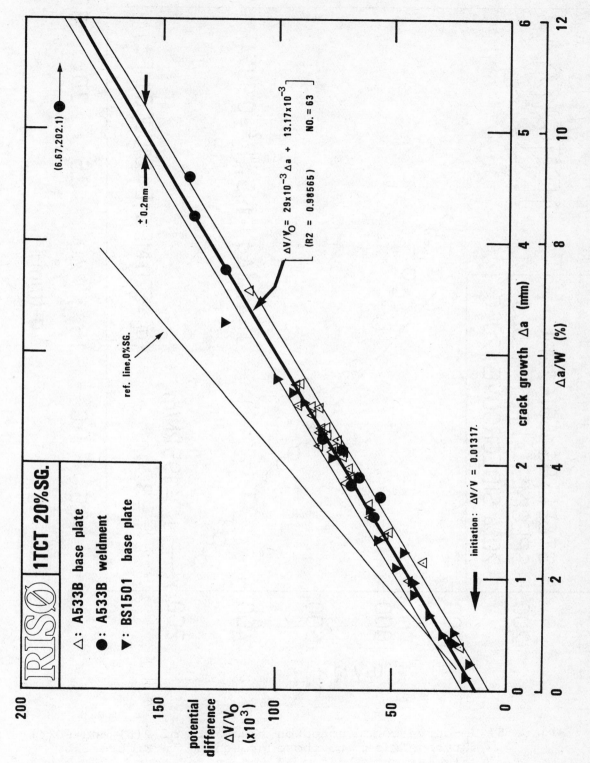

Figure 4b: As Figure 4a, but covering 20% side-grooved 1TCT specimens.

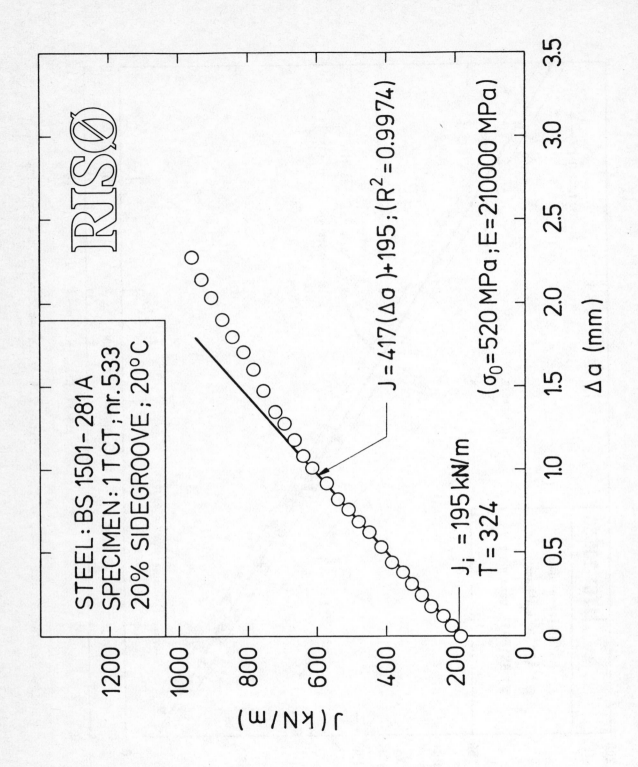

Figure 5: J_R-curve constructed on the basis of $P(\delta)$ and $PD(\delta)$ curves similar to those presented in Figure 2 by calculating values of J and Δa for each 1/10 mm specimen displacement. No blunting line is employed since blunting is excluded in the analysis of the fracture surfaces, and J_{Ic} can not be assesed. However, an initiation value J_i is obtained.

Session III

EXPERIENCE AND PROBLEMS WITH EXISTING TECHNIQUES

Chairman - Président

J.D. LANDES

(United States)

Séance III

METHODES COURANTES : EXPERIENCE ET DIFFICULTES

SESSION NO. 3: EXPERIENCE AND PROBLEMS WITH
EXISTING TECHNIQUES

Summary by the Chairman: J. D. Landes

Present experience with the test methods for ductile fracture
toughness show that there is a problem in developing reproducible
results between the various laboratories. The test methods being used
may not be explicit enough to guarantee that the exact test procedure
is being duplicated in different laboratories. It is difficult to
separate differences due to procedure from those due to material
variability. The test method for measuring an initiation value of
toughness (ASTM E 813-81) is judged to be unsuitable by the European
laboratories. Several American laboratories expressed more interest
in using an R curve approach to toughness evaluation for structural
application than in using an initiation value for this purpose. The
present proposed ASTM J_R curve method has not had enough exposure to
be adequately assessed. Nevertheless, many representatives at the
Workshop felt that they would need to use an initiation value for
design and therefore need to have an improved standard for initiation
testing.

There was general agreement on how the testing should be
conducted. Most laboratories used a standard type ASTM specimen such
as the compact toughness specimen. During the test load, displacement
and ductile crack advance are measured. From this a curve of a
mechanical characterizing parameter versus crack extension is
developed (R curve). The mechanical parameter most frequently used is
J. Almost unanimously a single specimen technique with an automatic
crack measuring technique is preferred (although duplication is
suggested to characterize material scatter).   The two automatic
methods used almost universally are unloading compliance (partial
loading) or potential drop (both AC and DC).

The greatest difficulty comes from lack of agreement on how
to analyse the R curve result to determine an initiation value of
ductile toughness. The only point of agreement is that the present
ASTM E 813 is inadequate. Special problems and points for development
raised during this session were:

1. The choice of initial crack length in the unloading
   compliance method can shift the R curve; a clearly defined
   method for determining this initial crack length is not
   available.

2. The present ASTM E 813 definition of the blunting line does
   not always appear to be satisfactory.

3. There is sometimes negative crack growth in the unloading
   compliance method which appears to inluence the R curve.
   (This may be linked to the choice of initial crack length.)

4. System errors such as transducer errors and nonlinearities
   and errors in A/D conversions can cause significant errors in
   the R curve.

5. The E 813 linear fit to the R curve points does not appear to follow the actual trend; a nonlinear fit appears to be more appropriate.

6. The E 813 method for selecting an initiation point is not a good measure of actual initiation.

7. The current size requirements do not appear to ensure size independence of the R curve. The E 813 method does not acknowlege the benefit of side grooving in terms of size effect.

8. The four points required in E 813 are not always adequate to characterize the R curve trend. Points selected for small Δa can give a lower J_{Ic} value than the case where points are selected for longer Δa.

Progress is being made in that solutions are being found for many of the problems with the test methods. Some ways suggested to improve the testing and analysis are:

1. Better fixturing to reduce friction and plastic deformation in the load application device can improve unloading compliance results and may eliminate negative crack growth.

2. In the unloading compliance method the results are improved if the load is allowed to relax before unloading. This may be time consuming and may be replaced by unloading twice at one displacement point and using the second unload to determine the compliance or by fitting only the lower portion of the unload curve with a straight line.

3. A better method for fitting the R curve points may improve the definition of initiation J. Present fitting methods are reversed; actually J is the independent variable and Δa the dependent variable in the test and the points should be fitted in that manner. Something other than a linear fit should be considered. Many possibilities were presented.

4. The test method and analysis of results should be responsive to the needs of the application of these results. One test method may not satisfy the needs of all application methodologies.

Some additional summary comments representing the opinion of the session chairman are as follows:

1. Despite the many problems presented at this workshop there is much general agreement on the testing for ductile fracture toughness. This includes choice of characterising parameter, methods for measuring crack extesion and the use of an R curve approach to the analysis of results. This agreement represents significant progress since the early days of ductile fracture testing.

2. Present methods of analysing the test results may not give adequate information for the application of the results. For example, the characterisation of initiation toughness does not relate to the maximum load bearing capacity of a structure. As discussed during the session, the analysis of test results must be made with a knowledge of how these results will be used for the evaluation of structural components.

3. The problem concerning lack of reproducibility of results between various laboratories is important. However, this problem must be viewed in light of what is needed for application and what can attained in real materials. Toughness characterisation by the LEFM K_{Ic} method could routinely result in $\pm$ 15% scatter. Translated to a J parameter this is equivalent to $\pm$ 30% scatter: Round robin test programmes where various laboratories used a single material and test method showed scatter comparable to this. Therefore this scatter may represent a realistic limit to reproductibility. The extreme sensitivity of an apparent initiation toughness value to the method of data analysis would suggest that the application (e.g., the load displacement curve of the structure) may not be very sensitive to the scatter.

4. In conclusion, significant improvements can be made in the methods for testing and analysing results from an international group such as this. The discussion from this workshop should be continued through periodic meetings. An opportunity to infuence the work of ASTM on the development of ductile fracture toughness test standards can be made through participation in the ASTM E24 Committee meetings. Participants from this workshop should try to attend E24 meetings and express their needs and opinions for developing more acceptable testing standards.

THE INFLUENCE ON INITIATION TOUGHNESS AND J_R OF VARIABILITY

IN INITIAL CRACK LENGTH PREDICTION FROM

UNLOADING COMPLIANCE MEASUREMENTS

E Morland and T Ingham

RNPDL, UKAEA

To be presented at an OECD/CSNI Workshop on "Ductile fracture test methods"
Paris, December 1-3 1982.

SUMMARY

The paper presents some results obtained at RNL when using the unloading compliance method to develop J_R curves. The RNL system is not yet fully developed but results to date have shown that systematic variations in elastic compliance prior to physical crack extension can have a pronounced influence on estimated initiation toughness values. The results suggest that reliable estimates of initiation toughness will only be obtained if the system can be refined so that initial crack length can be predicted to within say $\pm$ 0.05mm.

The RNL Unloading Compliance System

A schematic of the system is shown in Fig 1. Tests are performed on either a servo-hydraulic or a servo-electric screw driven machine. The testing clevises contain flat-bottomed pin-holes. Both the clevises and loading pins are manufactured from maraging steel to minimise indentation effects. Under-sized pins are used to minimise friction.

Specimens are unloaded by 10-15% of the attained load and 40 load-displacement voltage pairs are recorded during the unloading portion of an unload-reload sequence. Maximum sensitivity on the load signal is obtained by calibrating to ensure that the maximum load anticipated in any test provides the full 10v output of the load cell. A DC offset device provides compatibility with the -5v/+5v voltage range of the 12 bit A/D converter. A conventional clip gauge is used to measure load-line displacement and two separate amplifiers are used to monitor (a) displacement throughout the test, and (b) displacement during an unloading sequence. The amplifier used in the unloading sequence provides 10v output for approximately 0.3mm load-line displacement and is re-zeroed prior to each unloading event. This procedure provides a capability for resolving displacements of $\sim 7 \times 10^{-5}$mm during an unloading event. Typical elastic displacements for a 10-15% unload are $\sim$ 0.05mm-0.2mm. For a 40mm compact with a/W = 0.6, a 0.1mm change in crack length is equivalent to a 0.8% change in compliance ie a change in displacement of $\sim 4 \times 10^{-4}$-1.6 $\times 10^{-3}$mm which is easily detectable using the system.

The 40 load/displacement data pairs are acquired at an effective rate of 2 pairs/ sec during an unloading event and data are stored on floppy discs for post-test analysis. Specimens are not held prior to unloading to allow relaxation effects to stabilise and some non-linearity can occur in the early stages of unloading. Hard copy output of each unloading line is obtained after the test and the operator defines the region over which a linear regression analysis will be performed. Typically 70-80% of the data are analysed and correlation coefficients of 0.99995, or better, are achieved.

Results

Tests have been made on a number of pressure vessel materials known, from multiple specimen tests, to exhibit significantly different J_R behaviour. Good agreement is obtained between measurements of resistance to crack growth from multiple and single specimen tests but a problem remains in defining initiation toughness/values from unloading compliance data due to variations in compliance prior to crack initiation.

Figure 2 shows normalised crack length predictions for tests at 20°C and 288°C on an A533B-1 steel. Pre-yield variations are much greater at 288°C presumably due to greater friction/clip gauge stability effects. All pre-yield predicted crack lengths agree with the appropriate measured initial crack length $a_{(o)}$ to within 2.5%. The question arises, therefore, as to which value of $a_{(o)}$ should be used when constructing the J_R curve. The approach has been to define $a_{(o)}$ as the crack length predicted at general yield (ie at $\sim$ 0.9 P_L where P_L is defined according to E813:81). This method has provided good correspondence with multi-specimen J_R data but has limited appeal and would be totally unsatisfactory for less tough materials where crack initiation may occur prior to general yield.

Figures 3 and 4 illustrate the influence of the predicted $a_{(o)}$ value on the position of the J_R curve for the test data shown in Fig 2. The curves are translated along the Δa axis by the difference in predicted $a_{(o)}$. Each curve should be corrected for stretch zone growth occurring up to the load level used to predict $a_{(o)}$ but this correction will be small (less than 0.05mm at 0.9 P_L). Whilst transposition of the J_R curve to either the left or the right with choice of $a_{(o)}$ should have no effect on dJ/da, the change in position of the J_R curve will have a large effect on the estimation of initiation toughness. This effect is shown in Fig 5 where initiation toughness values have been defined as the intersection of the J_R power curve and a 0.1mm offset from a $J = 4\sigma_f \Delta a$ blunting line.

Similar behaviour has been observed during testing of a submerged arc weld metal where particular attention is being paid to the variation in elastic compliance prior to general yield. Figure 6 shows the normalised crack length vs. normalised load line displacement relationships for the first two tests in this programme. The first specimen tested, DA25, showed decreasing compliance prior to yield. This was largely eliminated in the second test, DA13, by liberally coating both the flat-bottomed clevis holes and loading pins with MoS_2 grease. However, although significantly different pre-yield behaviour was observed, all pre-yield estimates of crack length agreed, as previously noted, to within 2.5% of the measured initial crack length. In addition, unloadings prior to general yield, were repeated either three times (DA25) or twice (DA13) and variations in predicted crack length for any particular unloading location were within 0.05mm (0.0013 a/W) in both tests. This suggests that the different pre-yield behaviour in the two tests is not attributable solely to frictional effects. J_R curves which would be defined using initial crack length predictions for unloadings at $0.2 P_L$ - $0.9 P_L$ are shown in Figs 7 and 8 for specimens DA25 and DA13, respectively. The predicted final crack extensions are within $\sim$ 0.1mm (0.25% W) of the measured crack extension. Despite the good agreement between the predicted and measured final increments of crack growth, the pre-initiation J_R data do not adequately predict the amount of crack tip blunting expected in pressure vessel materials. This is particularly noticeable in Fig 8 where the compliance data for $J \lesssim 0.1$ MJ/m^2 provide a gross over-prediction of the amount of crack blunting expected to occur. The variation in initiation toughness values, which are defined essentially by post-yield J-Δa data, would be of the same order as shown for the test on A533B-1 steel (Fig 5).

Conclusions

1.   The RNL system appears to accurately follow the changes of compliance associated with physical crack extension.

2.   Closer examination of compliance changes prior to general yield has shown that any prediction of initiation toughness is highly dependent upon the initial crack length prediction and suggest that consistent predictions of Ji (J_{Ic}) will only be obtained if initial crack lengths can be predicted to within $\pm$ 0.05mm AND the changes in compliance monitored during crack tip blunting (rather than actual crack growth) provide J-Δa data which follows the known (anticipated) blunting relationship for the material under test.

SCHEMATIC DIAGRAM OF UNLOADING COMPLIANCE TEST FACILITY

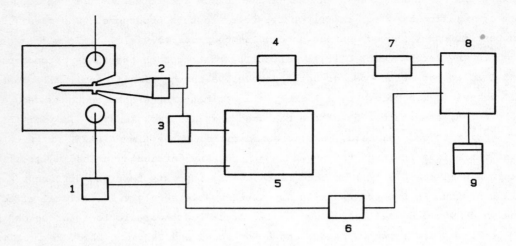

1) LOAD CELL
2) CLIP GAUGE
3) AMPLIFIER (×A1)
4) AMPLIFIER (×A2) NB: A2>>A1
   (incl. offset facility)
5) AUTOGRAPHIC RECORDER

6) DC OFFSET DEVICE
7) R-C FILTER
8) COMPUTER (WITH 12bit A-D CONVERTER)
9) FLOPPY DISC STORAGE

Fig. 1

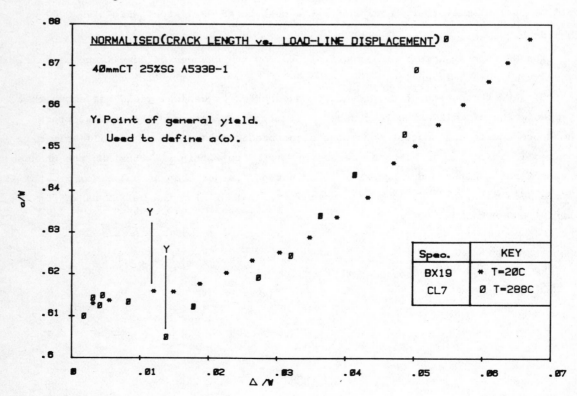

Fig. 2

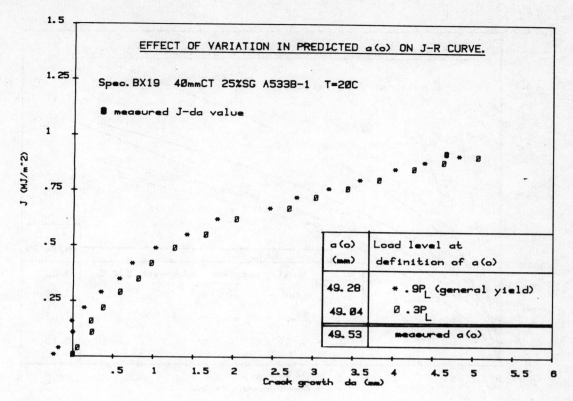

Fig. 3

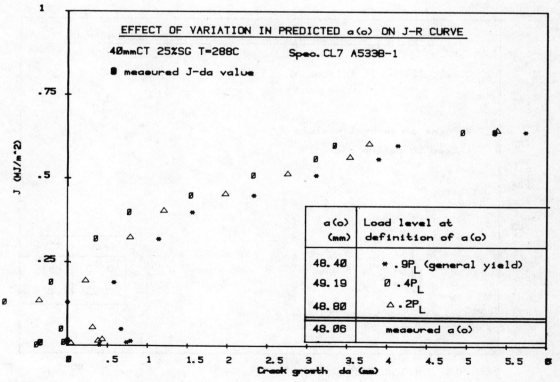

Fig. 4

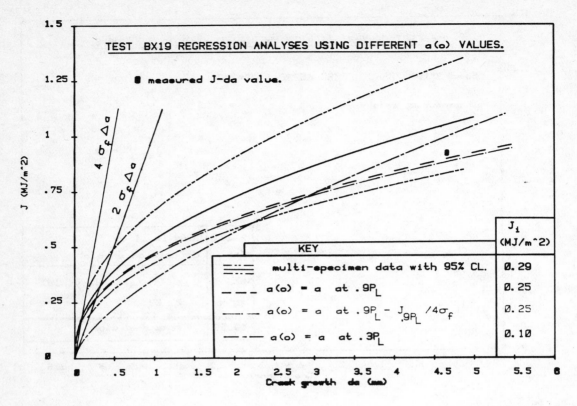

Fig. 5

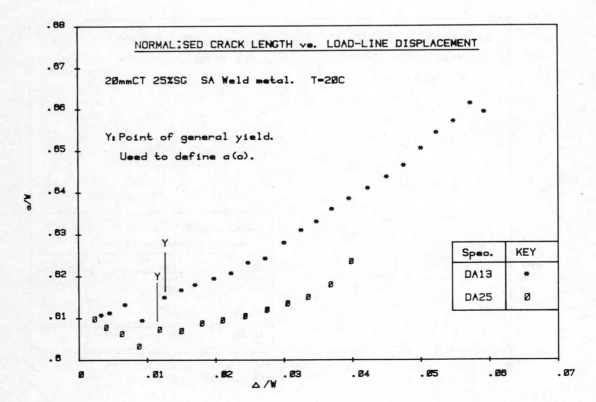

Fig. 6

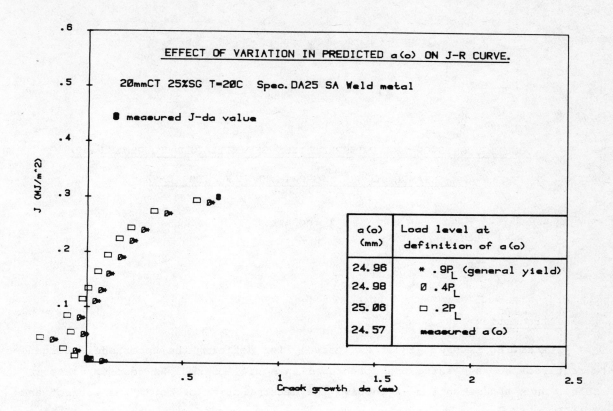

Fig. 7

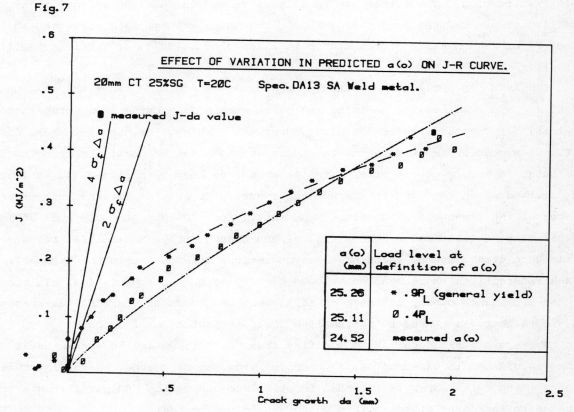

Fig. 8

SOME THOUGHTS ON THE INTERPRETATION OF INITIATION TOUGHNESS AND

RESISTANCE TO CRACK GROWTH FROM J_R TEST DATA

T Ingham

RNPDL

1.   Introduction

J_R testing is used almost exclusively for defining the upper shelf toughness data required when assessing the structural integrity of nuclear reactor pressure vessels. Such assessments must involve the determination of both "initiation" and post-initiation data. The former are usually expressed in terms of an equivalent K, and the latter are defined as either enhanced toughness values (ie J expressed in terms of K) or as a measure of resistance to crack growth (dJ/da or T for instability analyses).

The ASTM has published a standard method for determining "initiation" toughness (E813:81)[1] and a method for defining the plane strain J_{I-R} curve is in preparation[2]. The E813:81 "J_{Ic}" procedure defines a toughness value which, for high toughness materials in particular, can over-estimate the toughness at the onset of ductile crack growth. To the author's knowledge, no other material standards have been issued and variants on the J_{Ic} method are often used in the nuclear industry to provide alternative, closer estimates of the toughness at ductile crack initiation. Despite the plethora of methods for estimating "initiation" toughness values and consequently, the initial stages of resistance to crack growth, there has been no serious attempt within the nuclear industry to reach international agreement on methods of testing and interpretation. In view of the inter-dependence of any estimate of initiation toughness and initial resistance to crack growth there is a need to rationalise the interpretation of J_R data.

The aim of this paper is to identify those factors which, in the author's opinion, should be discussed before a concencus view can be reached on the interpretation of J_R data. The paper is intended to stimulate discussion and participants are encouraged to introduce others points which should be considered.

2.   J$_R$ Construction Methods

(a)   Linear J-Δa relationships

For linear analyses, a definition of "initiation" toughness and initial resistance to crack growth depends critically upon the crack growth (Δa) range and the assumed blunting line relationship.

For E813:81 Δa measurements, Δa ranges which have been used in practice are:-

(i)   The ASTM E813:81 range;

(ii)   A J controlled growth range eg Δa $\leq$ 6%b;

and blunting line relationships have been defined as:-

(i) $J = 2\Delta_f \sigma a$.

(ii) $J = 4\Delta_f \sigma a$.

(iii) Experimentally defined relationships.

A further method which could be considered would be to define the blunting line relationship using $J = 2/dn \cdot \sigma_f \Delta a$ where dn is defined in ref (3) as a function of ($\frac{\sigma_f}{E}$) and strain hardening exponent (n).

Blunting line relationships are not required when Δa measurements involve ductile crack extension only (ie excluding the stretch zone growth component). This approach, which has been used mainly in the UK, would not be suitable for interpreting unloading compliance data but could be more appropriate to the interpretation of potential drop data.

(b)   Non-linear curve fits

Non-linear curve fits provide a more accurate characterisation of resistance to crack growth over a larger Δa range than linear analyses.

Possible curve fits are:-

(i)   A power curve.

(ii)   A skewed hyperbola.

(iii) A best-fit polynomial.

Δa range:

In principle this method allows a more complete J$_R$ curve characterisation with no specific limit on Δa but for practical application it would be necessary to limit Δa in terms of the size requirements for J-controlled growth.

Definition of initiation:

Apart from perhaps the skewed hyperbolic fit which could attempt to simulate crack tip blunting, continuous non-linear curve fits pose a problem when defining initiation toughness. Methods which have been used to estimate initiation values from power curve analyses are:

The intersection of J_R with:-

(i) $J = 2\sigma_f\Delta a$.

(ii)   A 0.15mm offset from $J = 2\sigma_f\Delta a$.

(iii) $J = 4/3\sigma_f\Delta a$.

(iv)   A 0.1mm offset from $J = 4\sigma_f\Delta a$.

(v)   Experimental blunting line relation ships.

Questions requiring resolution are:

(a)   Are linear or non-linear fits more suitable for defining:

     (i)   Initiation toughness?

     (ii) Resistance to crack growth?

(b)   How is initiation best defined ie which blunting relationship should be used when Δa includes stretch zone growth?

(c)   What Δa range should be used for:

     (i)   Linear constructions?

     (ii) Non-linear constructions?

(d)   Are current size requirements adequate for defining specimen size independent values of:

     (i)   Initiation toughness?

     (ii) Resistance to crack growth?

(e)   In view of the difficulties in identifying ductile initiation in high toughness (ie high initial dJ/da) materials and bearing in mind the likely irrelevance of initiation to instability, could a J value at a specific low value of Δa (eg $\Delta a = 0.15$mm or 0.2mm) be used in lieu of a blunting line assumption?

3.   Determination of J

It is assumed that most workers use the J formulae corrected for crack growth proposed by ASTM in either the draft or final version of E813:81 and that alternative expressions which may be used are sufficiently well-documented not to require detailed consideration.

4.   Determination of Crack Growth (Δa)

(i)   Non-side grooved specimens

For multiple specimen testing, where Δa is measured directly, different crack growth averaging procedures should not influence estimates of initiation toughness but can have a pronounced effect on resistance to crack growth.

Δa is usually measured as a 9 point average (E813) but, when defining resistance to crack growth for non-side grooved specimens it may be more appropriate to use Δa values which are weighted towards the amount of growth occurring at the centre of the specimen.

(ii)   Side-grooved specimens

Side-grooving promotes growth along a uniform crack front and eliminates the problem noted above. The testing of side-grooved specimens would seem to be an essential requirement for J_R curve testing, but, by manipulation of the Δa averaging procedure in non-side grooved specimens, it may be possible to obtain equivalent data from both types of specimen.

(f)   What is the optimum side-groove depth (as a % of thickness)?

(g)   Is side-grooving really necessary? ie can a side-grooved J_R curve be estimated using data from non-side-grooved specimens and a modified Δa measurement.

(h)   Do other methods of estimating Δa (eg area measurments from photographs) offer improved accuracy over direct optical measurements of crack growth?

5.   Single Specimen Tests

Single specimen testing is sufficiently well-developed in some laboratories to be used as a viable alternative to multiple specimen tests. The principal single specimen methods are:

(i)   Unloading compliance.

(ii)   D-C potential drop.

(iii)   AC potential drop.

(iv)   Dual displacement measurements.

(v)   Back-face strain.

Factors requring resolution before a particular experimental method can be regarded as being suitable for adoption as a general test procedure are technique-specific and, are best detailed by workers most experienced with each technique. Experience at RNL with AC and PD methods has been limited to initiation detection only and no comments can be offered on the application of these methods for defining J_R data. More experience has been gained with the unloading compliance technique and the two key areas requiring resolution before the RNL system could be considered as a satisfactory replacement for multiple specimen testing are:-

(i)   Elimination of pre-yield variations in elastic compliance (ie the prediction of -ve crack growth).

(ii)   The development of stable, displacement gauges capable of operation at high temperatures ($\sim 300^{\circ}$C).

References

1.   ASTM "Standard test for J_{Ic}, a measure of fracture toughness". ASTM Book of Standards, part 10, pp.810-828, 1981.

2.   ASTM E24-08 Task Group Activity on J_{I-R} testing.

3.   Shih C F et al. Methodology for plastic fracture. Combined 9/10 Quarterly Progress Reports, GE Report SRD-79-031, Jan 1979.

On the Problem of "Negative Crack Growth" and "Load Relaxation" in Single
Specimen Partial Unloading Compliance Tests

B. Voss
Fraunhofer-Institut für Werkstoffmechanik, D-7800 Freiburg, W-Germany

1. Apparent negative crack growth

Figure 1 shows two $J(\Delta a)$-resistance curves of different steel specimens. The upper one shows apparent negative crack growth, well known to
arise sometimes in partial unloading compliance tests, the lower one does
not. One possible reason for negative crack growth indicated by a round
clevis hole in Fig. 1 [1] will be discussed and an alternative clevis design to avoid this error indicated by a flat bottom hole will be proposed.

The ASTM-standard [2] proposes flat bottom holes or roller bearings in
the clevisses for bolt loading of compact specimens to avoid friction effects. Roller bearings are difficult to handle and have some problems at
elevated test temperatures. Use of flat bottom holes instead may result in
negative crack growth. Figure 2 shows a compact specimen with a clevis with
flat bottom holes according to [2] (schematically). By loading the specimen
(Fig. 3) the loading bolts bend and may cause plastic deformation of the
flat surfaces of the clevis in the regions marked by arrows. Figure 4 shows
side views of the contact area of bolt and clevis

- without load
- with plastic indentation by vertical loading and
- the effect of additional rotation of the loading bolt, necessary to maintain mechanical equilibrium while the crack is opened by loading.

Figure 5 models this situation by loading the specimen (radius of the
holes r_s) by a bolt (radius r_b) rolling on a circular surface of the
clevis hole (radius r_c).

At a displacement V_{LL} measured in the load line position each half of the specimen near the load line is tilted relative to the crack plane by an angle α_s. Assuming a rotation centre in front of the crack tip (crack length a) on the ligament (size b) at a distance $\varepsilon \cdot b$ ($\varepsilon \approx .3$) α_s is approximately given by

$$\alpha_s \approx \tan \alpha_s = V_{LL}/(2(a + \varepsilon \cdot b)) \tag{1}$$

Because of this rotation of the bolt hole region of the specimen the lines of contact of the specimen, the bolt and the clevis respectively must move towards the crack tip by a distance Δx quantified below. The assumed load line for compliance evaluation and crack length calculation passes through the centre of the bolt hole in the specimen (dashed arrows in Fig. 5). But the real load line is at a distance Δx towards the crack tip and the measured force F_m or a difference ΔF_m during a partial unloading will be greater compared to the assumed load line position. This causes an under-estimation of the compliance and the crack length and consequently results in negative Δa-values before real crack growth.

Assuming free rolling of the contacting surfaces without any slip Δx is given by

$$\Delta x = r_s \cdot \sin (\alpha_s \cdot r_s/(r_s + r_c - 2 r_b)) \tag{2}$$

and the measured compliance C_m may be corrected for this error by

$$C_{corr} = C_m \cdot \frac{a + \varepsilon \cdot b}{a + \varepsilon \cdot b - \Delta x} \tag{3}$$

In practice the correction by eqn. 3 normally will not be applicable because r_c ($\approx r_b$) will be an unknown and not constant number if caused by plastic indentation. For flat bottom holes ($r_c = \infty$) follows from eqn. 2 $\Delta x = 0$. For reasonable assumptions (CT 25; $r_s = 6.25$ mm, $r_b = 6$ mm, $r_c = 6$ mm) the consequences of this model are demonstrated in Fig. 6 for a real experiment without negative crack growth by introducing the error by the invers of eqn. 3 resulting in a shape of the new J(Δa)-curve and $\Delta a_{min} = -0.25$ mm similar to Fig. 1. Obviously the value of $J_{IC} = 173$ kJ/m^2 is greater than the correct one $J_{IC} = 122$ kJ/m^2, but even correcting for the effect of negative Δa by shifting all points to positive Δa-values with the most negative point on the blunting line results in $J_{IC} = 147$ kJ/m^2 greater than the correct one. So apparent negative crack growth may result in non conservative J_{IC}-estimations.

To avoid this effect of possible plastic deformations sometimes disturbing results measured with flat bottom hole clevisses an alternative design was developed and tested successfully [3]. Figures 7 and 8 (comparable to Figs. 2 and 3) show that hardened inserts allow for tilting of the flat bottom planes if the loading bolts are bending. So line loads are applied instead of point loads (Fig. 3, arrows) thereby decreasing pressure and consequently minimizing risk of plastic deformation.

Clevisses of this type were used at temperatures up to 600°C without negative crack growth and with low scatter of crack length measurements.

2. Relaxation

Especially in tests at elevated temperatures (e.g. range of service temperature of light water reactors 300°C) time dependent effects in the specimen may cause problems for the crack length determination from the compliance measured in partial unloadings. Two examples of CT 48 specimens (B = 48 mm, W = 96 mm) will be discussed [1].

Figure 7 shows part of a load (F) vs. displacement (V)-diagram measured at 300°C, with load drops of about 5 % during constant displacement control up to 3 min. Though the nearly exponential load drop did not reach a (nearly) constant level during this time the crack length could be estimated from the compliance. The differences in crack lengths calculated from unloading and reloading compliance respectively showed an (apparent) crack growth of (0.2 $\pm$ 0.2) mm during the unloading cycles. In spite of these differences the final crack length agreed with the crack surface measurement within 0.2 mm.

Even without the 3 min relaxation time reasonable J_R-curves can be measured. Figure 8 shows one of several unloadings without relaxation followed immediately by a second unloading in a test with a displacement velocity of about 1 mm/min. Caused by the visible curvature of the unloading and reloading paths the differences of estimated crack lengths were 1.6 mm for the first cycle and .45 mm for the second one. The mean values differed by only 0.24 mm (0.02 mm to 0.29 mm for several pairs of unloadings) comparable to the "absolute accuracy" measured by the difference of .2 mm between the final crack lengths from the last single cycle unloading and the crack surface measurement for this test.

From these results typical for several more tests it may be deduced that reliable J(Δa)-curves may be measured. But as discussed in [1] and [4] relatively small errors in the initial crack length determination (definition of the point Δa = 0) may cause relatively great changes in the extrapolated J_{Ic}-value according to the ASTM-standard [2].

Literature

[1]   B. Voss: Proceedings of the 14. Sitzung des Arbeitskreises Bruchvor-
           gänge, Mülheim 1982, DVM Berlin (1982)

[2]   ASTM E 813-81, Annual Book of ASTM Standards, Part 10, Philadelphia
           (1981)

[3]   Patent applied for, P 32 34 472.4 (1982)

[4]   B. Voss, J.G. Blauel: Proceedings of the 4th ECF-Conference, Leoben
           1982, Vol. 1, edit. K.L. Maurer, F.E. Matzer, EMAS, UK (1982)

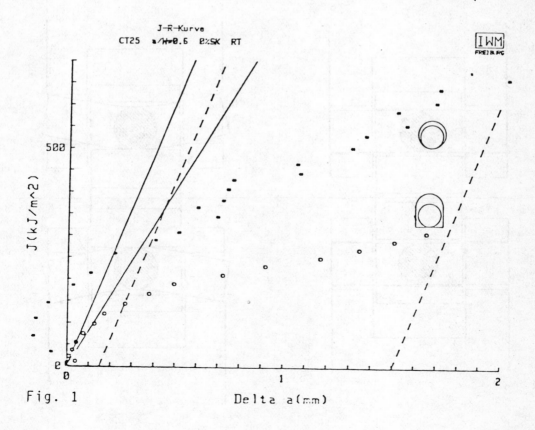

Fig. 1

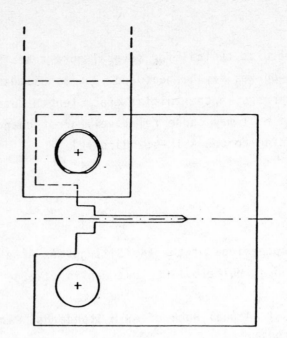

Fig. 2

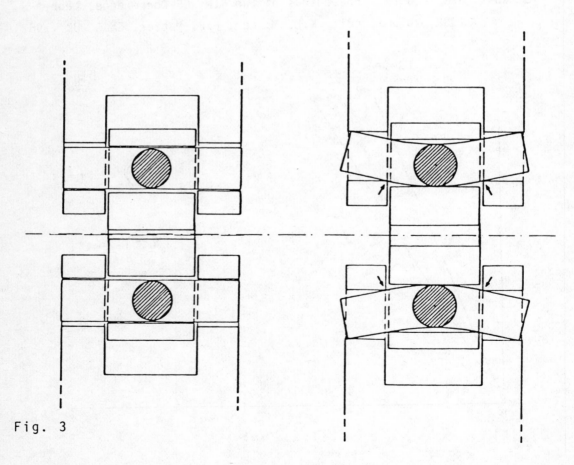

Fig. 3

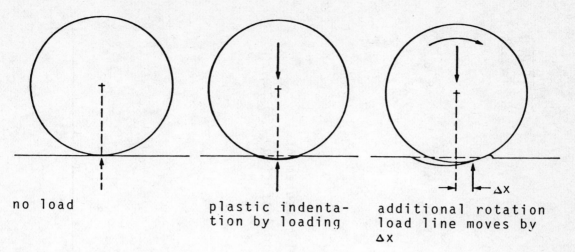

no load

plastic indenta-
tion by loading

additional rotation
load line moves by
Δx

Fig. 4

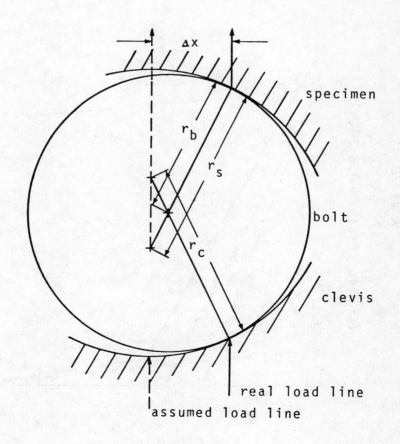

Δx

specimen

r_b

r_s

bolt

r_c

clevis

real load line

assumed load line

Fig. 5

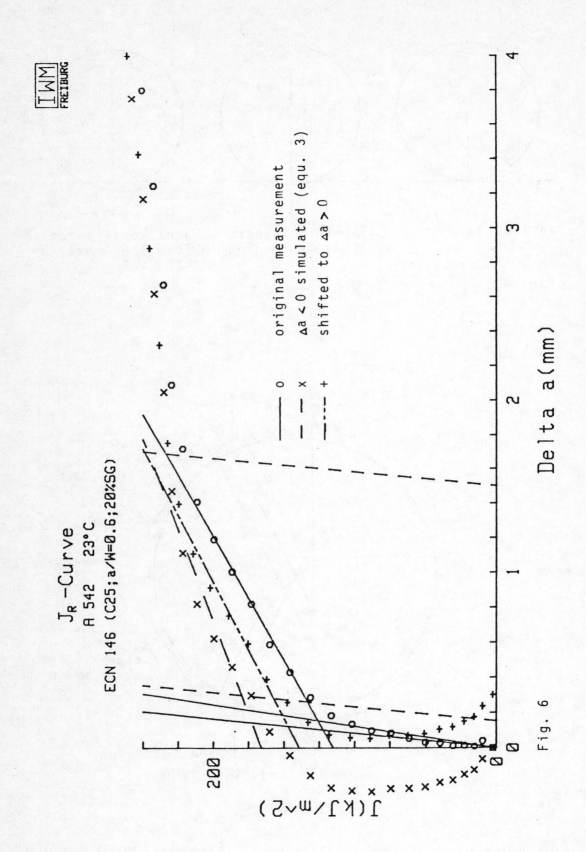

J_R-Curve
A 542   23°C

ECN 146 (C25;a/W=0.6;20%SG)

original measurement
Δa < 0 simulated (equ. 3)
shifted to Δa > 0

o
x
+

J(kJ/m^2)

200

Delta a(mm)

Fig. 6

IMI
FREIBURG

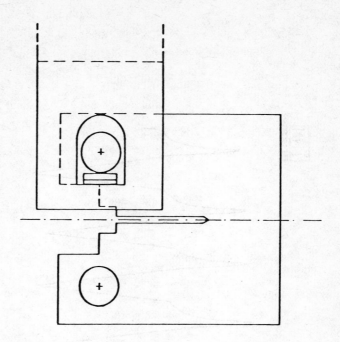

Fig. 7

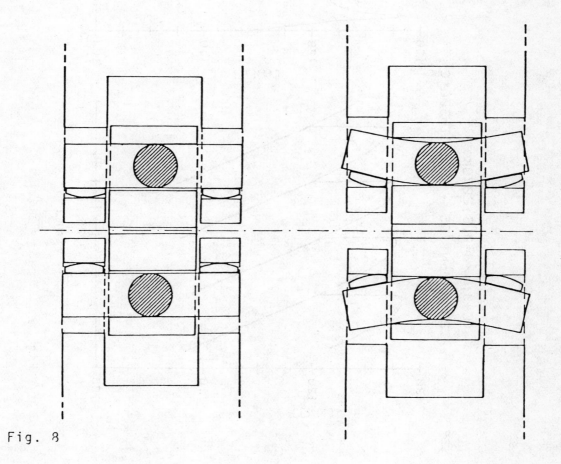

Fig. 8

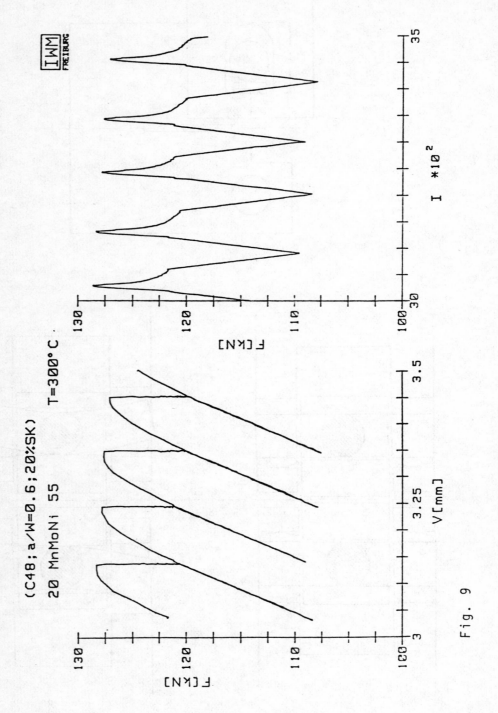

Fig. 9

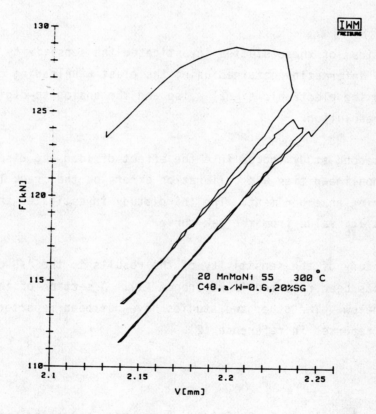

Fig. 10

Sensitivity of J-R Curve Results to Testing Variables

W. A. Van Der Sluys
R. J. Futato

In this presentation, the results of three different studies performed at B&W on the sensitivity of the J-R curve results to testing variables will be presented.

The first of these studies investigated the sensitivity of the measured crack length information obtained using the elastic unloading compliance technique to the electronic signal noise and the analog-to-digital (A/D) conversion resolution.

The second study looked into the effect of load and displacement transducer nonlinearities and calibration errors on the crack length and J values measured during an experiment. The third study investigates the difficulties in estimating a J_{Ic} value from the J-R curve.

The study of the sensitivity of the results to the A/D conversion resolution has been reported in reference (1). A section of that report is reproduced below. The other two studies have not been reported in the literature but will be reported in reference (2).

References:

1. W. A. Van Der Sluys, R. J. Futato, "Computer-Controlled Single Specimen J-Tests," presented at Second International Symposium on Elastic-Plastic Fracture Mechanics, Philadelphia, PA., October 6 - 9, 1981. To be published in Symposium proceedings.

2. R. J. Futato, J. D. Aadland, W. A. Van Der Sluys, A. L. Lowe, "A Sensitivity Study of the Unloading Compliance Single Specimen J-Test Technique," to be presented at the ASTM Symposium on User's Experience wtih Elastic-Plastic Fracture Toughness Test Methods, Louisville, Kentucky, April 1983.

Sensitivity Study (From Reference 1)

The generation of an entire J-R curve from a single specimen requires the remote detection of small amounts of crack extension with a high degree of confidence. Instrumentation for an automated procedure should therefore be selected based on a consideration of the effect of the measurement technique on the confidence of the results.

When the elastic unloading compliance measurement technique is used, crack lengths are determined from a knowledge of specimen compliance, calculated from load and displacement data obtained during a partial unload of the specimen at various points throughout the test. Consequently, any scatter or uncertainty in measured load or displacement causes a corresponding uncertainty in the calculated compliance and ultimately the calculated crack length. The scatter in measured data, due primarily to electronic signal noise and analog-to-digital (A/D) conversion in an automated test, must therefore be reduced to known acceptable levels.

The effect of A/D conversion and electronic noise on measured data can be modeled as shown in Figure 1. The A/D conversion process transforms a continuously varying signal into a set of integers varying linearly between zero and $\pm 2^{b-1}$ full scale, where b = A/D word length. (The highest order bit in the word is generally used to indicate signal polarity). The minimum resolvable signal change, U, is that causing a change of 1 bit in the A/D output, so that $U = (\text{full scale signal})/2^{b-1}$. The individual (dimensionless) data points generated by the conversion process are therefore a measure of the number of smallest resolvable units, U, composing a signal of any given time, and can be described by the expression

$$\text{A/D Output} = \text{Integer} \left[\frac{(\text{signal level})}{U} + .5 + \text{noise} \right] . \qquad (1)$$

The above expression allows for rounding between bits as well as for the superposition of noise on the true signal.

Using the above model, the effect of A/D conversion and transducer signal noise on crack length confidence can be estimated. Since in general

$C = C(E,B_e,a/w)^{1,3}$, the precision in a/w due to the precision of a given compliance measurement can be approximated by the expression

$$\Delta(a/w) = \Delta C \left(\frac{\partial C}{\partial(a/w)}\right)^{-1} \tag{2}$$

for small errors in C and a/w, ignoring changes in modulus or specimen thickness. Further, if a linear regression analysis can be used to calculate compliance, then the precision, ΔC, can be estimated statistically as the percent confidence interval of the compliance, C, given by the expression

$$\Delta C = \left(t_{(1-\alpha/2)} - t_{(\alpha/2)}\right)\frac{S_{y.x}}{S_x(n-1)^{1/2}} \tag{3}$$

where n = total number of points used in the compliance calculation, t = values of the t distributions for n-2 degrees of freedom, Sy.x = standard error of estimate, and Sx = variance of load distribution for a given displacement.[2] For the above equation, percent confidence = $100(1-\alpha)$. Using equation (1), the individual load and displacement data points measured during an unload, P_i and V_i, respectively, can be described by

$$P_i = U_P \text{ Integer} \left[R_P \frac{i}{n} + .5 + noise\right]$$

$$V_i = U_V \text{ Integer} \left[R_V \frac{i}{n} + .5 + noise\right] \tag{4}$$

where

$$R_p = \frac{\Delta P}{U_p} = \frac{\Delta P}{\text{load range}} \times 2^{b-1}$$

$$R_V = \frac{\Delta V}{U_V} = \frac{\Delta V}{\text{displacement range}} \times 2^{b-1} \qquad (5)$$

and ΔP and ΔV are the total change in load and displacement, respectively, during the unload. Therefore,

$$\Delta C = (U_V/U_p)\, \overline{\Delta C} = C(R_p/R_V)\overline{\Delta C}$$

where $\overline{\Delta C}$ is a dimensionless function of R_p, R_V, a/w, n, and noise, and the confidence interval in a/w corresponding to ΔC can be estimated by

$$\Delta(a/w) = (R_p/R_V)\ (C)\ (\overline{\Delta C})/\frac{\partial C}{\partial(a/w)} = \text{function } (R_p,\ R_V,\ a/w,\ n,\ \text{noise}) \qquad (6)$$

Consequently, at a given crack length ratio, the statistical confidence interval in a/w is dependent on the effective resolution, R_p and R_V, and the level of electronic noise.

Several assumptions were made for computational ease in the above analysis. These assumptions are: 1) the crack length vs compliance equations are exact, 2) the variance of the distribution of loads for a given displacement are equal for all displacements, and 3) the distribution of loads for a given displacement is a normal distribution. In addition, equation (3) used to define the percent confidence interval in compliance, ΔC, is strictly valid only in cases when the load values are known without error. However, since order of magnitude estimates of confidence are being sought, the errors due to these assumptions are considered to be negligible.

Confidence intervals calculated according to equation (6) at the 95 percent confidence level are shown in Figure 2 as a function of R_p and R_V for a/w = .6. The calculation assumes displacement measurement on the specimen load-line, and that approximately 100 data points are available for the compliance calculation. A constant noise level equal to eight times the digital resolution ($\pm$4 A/D output units) on both the load and displacement signals is also assumed. The noise spectra of the signals are independent and totally random with σ=1. Note that when R_p and R_V vary greatly, the smaller determines the size of the confidence interval. In fact, regions of total R_V-control and total R_p-control exist, corresponding to the horizontal and vertical regions of the curves, respectively. In these regions, increases in the effective resolution of the non-controlling parameter will have little, if any, effect on measurement confidence. Notice also the interchangability of R_p and R_V, not U_p and U_V, in determining confidence interval size.

Figure 3 shows the effect of slight changes in the noise level on the effective resolution required to produce a confidence interval $\Delta(a/w)$ = .001. As before, the peak noise level is assumed to be equal on both load and displacement, although the noise spectra of the two signals are independent and totally random. Results of the analysis reveal that a significant increase in effective resolution is required to offset relatively slight increases in electronic noise. Conversely, electronic noise will sharply limit the statistical confidence in a crack length measurement, to a degree even greater than that due to A/D conversion resolution alone. Consequently, care must be taken to maintain electronic noise at minimal levels throughout testing.

In designing a system for both automated data acquisition and test control, the need for high resolution data acquisition must be balanced against the need for short A/D conversion time. It is therefore desirable to use the shortest A/D word length that will supply the necessary resolution. Assuming R_p = 1500 and R_V = 550 at a $\pm$4 unit noise level will provide adequate statistical confidence for proper J-R curve generation, then for a typical J test in which maximum load = 47 kN (10000-lb) and maximum displacement = 4 mm (.16-in) with ΔP = 7 kN (1600 lb) and ΔV = .08 mm (.003 in.) during unloading, a 16-bit A/D can be used provided

$$\text{load range} \leq \frac{7 \text{ kN}}{1500} \times 2^{15} = 153 \text{ kN (34000 lb)}$$

$$\text{displacement range} \leq \frac{.08 \text{ mm}}{550} \times 2^{15} = 4.7 \text{ mm (.19 in)} \ .$$

However, since both R_p and R_V are less than 2^{11} (= 2048), then a 12-bit A/D will work as well, provided,

$$\text{load range} \leq \frac{7 \text{ kN}}{1500} \times 2^{11} = 9.5 \text{ kN (2100 lb)}$$

$$\text{displacement range} \leq \frac{.08 \text{ mm}}{550} \times 2^{11} = .3 \text{ mm (.012 in)} \ .$$

Therefore, while for a typical J test a 16-bit A/D converter can be used with the standard analog electronics for test control, a 12-bit A/D will provide equal statistical confidence if additional electronic signal amplification is used to expand the unload. In general, a $2^{15}/2^{11}$ (= 16-fold) increase in signal amplification is required to interchange 12-bit and 16-bit A/D converters, when scatter in load and displacement due to electronic noise is kept constant.

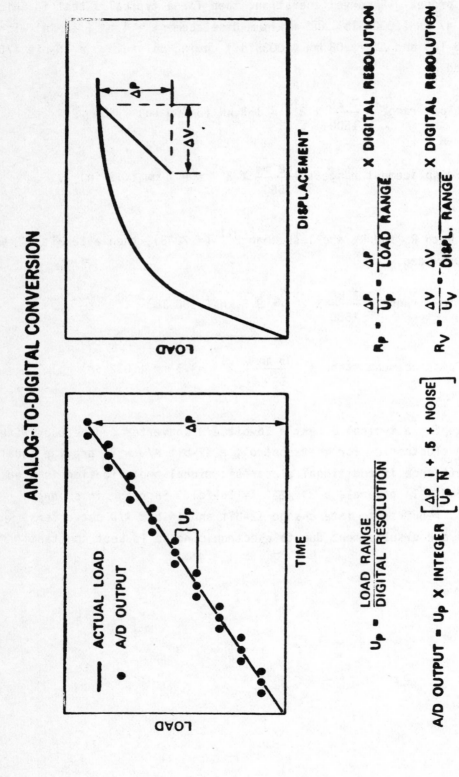

ANALOG-TO-DIGITAL CONVERSION

$$U_P = \frac{\text{LOAD RANGE}}{\text{DIGITAL RESOLUTION}}$$

$$R_P = \frac{\Delta P}{U_P} = \frac{\Delta P}{\text{LOAD RANGE}} \times \text{DIGITAL RESOLUTION}$$

$$R_V = \frac{\Delta V}{U_V} = \frac{\Delta V}{\text{DISPL. RANGE}} \times \text{DIGITAL RESOLUTION}$$

$$\text{A/D OUTPUT} = U_P \times \text{INTEGER} \left[\frac{\Delta P}{U_P} \left| \frac{1}{N} + .5 + \text{NOISE} \right. \right]$$

FIGURE 1 Model Showing the Effect of the A/D Conversion Process (Digitization) on the Load and Displacement Signal Resolution During an Unload.
U = Smallest Signal Change Resolvable by the A/D.

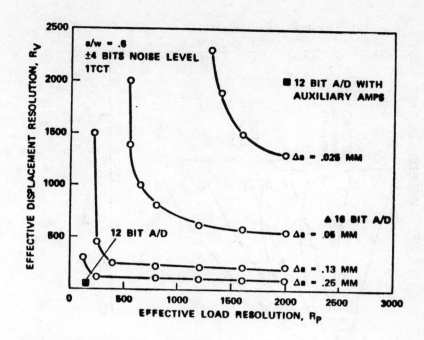

FIGURE 2   Effect of Load and Displacement Signal Resolution During Unload on Crack Length Measurement Confidence, $\Delta a/w$. $R_p = \Delta P/U_p$, $R_V = \Delta V/U_V$ for ΔP Load Change and ΔV Displacement Charge During the Unload. The Crack Length Confidence Attainable Using 12 Bit A/D Converters, 16 Bit A/D Converters, and 12 Bit A/D Converters With Additional Amplification of the (Analog) Unload Signals Before Digitization are Also Shown.

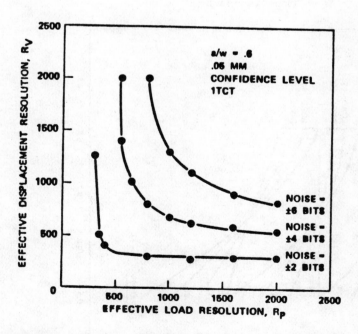

FIGURE 3   Effect of Noise on Signal Resolution Required to Achieve a Crack Length Confidence Interval of .05 mm Using a 1T Compact Fracture Specimen.

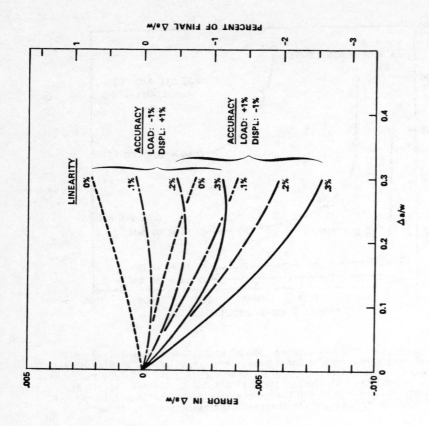

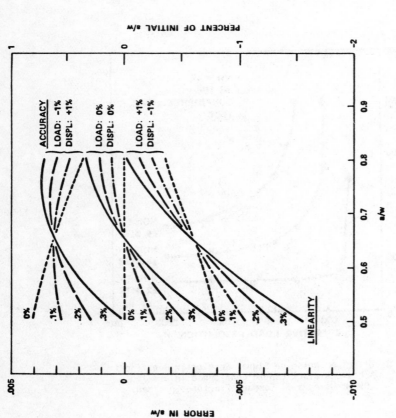

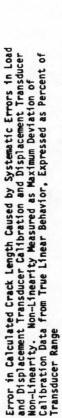

Error in Calculated Crack Extension Caused by Systematic Errors in Load and Displacement Calibration and Displacement Transducer Non-Linearity. Non-Linearity Measured as Maximum Deviation of Calibration Data From True Linear Behavior, Expressed as Percent of Transducer Range

Error in Calculated Crack Length Caused by Systematic Errors in Load and Displacement Transducer Calibration and Displacement Transducer Non-Linearity. Non-Linearity Measured as Maximum Deviation of Calibration Data from True Linear Behavior, Expressed as Percent of Transducer Range

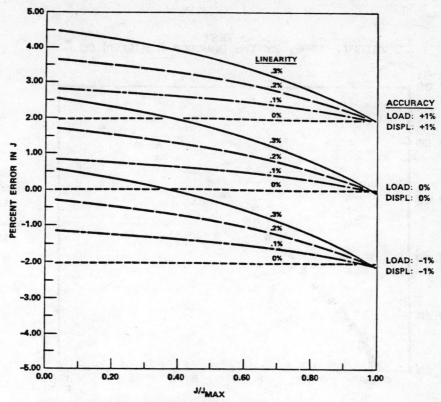

Percent Error in J Caused by Systematic Errors in Load and
Displacement Calibration and Displacement Transducer Non-Linearity.
Non-Linearity Measured as Maximum Deviation of Calibration Data From
True Linear Behavior, Expressed as Percent of Transducer Range

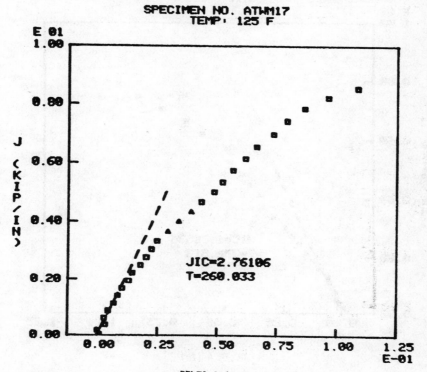

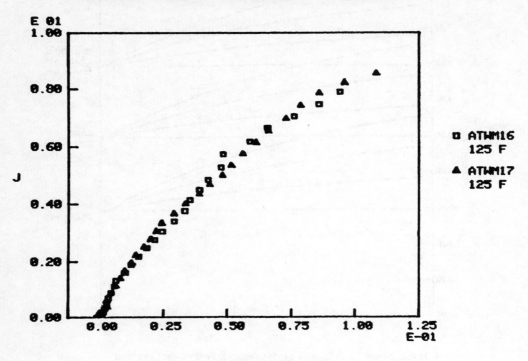

SPECIMEN NO. ATWM16
TEMP: 125 F

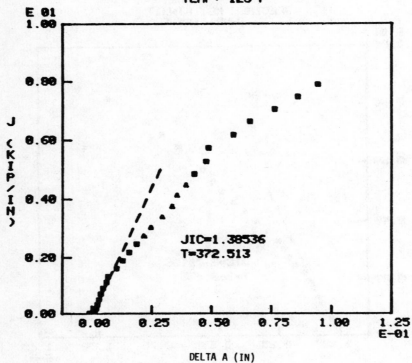

JIC=1.38536
T=372.513

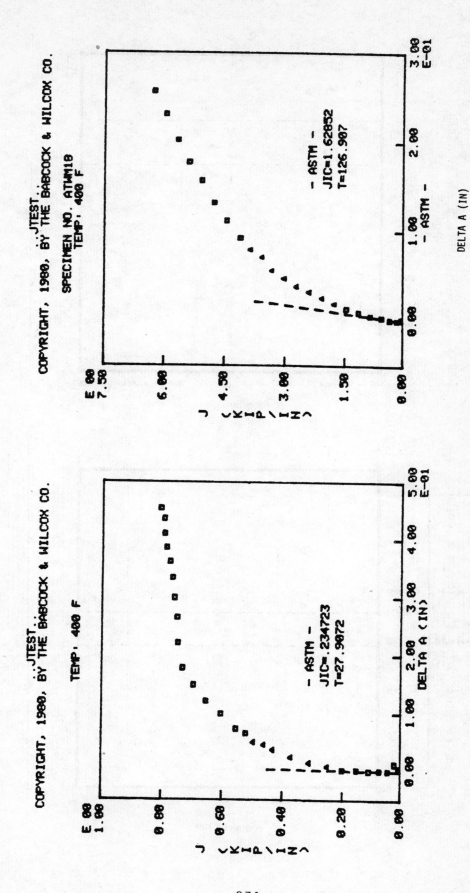

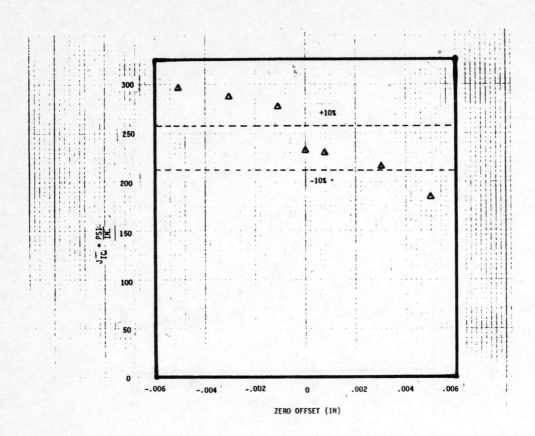

ZERO OFFSET (IN)

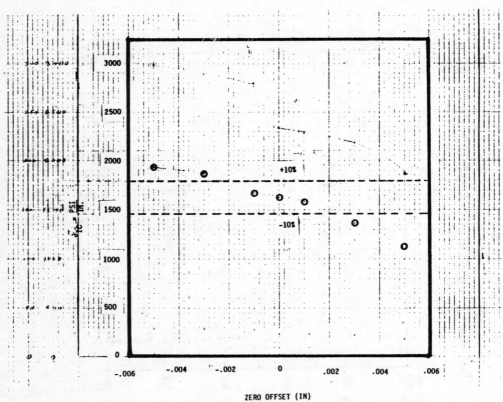

ZERO OFFSET (IN)

ROTATION CORRECTIONS FOR STAINLESS STEEL SPECIMENS
IN THE UNLOADING COMPLIANCE METHOD

B. HOUSSIN

FRAMATOME, PARIS-LA-DEFENSE, FRANCE

ABSTRACT :

Some difficuties encountered in the determination of tearing
resistance R-curves of stainless steels by the unloading compliance method are reported. Due to the large displacements exhibited
by the specimens, rotation corrections are needed but are insuffi-
cient to account for the discrepency observed with the reference
heat tinting method.

BACKGROUND

In the partial unloading method {1} small reversed displacements are periodically made during the recording of the load-displacement curve of a compact tension (CT) type specimen.

The compliance of the specimen is measured from the slope of the small elastic variation in load (ΔP) versus displacement ($\Delta\delta$) during each unloading-reloading event.

The amount of stable crack extension Δa is deduced from the change in crack length calculated using the elastic relation between crack length and compliance {2} {3} .

The major two difficulties of the method are first, to measure accurately the slope of the small elastic unloading and second, to convert properly this quantity or its change into crack length (or change in crack length).

The compliance-crack length relationships {2}{3} have been established using small displacement and elastic analysis. These assumption are widely violated in elastic-plastic fracture tests, due to the important change in specimen geometry and yielding of the remaining ligament. Corrections are needed when a large opening of the specimen occurs especially during tests on very tough and moderate strength materials such as austenitic stainless steels. The importance of the rotation correction proposed by LOSS {4} to evaluate properly the J Resistance curve was examined for austenitic stainless steels.

The mechanical properties of the stainless steel investigated and the design of the 50mm thick CT specimens (20% side-grooved) are reported in the paper of P. BALLADON {5} presented at the present conference.

1. MEASUREMENT OF THE UNLOADING SLOPE

The monotonic load-displacement curve of the specimen obtained by an autographic method is used for calculating at each partial unloading the J value in accordance with the ASTM E 813-81 {6} procedure.

The small variations in load (ΔP) <u>vs.</u> displacement ($\Delta\delta$) of the unloads are recorded on a second x-y recorder using appropriate magnification factors to obtain a pen travel of about 80mm in each axis direction in the linear part of the unload-reload curve (fig.1).

The first part of the unloading, where relaxation effects occur leading to a non linear variation of load vs. displacement, is not taken into account, such as the end of the reloading which also exhibit a non linear behaviour.

Sometimes, narrow hyteresis loops were observed, in this case a mean value of the slope was considered.

The reading errors induce an uncertainty in the determination of the unloading slope $C = \Delta\delta/\Delta P$ estimated to be less than 1%. However, the total error in the unloading slope was estimated to be 1% to also account for the linearity of the load cell and displacement gauge.

The relationship between crack length extension and change in compliance proposed in {1} reduces for a/w = 0.65 to the following expression :

$$\frac{\Delta a}{a} = 0.24 \frac{\Delta C}{C}$$

The absolute error on the amount of crack extension Δa estimated for the specimen used, is : $\Delta (\Delta a) = 0.16$ mm

The scatter exhibited by the data around each of the three mean J-Δa resistance curves plotted on figure 2 supports this evaluation. The differences between the three mean J-Δa curves are attributed to the intrinsic variability in toughness of the material.

Nevertheless, large differences (40%) are observed for the final crack extensions obtained by the compliance method when compared to the values directly measured by the heat tinting method (fig. 2). Corrections are required to reach a better agreement with the heat tinting method considered as the reference method.

2. ROTATION CORRECTIONS

Yielding of the ligament of the compact tension specimen causes rotation of the loading arms which reduces the moment arm of the applied load and the clip gage displacement (fig.4). This results in a measured elastic compliance ($\Delta\delta_m/ \Delta P_m$) which is less than it would be ($\Delta\delta_c/ \Delta P_c$) for a specimen having the virgin geometry of the undeformed CT specimen.

A correction valid for small rotation angles was proposed in 1977 by DONALD and SMITH and reported by CLARKE and LANDES {7} . Some months later LOSS {4} published a general formulation valid up to large rotation angles. Minor corrections were introduced by MERCKLE and recently published {8} .

This formulation is given in figure 4, using the notations summarized in figures 3 and 4.

Attention is paid to the incorrect simplified relationships reported in the literature {9} which results from an error between the total displacement δ_m and the value $V_m = \delta_m/2$ which is used in the calculation made on half the specimen.

The general formulation established by LOSS   :

$$\Theta = \sin^{-1}\left[(D + \delta_m/2) / (D^2 + R^2)^{0.5}\right] - \tan^{-1}(D/R)$$

$$\frac{\Delta\delta c}{\Delta\rho c} = \left(\frac{\Delta\delta m}{\Delta\rho m}\right)\left[\cos\Theta - \frac{H}{R}\sin\Theta\right]^{-1}\left[\cos\Theta - \frac{D}{R}\sin\Theta\right]^{-1}$$

$$\frac{\Delta\delta c}{\Delta\rho c} = Fc\left(\frac{\Delta\delta m}{\Delta\rho m}\right)$$

may be approximated for small angles of rotation using :

$$\sin\Theta \sim \frac{\delta_m/2}{R} = \frac{\delta m}{2R}$$

$$\cos\Theta = (1 - \sin^2\Theta)^{0.5} \simeq 1 - \frac{\delta m^2}{8R^2}$$

$$\left[\cos\Theta - \frac{H}{R}\sin\Theta\right] = 1 - \frac{H\delta m + \delta m^2/4}{2R^2}$$

$$\left[\cos\Theta - \frac{D}{R}\sin\Theta\right] = 1 - \frac{D\delta m + \delta m^2/4}{2 R^2}$$

loading to :

$$\frac{\Delta\delta c}{\Delta\rho c} = \frac{\Delta\delta m}{\Delta\rho m}\left[\frac{1}{(1 - \frac{x}{R})(1 - \frac{y}{R})}\right]$$

$$x = \frac{H\delta m + \delta m^2/4}{2 R}$$

$$y = \frac{D\delta m + \delta m^2/4}{2 R}$$

The expression of x and y are less than half the values given in {9} which over estimate the correction.

In fact, these simplified relationships are not recommended for stainless steels where large rotation angles occur during the test.

The more general LOSS formulation was adopted, with the usual definition for the distance of the rotation center to the load displacement line :

$$R = a + r (w - a)$$

It is often argued that the incertainty in the location of the center of location induces significant errors in the correction.

This fact was examined using constant r values of 0.3, 0.5 and 0.75 for the specimen tested to the largest displacements ($\delta_m = 11$ mm). The uncorrected final crack extension Δa_{cu}^f and the corresponding values of the compliance correction factor F_c and corrected final crack extensions Δa_{cc}^f are given for the different values of r, in the following table :

r		0.3	0.5	0.75
Δa_{cu}^f (mm)	7.0			
Δa_{cc}^f (mm)		8.43	8.22	8.06
F_c		1.1214	1.1042	1.0867

The choice of the location of the rotation center is demonstrated to induce variations which are an order of magnitude less than the correction on Δa_{cu}^f.

The realistic value r = 0.5 currently adopted was taken for all the calculations. The corrected resistance curve of the same specimen is presented in figure 5. At the end of the test the specimen exhibits a half rotation angle θ more than 7°.

The uncorrected and corrected final crack extending of the three specimens are compared in figure 6 to the physical crack extension (Δa_p).

Despite an increase of about 20% of the uncorrected crack extension, the corrected compliance crack extension is found to be about 20% less than the physical crack extension determined using the reference heat tinting method.

The compliance relationship might be improved by performing elastic-plastic calculations on side-grooved CT specimens in the same material to account accurately for the important change in geometry.

Nevertheless, the corrected Resistance curve could be empirically ajusted to fit the final crack extension when estimated data are needed.

CONCLUSION

The main conclusions drawn from the determination of the J-Δa Resistance curves of stainless steel by the compliance method are :

- large rotation angles exhibited by CT specimens in stainless steel require a correction of the measured change in compliance during partial unloadings.

- the compliance correction factor reaches 1.10 for half rotation angle θ of about 7°.

- corrections on final crack extension are about 20% of the uncorrected value.

- Nevertheless, the corrected final value of the crack extension was found about 20% less than the physical extension determined using the heat tinting method.

REFERENCES

{1} G.A. CLARKE, W.R. ANDREWS, P.C. PARIS and G.W. SCHMIDT,
Mechanics of Crack Growth, ASTM STP 590, (1976), pp.27-42.

{2} A. SAXENA and S.J. HUDAK Jr.,
Int. Journ. of Fracture, 14 (1978) pp. 453-468

{3} C.F. SHIH, H.G. DELORENZI and W.R. ANDREWS,
Int. Journ. of Fracture, 13 (1077) pp.544-548

{4} F.J. LOSS,
"Structural Integrity of Water Reactor Pressure Boundary Components ",
NRL Memorandum Report 3782, (1078)

{5} P. BALLADON,
" Comparison of unloading Compliance and Interrupted Test methods :
effect of specimen size and Notch effect ", CSNI Workshop on ductile
fracture test methods, PARIS, 1-3 December 1982.

{6} ASTM E 813-81
" J_{Ic} a measure of fracture toughness ", Annual Book of ASTM Standards,
Part 10, (1982) pp. 822-840.

{7} G.A. CLARKE and J.D. LANDES, in
" Toughness Characterization and Specifications for HSLA and Structural
Steel ", MANGONON Ed., The Metallurgical Society of AIME, (1979) pp.79-111.

{8} J.G. MERKLE, in
" Resolution of the Reactor Vessel Materials Toughness Safety Issue ",
Report NUREG - 0744, vol.III, Appendix C, (1981).

{9} G.A. CLARKE,
Fracture Mechanics, ASTM STP 743, (1981) pp.553-575.

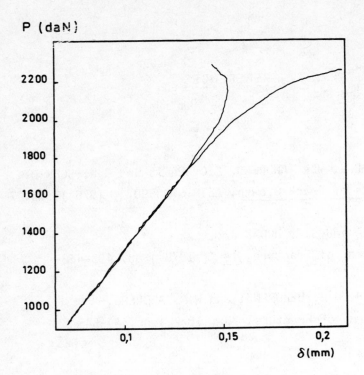

FIGURE 1 : TYPICAL UNLOADING ON STAINLESS STEEL

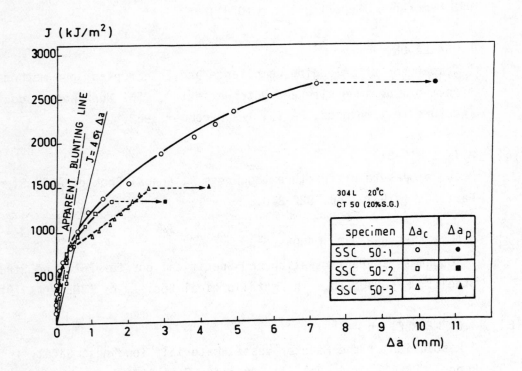

FIGURE 2 : J-Δa RESISTANCE CURVE OF THREE STAINLESS STEEL
SPECIMENS OBTAINED BY THE UNLOADING COMPLIANCE
METHOD.

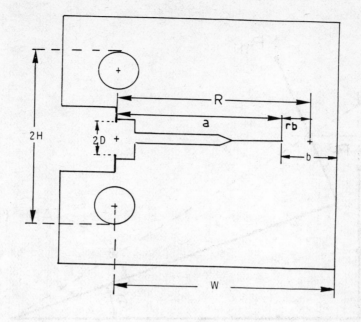

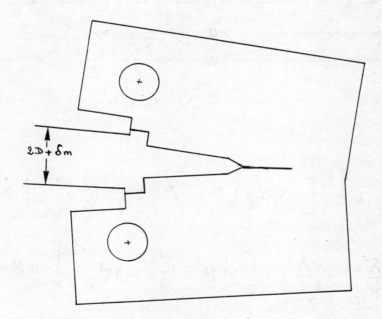

FIGURE 3 : DEFINITION OF DIMENSIONS USED IN THE CORRECTION
ROTATION EXPRESSION.

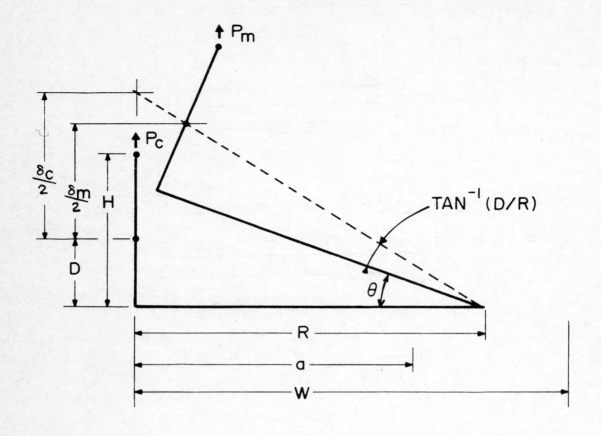

$$\theta = \text{SIN}^{-1} \left[(\delta M/2 + D) \middle/ (D^2 + R^2)^{0.5} \right] - \text{TAN}^{-1} \left(\frac{D}{R} \right)$$

$$\left(\frac{\delta}{P} \right)_C = \left(\frac{\delta}{P} \right)_M \left[\cos\theta - \frac{H}{R} \sin\theta \right]^{-1} \left[\cos\theta - \frac{D}{R} \sin\theta \right]^{-1}$$

FIGURE 4 : CORRECTION ROTATION EXPRESSION {8}

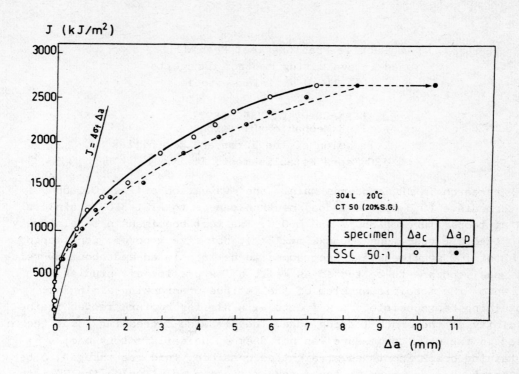

J (kJ/m^2)

$J = 4\sigma_f \, \Delta a$

304 L   20°C
CT 50 (20%S.G.)

specimen	Δa_c	Δa_p
SSC 50-1	○	●

Δa (mm)

FIGURE 5 : EFFECT OF ROTATION CORRECTIONS ON THE J-Δa
RESISTANCE CURVE.

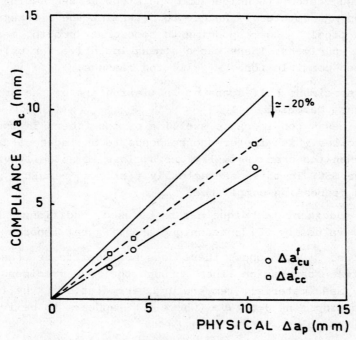

COMPLIANCE Δa_c (mm)

$\simeq -20\%$

○ Δa^f_{cu}
◻ Δa^f_{cc}

PHYSICAL Δa_p (mm)

FIGURE 6 : COMPARISON OF UNCORRECTED AND CORRECTED FINAL CRACK
EXTENSIONS OBTAINED BY COMPLIANCE METHOD TO THE
PHYSICAL CRACK EXTENSION (HEAT TINTING METHOD)

Ductile Fracture Test Methods
used in evaluating neutron damage to
AISI 316 H stainless steel

J. BERNARD   , G. VERZELETTI
Applied Mechanics Division
Commission of the European Communities
JRC Ispra Establishment, Italy

The FM research in JRC concerns mainly the evaluation of low (0.1 dpa) to intermediate (0.3 dpa to 1 dpa) neutron damage to AISI 316H stainless steel at temperature of 350°C and 550°C. The test specimens are relatively small sized 3PB specimens (80mmx20mmx15mm), not size grooved. Considering that about 100 specimens are programmed to be irradiated and about 50 reference specimens are to be tested as well, a few preliminary studies have been run on the specific problem of the fatigue precracking aiming at:

1) obtaining reproducible crack fronts with similar rupture facies in the vicinity of the crack leading edge. The stepwise decreasing loads technique used in the usual procedure was not deemed the most satisfactory one;
2) obtaining crack fronts as straight as possible. This was thought to be an asset when determining the crack initiation point using the PD method.

These goals have been reached in a satisfactory way using a smoothly decreasing cyclic load amplitude controlled in such a way that the deflection at the load point of the specimen is kept constant. Figs. 1,2 show the crack fronts obtained using the suggested technique and the stepwise decreasing load technique. All tests are run using computer technique to determine the J integral, J_{ic} at initiation and a final J_f corresponding to some crack growth, measured by heat-tinting the specimens. There was a strong incentive for using the SSCT but this has not been possible for the following reasons:

1) rotation of the specimens half-spans at initiation are of the order of 8° to 10° for the base material
2) this fact determines a considerable swelling of the fibers in the vicinity of the central roller which in turn corresponds to an increase of the ligament size even when the crack actually grows. This is actually visible on fig. 1   . The SSCT in this case actually yields as result a practically constant or even decreasing compliance.

Since the multiple specimens technique had to be used, additional information concerning the J_{IC} values by PD dc technique are of great importance.

As the graphs of Figs. 3, 4, 5 show, there does not seem to be constantly a good coeherence between regression lines through the four specimens results available for each case (material, temperature, irradiation), the intercept of those with a blunting line $J = \alpha \bar{J} \Delta a$ (where $\alpha = 2$ appears to be irrelevant in most cases) and the values of J_{IC}.

In our view this can be described to the fact that the determination of some stretch zone is practically impossible considering also that no SEMIcroscopy can be run on the irradiated material (as yet, at the JRC).

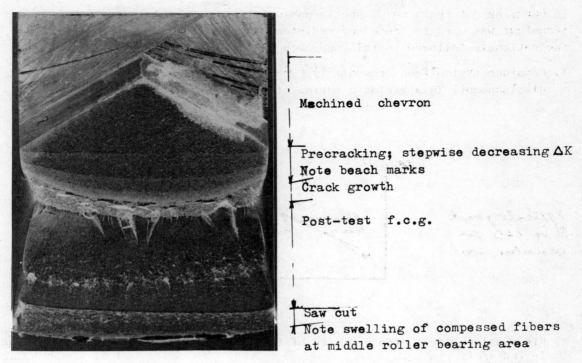

Machined chevron

Precracking; stepwise decreasing ΔK
Note beach marks
Crack growth

Post-test f.c.g.

Saw cut
Note swelling of compessed fibers
at middle roller bearing area

FIG. 1 3PB: precracking by decreasing ΔK *stepwise*.

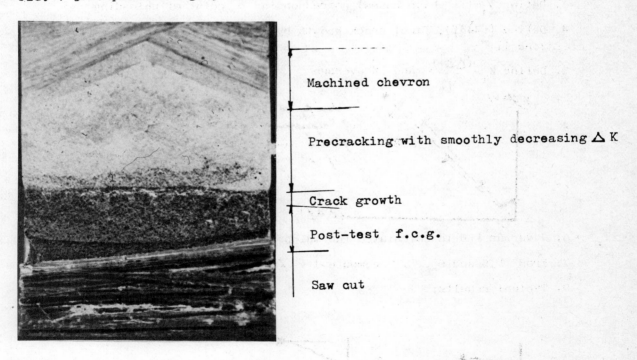

Machined chevron

Precracking with smoothly decreasing ΔK

Crack growth

Post-test f.c.g.

Saw cut

FIG. 2 3PB: precracking at constant middle point deflection

In carrying out tests on CT specimens in AQ42 ferritic steel, a computer technique was used to trace the resistance curves using the PDdc information, the rationale followed in this approach is as follows:

1. Consider typical PDdc measurements $[\varphi(\mu V)]$ in function of i.e. load-line displacement. This yields a series of graphs

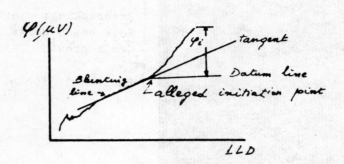

Typical graph φ vs. LLD for stainless steel.

2. Fix a datum line in the vicinity of the change in curvature convex-concave of the φ vs. LLD curves.

3. Define φ_i (i = 1..n cases) corresponding to point of unloading

4. Define $(\Delta a_f)_i$; final crack growth by heat-tinting, excluding the stretch zone i.

5. Define $K_i = \dfrac{(\Delta a_f)_i}{\varphi_i}$ and $\bar{K}$ average

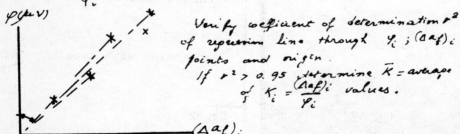

Verify coefficient of determination r^2 of regression line through φ_i ; $(\Delta a_f)_i$ points and origin.
If $r^2 > 0.95$ determine $\bar{K}$ = average of $K_i = \dfrac{(\Delta a_f)_i}{\varphi_i}$ values.

6. J versus LLD is calculated (normal practice).

7. From 1, 5 and 6 ., compute J vs Δa with $\bar{K}$ value.

8. Typical results:

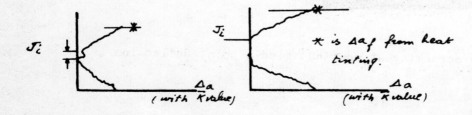

$*$ is Δa_f from heat tinting.

9. Conclusions: As far as the definition of J_{Ic} is concerned the method is not sensitive to the scaling factor $\overline{K}$.

As far as the obtained J-R curve is concerned, it is meaningful if $(\Delta a_f)_i$ by heat tinting corresponds closely to the $(\Delta a_f)_i$ obtained with $\overline{K}$.

Typical results are shown on Fig. 6. In our view this method is worth to be checked on the test results of the 48 unirradiated 3PB speciemns. Indeed in this case the fact that maximum crack growths to be taken into consideration for J controlled growth are only 700 um at the most, a continuous information on the J resistance curve from initiation to final crack length is most useful. In order to fully endorse this methodology, fractographic examination of stretch zone width and amount of crack growth should confirm the dcPD results. This work is actually in progress at JRC on unirradiated material.

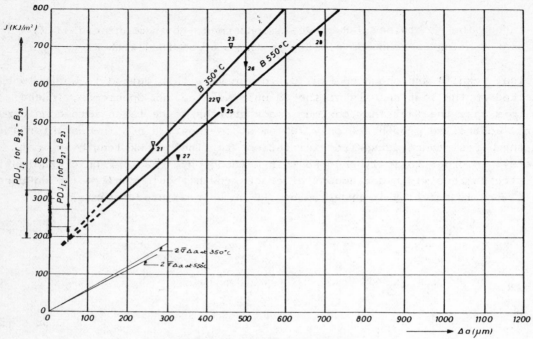

Fig.3 BASE MATERIAL ; 78 DAYS AGING IN Na

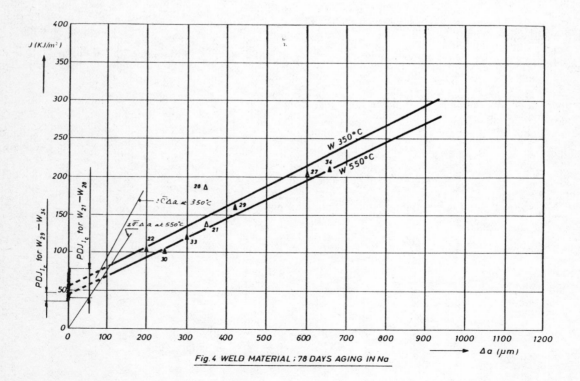

Fig.4 WELD MATERIAL ; 78 DAYS AGING IN Na

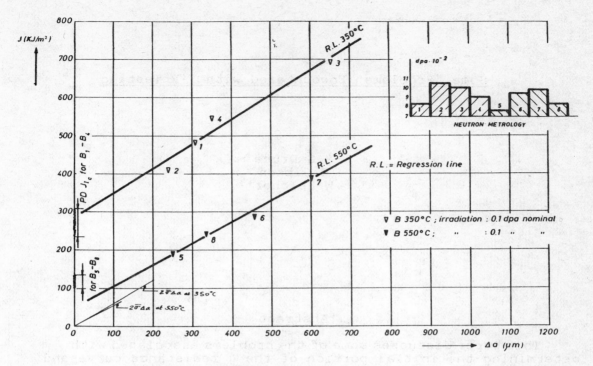

Fig.5 BASE MATERIAL ; 1 CYCLE IRRADIATION

J - RESISTANCE CURVES

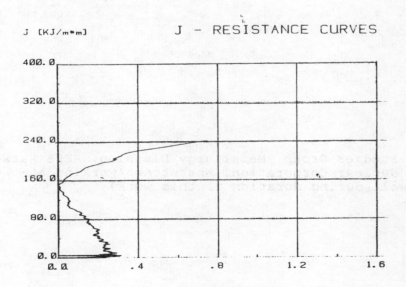

FIG. 6 Typical PDdc J-resistance curve for AQ-42 CT specimen

Some "Problems" Encountered with J-R Testing

S G Druce [+]
G Gibson [+]
W Belcher [++]

Abstract

This paper discusses some of the problems associated with determining the initial portion of the J resistance curve and defining the onset of ductile crack initiation. The ASTM J_{IC} testing standard E813-81 is used as a basis for evaluating multi-specimen data obtained from several ferritic steels. Deficiencies in this procedure are highlighted and some amendments proposed.

[+] Fracture Studies Group, Metallurgy Division, AERE Harwell
[++] National Nuclear Corporation, Whetstone (working at
     AERE Harwell during duration of this work)

1   Introduction

At Harwell, the interrupted loading multi-specimen test method has been used regularly over the past 4 years to determine the onset of ductile initiation and the resistance to subsequent ductile crack growth in several ferritic steels including ASTM A508, A533B, A542 and C-Mn steels. Usually some 6-9 specimens are tested to define the resistance curve up to 2-3mm ductile crack extension. However on occasions as few as 4 and as many as 21 specimens have been used.

Experimental programmes determining J-R data generally have at least one of the following objectives

(i)   To quantify the intrinsic material toughness (as for example in examining microstructural effects)

(ii)   To determine fracture mechanics data for subsequent use in structural integrity assessment

(iii) To further our understanding of elastic plastic fracture mechanics, EPFM (as for example in the study of geometric effects, side-grooving, or loading compliance)

These objectives impose different requirements on the testing method and what constitutes a "problem" depends very much how the data derived from the test is to be used. For example, comparison of the intrinsic toughness of materials requires a testing procedure allowing consistency with other workers but without necessarily the need to define an "initiation" criterion. However, several integrity assessment procedures currently use an "initiation" concept and therefore require some definition of this point. If J at "initiation" is being determined to derive a value of K_{IC} for subsequent use in linear elastic fracture mechanics (LEFM) integrity analyses then consistency with test methods determining the LEFM K_{IC} value is required. Such an approach would not intentionally recognise the apparent increase in toughness from crack extension. Alternatively in other integrity assessment methods some small amount of crack extension say, $\approx$2mm, may be allowed on the basis that J quantifies the stress and strain field remote from the immediate crack tip vicinity for restricted amounts of crack growth. In these cases the definition of "initiation" need not correspond with K_{IC} procedures and tends to be less important as the beneficial effects of increasing toughness with small increments of crack growth will be included in the integrity assessment method. However the specimen geometry and size must be chosen to ensure 'J control' for the required amount of crack extension. Finally, there is another class of integrity assessment procedures which totally disregard the initiation of ductile crack extension but predict the onset of unstable ductile crack extension which may only occur after large amounts of crack growth. Under these conditions the physical interpretation of J is not clear as it no longer quantifies the strain field ahead of the immediate crack tip vicinity but may represent the energy absorption term in an energy balance. For this type of analysis to remain conservative the constraint of the specimen being tested must be greater than that of the structure but need not satisfy the generally more onerous conditions for 'J control', or plane strain.

In view of these varied and sometimes conflicting requirements, it is unlikely that a single testing procedure will satisfy all applications. In the authors' opinion there is a pressing requirement for standardization in test methods for each application. Furthermore the acceptability of any testing standard for use in integrity assessments must be evaluated in conjunction with the assessment procedure using experimental data encompassing structural as well as specimen tests.

In the remaining sections of the paper various aspects of testing are discussed with reference to defining a testing procedure concerned with deriving data for comparing materials or for use in integrity assessments adopting an "initiation" concept followed by small amounts of 'J controlled' crack growth. Only ductile crack extension is considered and before small specimen data is applied to large structures, additional testing may be necessary to ensure that a fracture mode transition will not occur. Currently the only formally published J integral testing procedure addressing this topic is ASTM E813-81[1] and this is primarily designed to determine a toughness value at the initiation of ductile crack growth that can be used as a conservative estimate of the plane strain fracture toughness K_{IC}. The requirements and procedures of E813 are therefore taken as a basis for further discussion.

2 Specimen Size Effects

E813 specifies that for valid J_{IC} measurement the specimen net thickness (B_{net}), ligament (W-a) and crack length (a) all exceed 25 J_{IC}/σ_{flow} and that for all data points used in the linear regression B_{net}, W-a, a $>15J/\sigma_{flow}$ where σ_{flow} is the mean of 0.2% proof stress and ultimate tensile strength. For mild steel where $J_{IC} \simeq 0.1$ MNm^{-1} and $\sigma_{flow} \approx 390$ MPa this effectively imposes a minimum specimen thickness of ≈ 6mm, and for a standard 37.5mm thick specimen allows J values up to ≈ 1 MNm^{-1} to be included in the regression analysis. A systematic study by one of the authors of the effects of specimen thickness, width and degree of side grooving on ductile initiation and subsequent propagation has been reported [2] where geometry and in particular specimen thickness and side grooving are shown to influence the level of toughness despite satisfying the above size requirements for valid J_{IC} determination. Fig 1 shows a typical set of data for one geometry analysed to give an initiation value $J_{\Delta a=o}$, and ductile crack growth resistance value dJ/da. Fig 2 illustrates the effect of various specimen thicknesses, all exceeding the E813-81 minimum on $J_{\Delta a=o}$ and dJ/da for a constant specimen width and a/w = 0.55. Fig 3 shows the effect of increasing the degree of side grooving in an otherwise constant geometry. Clearly dJ/da is markedly influenced by both thickness and side grooving decreasing with increasing constraint. Any effect on initiation is less marked and obscured by the relatively large statistical scatter associated with the linear regression. However a small effect of thickness on $J_{\Delta a=o}$ is consistent with measurements of the critical stretch zone width (SZW) taken from the central portion of the specimen using scanning electron microscopy and shown in Fig 4. In taking numerous measurements at different locations, the statistical confidence in the mean measurement can be substantially improved compared with the limited number of data points used in the ASTM linear regression procedure.

These results indicate that even for small amounts of crack extension geometry independent J-R behaviour is not assured by the current E813 size requirements. However similar exercises (3,4) conducted using PWR PV steels have repeatedly indicated much less sensitivity to geometry. In conclusion the J/σ_{flow} scaling parameter does not always appear to define adequately size requirements for geometry independent behaviour.

3 Blunting Line Formula

In E813-81 J_{IC} is defined by the intersection of the regression line through the valid data points with an idealised blunting line given by

$$J = 2 \, \sigma_{flow} \, \Delta a$$

For low toughness materials, where the slope of the R curve is shallow, J_{IC} varies little with variation in the blunting line proportionality constant over the range 1.5-5.0. However for tough materials with steeply rising R curves J_{IC} can vary by a factor of two or more depending on the exact value of the proportionality constant. It is therefore important that the value used closely relates to the observed amount of crack tip blunting, if an estimate of "true initiation" is required

Fig 5 shows data for a high toughness pressure vessel steel tested at 290°C. The intersection of the E813 blunting line with the regression line occurs at a crack extension of 0.44mm whereas the measured critical stretch zone width is 0.097mm. This overestimation of the amount of blunting has resulted in significant amounts of ductile crack extension occuring prior to the defined "initiation" point and therefore results is an unduly high value of "initiation" toughness. The occurence of true ductile crack extension prior to "initiation" is particularly important if the data from small scale specimens is to be applied to larger structures where, in the transition temperature regime, brittle fracture may occur at loading less than that corresponding to J_{IC}.

4 Regression Analysis Procedure

The ASTM procedure uses a linear regression to valid data points within in limited Δa range treating crack extension as the controlled variable and J as the dependent variable. A major criticism of this method is that it attempts to fit a line to non-linear behaviour and therefore the result is subject to the chosen Δa

range and the exact positioning of data points within that range. In high toughness materials this can lead to a factor of 2 difference in J_{IC}. An additional criticism is that the procedure is physically incorrect in assuming crack extension to be the controlled variable and therefore by implication the greatest variability or error in measurement to occur in the level of J. In practice, J is the controlled variable and can be measured with good precision. Crack extension is the dependent variable and exhibits a large specimen to specimen variability of up to a factor of 2.

In an attempt to identify an improved procedure several alternatives have been compared using the data set shown in Fig 6 analysed in accordance with ASTM E813, yielding $J_{IC} = 0.39$ MNm^{-1}. The alternative procedures are given in Table 1, and all have treated J as the controlled variable. Procedures (1) and (2) are linear regressions, included for comparison, (1) using all the data points and (2) using only those within the ASTM exclusion limits. Procedures (3)-(6) are all power law expressions using all the data points: (3) is a simple power law assuming equal weighting to each point; (4) combines linear and power law terms allowing each of the 3 variables to vary to produce the best overall fit; (5) also combines linear and power law terms but specifies the linear terms so reducing the number of free variables to 2; (6) is a power law including a offset. Procedure (7) is to fit a second order polynomial.

Procedures (3-7) were chosen as being non-linear expessions with up to 3 free variables. The combination of linear and power law term were included in procedures 4 and 5 to model the possibility of an initial linear blunting line, dominated by power law behaviour at higher values of crack extension. Procedure (6) included an offset to the power law term to model power law behaviour after some finite amount of crack extension. Clearly there are other possible non-linear expressions which could be used and preference for any one procedure would be greatly enhanced by a physical interpretation in terms of the micro-mechanisms of ductile fracture. In the absence of suitable models the alternative procedures may be evaluated only in terms of the statistical accuracy of the fit. Table 1 gives the residual sum of squares and variance ratio resulting form a least squares analysis to find the optimum parameter values for each procedure. A decreasing value of variance ratio indicates a statistical improvement in the way in which the expression fits the data. Also given in Table 1 are the J values at a crack extension equal to the stretch width as measured on the fracture surface, ie "true initiation".

Table 1 shows that all the non-linear procedures provide a considerable improvement in the degree of fit to the data as compared to the linear methods. The best fit is obtained from procedure 6 which combines a power law with an offset. For the set of data examined the offset providing the best least squares fit was numerically equal to the measured stretch zone width with the consequence that Ji from the above definition would be zero. This extreme case emphasises the general point that "true initiation" occurs at a low value of J and would represent an unduly conservative approach for design or integrity assessment. Procedure 6 has three free variables whereas the simple power law, procedure 3, only has two and provides the next best fit. Procedure 3 is therefore recommended for data analysis when there are only a few data points as commonly found in multi-specimen testing. Fig 7 shows a simple power law regression to the same data as given in Fig 6. The improvement in the degree of fit to the data is particularly noticeable at low Δa values, that is in the region where any "initiation" value will be defined.

The question of what can be considered a suitable "engineering" definition of ductile crack initiation is beyond the scope of this contribution and as discussed in the introduction should be evaluated in conjunction with a particular failure assessment procedure. One possible approach favoured by the present authors is to define initiation at an arbitrary amount of total crack extension specific to a class of materials. For example, 0.2mm might be a reasonable value for medium strength ferritic steels as in relative terms this is only a very small amount of crack extension, and yet sufficiently large to readily allow the determination of Ji without back extrapolation of the data points. Alternative approaches have been suggested; for example Loss[5] has proposed a 0.15mm offset to a $2 \sigma_{flow}$ Δa blunting line defining Ji at the point of intersection with a power law regression to the data. Further evaluation is required before the optimum definition may be defined.

5   <u>Statistical Confidence in J_{IC}/Ji Data</u>

     E813-81 specifies that a minimum of 4 valid data points are required to determine J_{IC} and does not require any statement of the statistical accuracy of determination. Unfortunately many materials exhibit sufficient specimen to specimen variability to provide only a very low confidence in the linear regression to a small number of data points. Fig 8 illustrates this point using data obtained from an A542 steel thought to be of good commercial quality. An ASTM analysis of the four circled points yields a valid $J_{IC} = 0.18$ MNm^{-1}. However to within 95% confidence the actual value lies in the range 0-0.34. Clearly such uncertainty is unacceptable for applications to components requiring demonstration of a high degree of integrity and whatever the application some assessment of confidence is desirable. Furthermore for materials exhibiting large specimen to specimen variability it is necessary to use a larger number of data points than currently recommended in order to adequately quantify Ji.

6   <u>Summary</u>

     J-R curve testing is used for a variety of purposes each with specific requirements. It is unlikely that a single procedure will be adequate for all applications. Testing procedures should be evaluated in the context of how the data so derived is to be used. In the case of integrity assessment this includes an appreciation the fracture assessment method.

     Data from a variety of ferritic steels has been examined using the current ASTM J_{IC} testing procedure as a basis for discussion. Deficiencies of the procedure have been identified with respect to specimen size requirements, the blunting line formula, the regression analysis procedure and the absence of any assessment of accuracy. In particular, there is a requirement for improved analysis procedures for both fitting the experimental data and defining an acceptable engineering definition for initiation.

References

1   ASTM E813-81. Standard Test for J_{IC}, A Measure of Fracture Toughness.

2   S G Druce, "Effect of Specimen Geometry on the Characterisation of Ductile Crack Extension in C-Mn Steel" in Advances in Fracture Research Ed D Francois pp843-854 Pergamon Press 1980

3   D A Davies, M G Vassilarus, J P Gudas "Specimen Geometry and extended crack growth effects on J_I-R Curve Characteristics for HY 130 and ASTM A533B Steels". Presented 2nd International Symposium on Elastic-Plastic Fracture Mechanics, Philadelphia October 1981. To be published by ASTM.

4   T Ingham, G Wardle, J Bland, "Influence of Specimen Size on the Upper Shelf Toughness of SA 533B Class 1 Steel". To be presented at SMIRT 7, 1983

5   F J Loss Ed, "Structural Integrity of Water Reactor Pressure Boundary Components, Annual Report, Fiscal Year 1979", NUREG/CR-1128 NRL Memorandum Report 4122, December 1979

TABLE 1

COMPARISON OF ALTERNATIVE ANALYSIS PROCEDURES

(cf ASTM J_{IC} = 0.39 MNm^{-1})

Procedure	Sum of Residual Squares $\sum_{I=1}^{N}(Yi-\bar{Y})$	Variance Ratio	J_i J at Δa = SZW MNm^{-1}
1 Linear - All data points $\Delta a = AJ + B$	0.152	0.0080	0.30
2 Linear - ASTM Valid Data $\Delta a = AJ + B$	0.065	0.0034	0.27
3 Power Law-Equal Weighting All data $\Delta a = AJ^B$	0.0391	0.0021	0.11
4 Free Linear + Power Law $\Delta a = AJ + BJ^C$	0.0388	0.0022	0.10
5 Forced Linear + Power Law $\Delta a = \dfrac{J}{4\sigma F} + AJ^B$	0.0430	0.0023	0.15
6 Power Law - With offset ($\Delta a = A' + BJ^C$)	0.0327	0.0018	0
7 Polynomial $\Delta a = A + BJ + CJ^2$	0.0399	0.0022	0.16

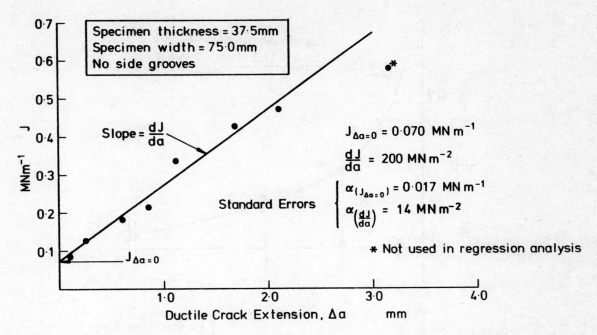

Fig 1   Typical experimental data set analysed to give initiation $J_{\Delta a=0}$, and crack growth resistance, dJ/da

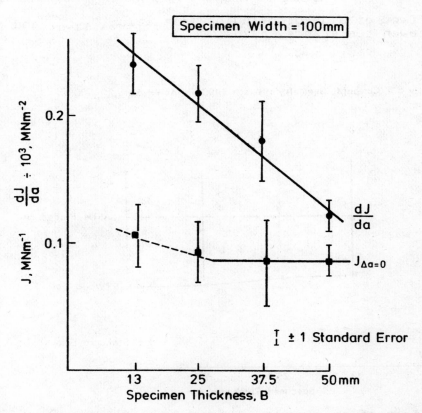

Fig 2   Effect of specimen thickness on initiation $J_{\Delta a=0}$, and crack growth resistance, dJ/da

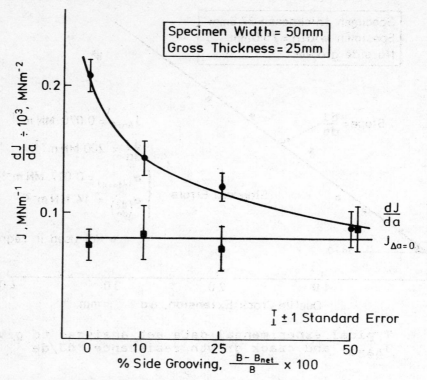

Fig 3   Effect of side-grooving on initiation $J_{\Delta a=0}$, and crack growth resistance, dJ/da

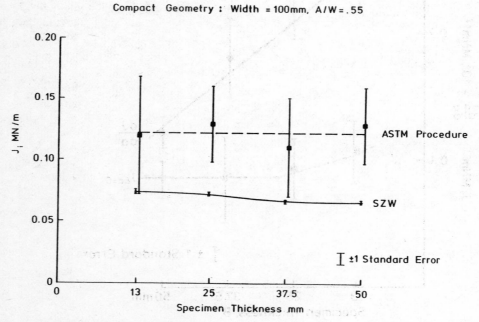

Fig 4   Effect of specimen thickness on initiation J derived from the ASTM procedure and critical stretch zone width (SZW) measurements

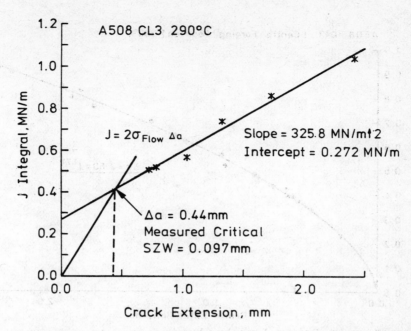

Fig 5   R Curve for A508 Class 3 at 290°C. Comparison of ASTM Blunting line with measured critical stretch zone width (SZW)

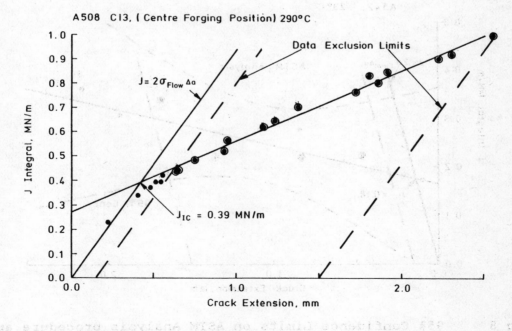

Fig 6   ASTM Data Analysis procedure applied to A508 Class 3 data

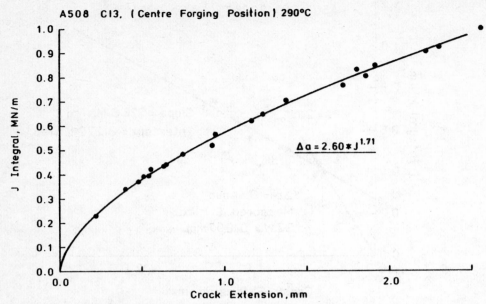

Fig 7   Simple Power Law Analysis procedure applied to A508 Class 3 data

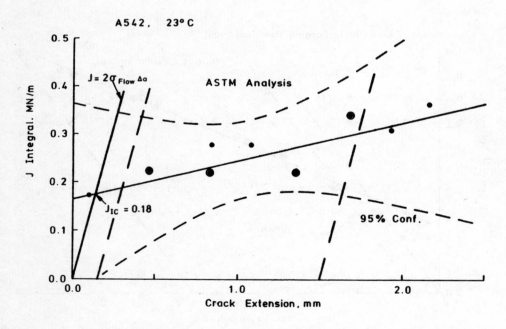

Fig 8   95% Confidence Limits on ASTM Analysis procedure applied to A542 data

PROBLEMS IN MEASURING CRITICAL J-INTEGRALS WITH
THE CURRENT ASTM E813 STANDARD METHOD

by

P. Uggowitzer, M. O. Speidel
Swiss Federal Institute of Technology, ETH
Institute of Metallurgy, ETH-Zentrum
CH-8092 Zurich, Switzerland

SUMMARY

Problems are highlighted concerning the present ASTM E813
for elastic-plastic fracture toughness testing using the J-Integral method.
Extensive investigations at the Institute of Metallurgy, Swiss Federal
Institute of Technology (ETH) Zurich, Switzerland, with a fully computerized,
single-specimen, partially-unloading technique show that for optimum and
accurate measurements of a critical J, three conditions must be met: First,
a reasonably large number of data points for the R-curve must be generated.
Second, these points should be evenly distributed up to $\Delta a = 1.5$ mm. Third,
the R-curve should be fitted to a polynom rather than a straight line.

I. Introduction

More and more, critical J-Integral measurements (J_{IC}) are used to characterize the fracture toughness of relatively tough and ductile materials. The reason for the popularity of the J-Integral measurements is due to the fact that they permit to obtain fracture toughness data with specimens of moderate size. However, the drawbacks of the J-Integral method include an incertitude of the proper test evaluation and therefore no total general acceptance of the method as a standard test. Although an ASTM standard test method has been developed (ASTM - E813, Ref. 1), its validity has been questioned and we cannot be sure that this standard even measures a materials property which is independent of procedural changes permitted by the standard. Moreover, it is by no means sure that the J-Integral measurements obtained according to ASTM-E813 are independent of the specimen dimensions as would be required if it was a materials property only.

II. Effects of procedural changes permitted by ASTM-E813

Figure 1 presents the crack resistance curve (R-curve) for a steam turbine rotor steel whose composition and mechanical properties are listed in table 1. The J-Integral and crack length data have been obtained with a fully computerized, single-specimen, partially-unloading testing technique. The tests were run at ambient temperatures, the CT specimens having no side-grooves and an a/w ratio of 0.6. A total of 30 data points have been measured and are presented in Fig. 1. The square points are those permitted by ASTM-E813. The data point represented by the crosses in figure 1 are not used in the evaluation of a critical J-Integral (J_{IC}) according to ASTM-E813 because they are situated outside the excluding lines at $\Delta a = 0.15$ mm and $\Delta a = 1.5$ mm. According to ASTM-E813, the intercept of a linear regression line of the not excluded data points with the so-called blunting line determines the critical J-Integral. According to figure 1, this J_{IC} equals 189 kJm^{-2}.

However, according to ASTM-E813 we are also permitted to evaluate a critical J-Integral with a smaller (sub-) group of data points. This is illustrated in figures 2, 3 and 4. For the first one of these different evaluations we use only the first ten data points closest to the blunting line (square points 12 to 21 in figure 2). The intercept of a linear regression line of these ten points with the blunting line results in J_{IC} = 152 kJm^{-2}, a strikingly different value from the result in figure 1, although both J_{IC} results were evaluated within the limits of ASTM-E813.

If, however, only the twelve valid data points farthest from the blunting line are used for the regression (square points number 18 through 29 in figure 3), the J_{IC}, still according to ASTM-E813, turns out to be 229 kJm^{-2}.

Yet a different evaluation of J_{IC} is shown in figure 4. In this case we have used only the four data points 17, 21, 25 and 29 which are not clustered. In this case, J_{IC}, evaluated according to ASTM-E813, is 236 kJm^{-2}.

We conclude that the best evaluation of critical J-Integral data should be based on a very large number of data points with equal distribution between the exclusion lines, if, indeed, the ASTM-E813 standard is to be used. The single-specimen, partial-unloading technique appears to be the best method to fulfill these requirements.

III. Polynomial instead of linear regression

It is obvious that the experimentally measured J-values between the exclusion lines Δa = 0.15 mm and Δa = 1.5 mm in figure 1 are much better represented by a convex curve than by a straight line. We have approximated the location of these valid data points by a polynom of the second order ($A_0 + A_1 x + A_2 x^2$). This is the dashed line in figure 1. Note that unvalid data points below Δa = 0.15 are also well approximated by the polynom. It is obvious that the J_0 value corresponding to the crack initiation is much more accurately given by the intersection of the polynom with the blunting line than by the intersection of the linear regression line with the blunting line.

We conclude that J_0 integrals corresponding to crack initiation would best be determined using a suitable polynom (Ref. 2, 3) rather than linear regression (Ref. 1).

IV. Effect of specimen size

A comparison of figures 1 and 5 permits the conclusion that there is a marked influence of specimen size (CT-1 versus CT-2) on the critical J-Integral value, no matter whether measured by the ASTM-E813 method or measured by the polynomial regression using the exclusion line $\Delta a = 0.15$ mm.

A specimen size effect on J_{IC}, measured by the multi-specimen technique, using the ASTM-E813 procedure, has been observed earlier (Ref. 4). There, however, it was concluded that the specimen size effect would disappear if a real J_o, corresponding to the real crack initiation was measured. In our measurements (figures 1 and 5) this specimen size effect does not disappear if we use the polynomial regression with the exclusion line $\Delta a = 0.15$ mm. It might however disappear if data points below the exclusion line $\Delta a = 0.15$ mm are incorporated in the polynomial regression.

REFERENCES

1. ASTM-E813-81 Standard Test for J_{IC}, a Measure of Fracture Toughness, Annual Book of ASTM Standards, Part 10 (1981)

2. Loss, F.J., Menke, B.H, Gray, R.A., and Hawthorne, J.R. (1979) Proceedings of the US Nuclear Regulatory Commission, St. Louis, Missouri, USA, Sept. 25-27.

3. Sigrist, P., Uggowitzer, P, and Speidel, M.O., in preparation.

4. Keller, H.P. (1979) Forschungsbericht der Deutschen Forschungs- und Versuchsanstalt für Luft- und Raumfahrt, DFVLR-FB 79-03, Köln, Germany.

TABLE 1 : Chemical Composition and Mechanical
Properties

Chemical composition:

C	0.29	Ni	2.79
Si	0.09	Cr	1.58
Mn	0.28	Mo	0.38
S	0.008	V	0.10

Mechanical properties:

$R_{p0.2}$ = 780 MNm^{-2}

R_m = 920 MNm^{-2}

A = 17 %

Z = 55 %

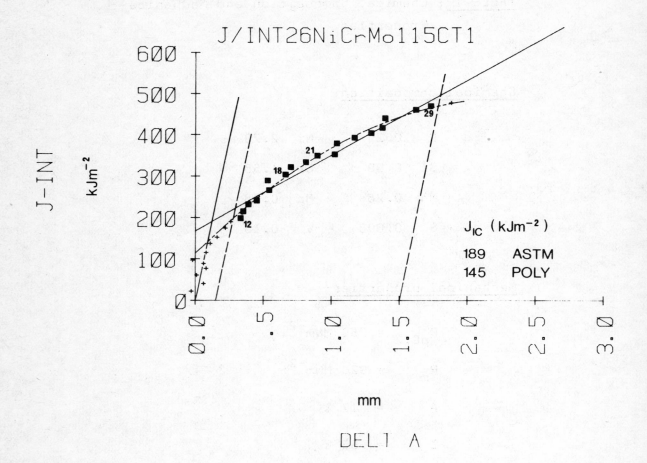

Figure 1 : J-R-curve obtained from CT-1 specimen
with the single-specimen partial unloading method. J_{IC}-values are evaluated
by linear regression (ASTM) or polynomial regression (POLY).

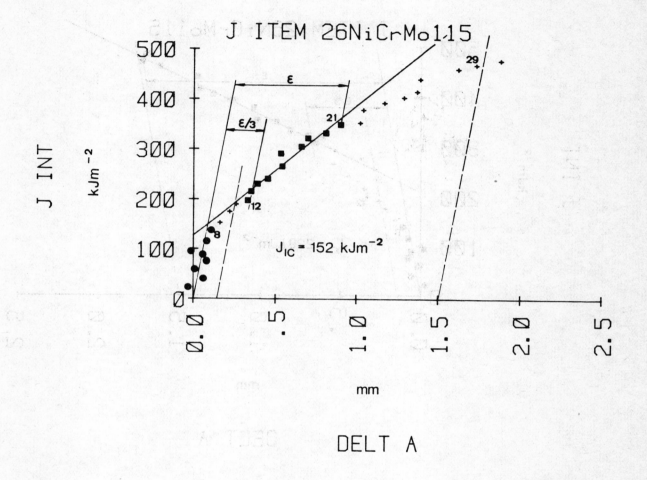

Figure 2 : J-R-curve with a sub-group of data
points (assuming that no unloading
was done for points 22 to 29).

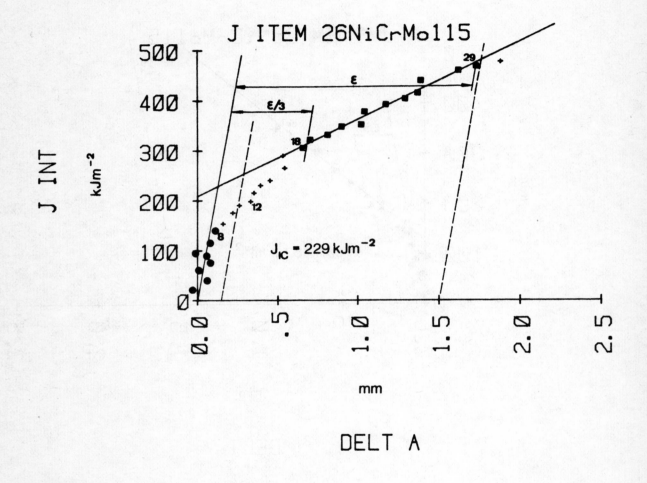

<u>Figure 3</u> : J-R-curve with a sub-group of data
points (assuming that no unloading
was done for points 12 to 17).

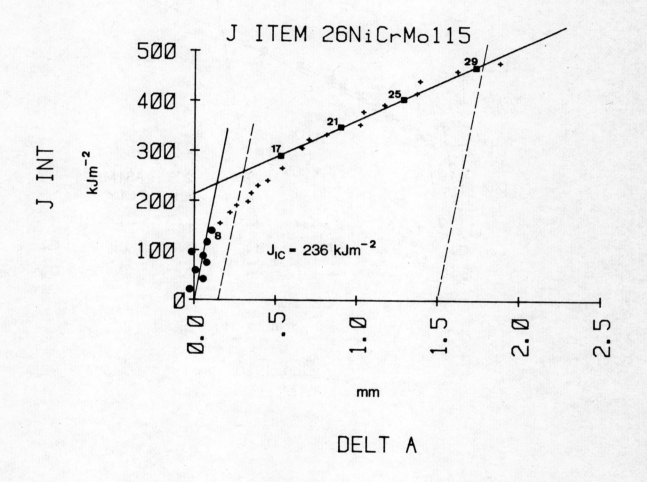

Figure 4 : J-R-curve with only 4 equally
spaced points.

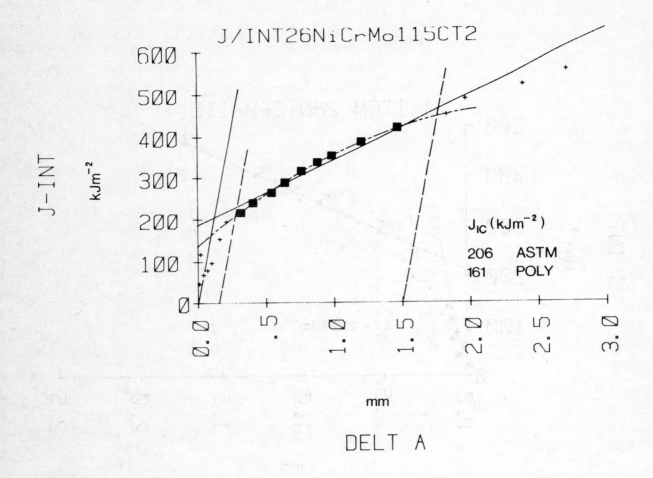

Figure 5 : J-R-curve obtained from CT-2 specimen.
J_{IC}-values are evaluated by linear
regression (ASTM) or polynomial regression (POLY).

APPLIED J-INTEGRAL VALUES IN TENSILE PANELS

D. T. Read
Fracture and Deformation Division
National Bureau of Standards
Boulder, CO 80303

ABSTRACT

The J contour integral has been applied widely in characterizing the fracture toughness of metals. In addition, the applied J-integral can be used to characterize the driving force for fracture in structures. This use of the J-integral requires knowledge of the dependences of the applied J integral on stress, strain, and crack size. Results from an experimental study of the applied J-integral as a function of strain in tensile panels are discussed in this paper. The main point of the paper is that the character of the relationship, between applied J-integral and applied stress and strain, changes with crack length. This point is illustrated by previously reported experimental data. Variables controlling the applied J integral are associated with the three observed deformation patterns: stress controls J for linear elastic strains and contained yielding, displacement controls J for net section yielding, and strain controls J for gross section yielding. Experimental results of this study show that the applied J-integral value for a given strain level is much higher for net section yielding than for gross section yielding. The yielding pattern obtained was found to depend on the ratio of crack length to specimen width. The transition from gross to net section yielding in the material tested occurred as the ratio of crack length to specimen width was raised from about 1 to about 5 percent. The significance of this transition is two fold; first, a strong increase in applied J values is produced by a small increase in crack size; second, this effect occurs at crack sizes that are small compared with specimen width.

Key Words: fracture mechanics, elastic-plastic; yielding; contained; net section; gross section; pattern; fitness-for-service.

This report was prepared as part of the Fracture Control Technology program under the sponsorship of Dr. H.H. Vanderveldt, Naval Sea Systems Command (SEA 05R15). The effort was directed by Mr. John P. Gudas, David Taylor Naval Ship R&D Center, under Program Element 62761N, Task Area SF-61-544-504.

INTRODUCTION

The J contour integral [1] has been applied widely in characterizing the fracture toughness of metals [2-4], but its application to evaluation of the durability of structures containing cracks has been delayed because dependences of the applied J-integral on stress, strain, and crack size have not been generally well established. In response to this need, a technique for measurement of the applied J-integral in tensile panels of HY-130 steel as a function of applied stress, displacement, and strain has been applied to the elastic-plastic strain range [5]. This technique is applicable to short, as well as long, cracks, unlike the J-integral evaluation techniques commonly used in material toughness testing. In this paper, experimental results on the behavior of the applied J-integral are discussed, and conclusions are drawn about the factors that control the applied J-integral. Two concepts central to the development of the arguments in this paper, controlling variables for J and deformation patterns, are now introduced:

The use of J as a fracture-characterizing parameter requires that J be related to structural loading. To formulate such relationships, the structural loading must be appropriately characterized. Some variables that might be used to characterize structural loading are remote stress, remote strain, strain over some gage length, and applied displacement. This choice of a characterizing parameter is not necessary in linear elastic fracture mechanics (LEFM) because the laws of linear elasticity enforce consistent relationships among the variables. For example, the stress and strain, are related through the material elastic constants. The LEFM stress intensity factor is related unambiguously to either remote stress or strain. Widespread yielding invalidates the laws of linear elasticity and also the fixed relationships among the loading variables. Therefore, one may not simply assume that some variable, for example stress, is the appropriate controlling variable for J in yielding fracture mechanics; the appropriate controlling variable must be identified. For some materials power-law hardening occurs over the entire strain range of interest, and so the various loading parameters may scale with one another [6]. In HY130, however, the stress-strain behavior is nearly elastic-perfectly plastic, so that deformations above yield can occur at essentially constant stress. This behavior prevents reliable scaling of deformations against applied stresses. The criterion for a proper J-controlling variable characteristic of the deformation state is that the J-integral must be a single-valued function of the controlling variable. Variables that change while the J-integral remains constant or that remain constant while the J-integral changes must obviously be rejected. A major argument herein is that different variables control J at different stress and strain levels and crack lengths for HY130 and possibly for many other materials.

A convenient classification of stress levels and crack lengths is by deformation patterns [7,8]. Figure 1 depicts deformation patterns characteristic of different deformation levels in a cracked specimen: linear elastic, contained yielding, net section yielding, and gross section yielding. Each of these patterns was observed in the present study and was important for interpreting the results. In this paper the appropriate controlling variables for the applied J-integral for the different deformation patterns are identified and formulas for estimating J for each region are discussed.

EXPERIMENTAL TECHNIQUE

The experimental technique used for measurement of the J-contour integral for elastic-plastic conditions was described in detail in a previous paper [5]. The materials tested was HY-130 steel, a 5-Ni, 0.4-Cr, 0.5-Mo steel with a yield strength of 933 MPa, an ultimate strength of 964 MPa, and very low strain hardening. Briefly, contours enclosing crack tips in single-edge-cracked, center-cracked, and surface-cracked specimens of HY-130 steel were instrumented with strain gages and linear-variable-displacement transducers (LVDTs) to measure the integrand terms of the J-integral. The integral was then evaluated numerically. Pin-loaded specimens with a gage section of 300 mm x 90 mm x 10 mm were used. Cracks were simulated by saw-cut notches of 1-, 2-, 4-, 7-, and 20-mm lengths. Some notches were sharpened by fatigue. Load was measured using the built-in load cell of the 1 MN servohydraulic test apparatus. Displacement over the gage length was obtained from the LVDT outputs. Remote strain was obtained from the strain gages near the end of the gage section. The experimental uncertainty in the J-integral in the plastic range was estimated at ± 10%.

RESULTS AND DISCUSSION

Measured values of the J-integral are plotted against gage length strain in Fig. 2. Gage length strain is defined as the change in the length of the gage section divided by its length. For most crack lengths, the J values assume a parabolic dependence on strain for low strain values and a linear dependence at higher strains. For a given strain level, J increases with crack length. In short, the general trends of the data in this figure are in accord with prior observations [9,10].

Appropriate controlling variables for J for the different deformation patterns are now considered.

Linear Elastic Strains and Contained Yielding

First, consider loads well below the nominal yield load of the cracked specimens. For such loads, the plastic zone at the crack tip is small compared with the length, width, and thickness of the specimen, so a linear elastic treatment is appropriate. For this situation, the elastic strain (given by σ/E where σ is the nominal remote stress and E is Young's modulus), the gage length strain, and the remote strain are all approximately the same except near the crack tip because linear elasticity holds throughout the specimen. Nominal stress is commonly used as the independent variable in linear elastic fracture mechanics [11]. In the linear elastic region it is clearly most convenient to regard nominal stress as the controlling parameter. Experimentally derived values of the applied stress intensity factor, K, are plotted against nominal stress in Fig. 3. The stress intensity factor data in this figure were calculated using measured J-integral values and the equation

$$K = \sqrt{(J\ E)} \qquad\qquad (1)$$

The experimental data of Fig. 3 clearly show the extent of the linear elastic regime; the stress intensity is proportional to the stress up to about 700 MPa, which is about 75% of the material yield strength. The experimentally measured K values are in agreement with predicted values from linear elastic fracture mechanics within the experimental uncertainty. For the elastic strain region, linear elastic relationships between applied stress σ, and stress intensity factor, K, for several crack configurations may be used. These are conveniently collected in Ref. [11].

As the load is further increased, the plastic zone expands rapidly outward from the crack tip. As long as the plastic zone remains sufficiently small, nominal stress remains the appropriate J-controlling parameter.[12]. This situation is termed contained yielding.

Net Section Yielding

The HY-130 material used in the present study has very little strain hardening, so a definite limit load is reached when the remaining ligament adjacent to the crack yields. The plastic zone penetrates to a free boundary forming a deformation band (Fig. 1). Upon further imposed displacement, the load no longer increases but the J-integral continues to increase because the crack continues to open. A plot of experimentally measured J-integral values vs. stress, Fig. 4, illustrates this behavior. After yield, nominal stress is no longer the J-controlling variable because J increases while stress remains constant. For relatively deep cracks, the remote strain becomes constant along with the nominal stress, because all the imposed deformation is transmitted to the deformation bands while the specimen ends move apart as rigid bodies. Therefore remote strain is also ineligible for consideration as the controlling parameter for J for this case. The gage length strain, that is, imposed displacement divided by some gage length, increases with the J-integral, but it depends on the gage length and so is not unique. The displacement, that is, the change of the gage length from its original value, increases along the width J and does not depend strongly on the choice of gage length, so it is a possible J-controlling variable for deep cracks. This choice is supported by relationships given by Rice, Paris, and Merkle [13] for the J-integral in specimens with deep cracks; in these relationships, J is proportional to the imposed displacement in tensile specimens:

$$J = -\frac{1}{B} \cdot \int_0^V dP/da \, dV. \qquad (2)$$

where V is imposed displacement at the load point, P is the load, and a is crack length. This relationship applies for both elastic and plastic deformation. In plasticity, P becomes the limit load. For the case of single-edge-cracked specimens pin-loaded in plane stress, the limit load is [14]

$$P = \sigma_f \, WB \, \{[(1-a/W)^2 + (a/W)^2]^{1/2} - a/W\} \, . \qquad (3)$$

Here σ_f is the flow stress and W the specimen width. Applying (2), assuming no strain hardening, the plastic part of J is found:

$$J_p = \sigma_f \, \{ \frac{1 - 2a/W}{[(1-a/W)^2 + (a/W)^2]^{1/2}} + 1\}V_p \, , \qquad (4)$$

where J_p is the part of J associated with plastic deformation and V_p is the total displacement less the elastic part. Using (4), the normalized rate of change of J with imposed displacement, M, given by $M = \Delta J/\sigma_y \, \Delta V$, is

$$M = \frac{1-2 \, a/W}{[(1-a/W)^2 + (a/W)^2]^{1/2}} + 1 \, , \qquad (5)$$

This M value is applicable for plastic strains. The flow stress for the limit load expression may be taken as the average of yield and ultimate strengths. This value is accurately known for HY-130 because its yield and ultimate strengths are nearly the same. Experimental data are shown in Fig. 5. In this figure, experimentally determined values of the quantity M are plotted against normalized crack length. The net section yield theory prediction (5) is shown as a broken line.

The noteworthy feature of this comparison is that the experimental data
fall away from the theoretical predictions of a/W at values less than
about 0.1. Similar behavior was previously observed for a C-Mn ship
steel [9]. Equation 2 with conventional limit load expressions usually
gives finite J values at a = 0. But a nonzero J value is nonsensical in
the absence of a crack. Therefore, at some crack length, J must deviate
from the conventional predictions and approach zero as crack length approaches
zero. The data fall approximately on a straight line through the origin,
although the actual behavior is probably more complex. Using the
slope of this line, g, dependence of J on displacement in this short-crack
net-section-yielding region, J_{sn}, can be described approximately by

$$\Delta J_{sn} = g \, a \, \sigma_f \Delta V/W \qquad (6)$$

The full value of J in this region is given by

$$J = J_e + g \, a \, \sigma_f (V-V_e)/W \qquad (7)$$

where J_e and v_e are the J-integral and displacement values at yield.
From the data of Fig. 5, the constant g was evaluated to be about 15.
The main point is that the experimental data show that J goes to zero as crack
length goes to zero, in contradiction to the simple theory of Eq. 5.
Therefore, a different theory must be sought.

In summary, in net section yielding for both long and short cracks,
imposed displacement controls the applied J value, but the simple theory is
inadequate for short cracks.

Gross Section Yielding

For very small crack lengths, the plastic strain is no longer concentrated by the crack tip in the remaining ligament but instead spreads throughout the whole specimen. The corresponding deformation pattern is gross section yielding. Strain is the controlling variable for the gross yielding region [15-17]. The transition from net section yielding to gross section yielding is illustrated in Fig. 6. This figure presents a plot of displacement contributed by strains associated with the crack (abbreviated crack-related displacement) against remote strain. The crack-related displacement is the total imposed displacement less the product of remote strain times gage length. Figure 6 shows that for the deeper cracks the remote strain remained constant as the crack-related displacement increased, as would be expected for net section yielding. But for the 1-mm crack the remote strain continued to increase and the crack-related displacement showed no steep rise at yield as in the net section yielding case, indicating that the entire specimen was yielding, behavior characteristic of gross section yielding. From this figure, it can be seen that the boundary between net section yielding and gross section yielding for the HY-130 material used in this study is located somewhere between the 1-mm crack and the 2-mm crack. It will be rounded off to a/W = 0.01 for purposes of the next section. Figure 2 shows that the measured J-integral values for the 1- and 2-mm cracks were much less than for the longer cracks. This illustrates the fact that J for the net section yielding is much higher that for gross section yielding. Relationships between J and strain for gross section yielding have been proposed previously [15-17]. These are in general agreement with one another but have not been critically evaluated in this study.

Deformation Behavior Map

It has been shown that the proper controlling parameters for J are stress, imposed displacement, and remote strain, and that each rules in a specific region of applied stress or strain and crack size; a behavior map depicting the location of these regions is shown as Fig. 7. Its ordinate is a logarithmic scale for normalized crack size, a/W. Its abcissa is normalized stress, σ/σ_y, up to $\sigma/\sigma_y=1$, and then it divides into separate scales for long and short cracks. For long cracks, the second part of the abcissa is displacement, and for short cracks, it is normalized remote strain, $\varepsilon_{rem}/\varepsilon_y$. For low nominal stresses, the deformation is linear elastic for all crack sizes. With increasing stress, specimens with long cracks yield first, and eventually as $\sigma/\sigma_y \to 1$ all specimens yield. The curve extending from the upper left corner of the graph down to $\sigma/\sigma_y = 1$, a/W $\approx$ 0 is the boundary between linear elastic behavior and contained yielding. The nearly parallel curve forms the boundary between contained yielding and extensive yielding. Extensive yielding is divided by a horizontal line into gross section and net section yielding. Here this line has been placed at a/W = 0.01. High strain hardening or yield elevation by constraint might raise this line. For small crack lengths (the gross section yielding region) the controlling parameter for J, namely, remote strain, is given on the lower abcissa. For larger cracks (net section yielding), the controlling parameter for J, namely displacement, is given on the upper abcissa. Within the net section yielding region, the short crack region is separated from the remainder by another horizontal line. Displacement is still the controlling parameter for this region, but the crack-size dependence is different. Deformation maps like this could be used to depict the effect of material, part geometry, or deformation type (for example, residual vs applied stress) on the formation of the different deformation patterns.

The large behavior difference observed between gross and net section yielding explains why one set of J-vs-displacement formulae, of the type given by Ref. [13], have been used in material toughness tests of deeply notched specimens, whereas an entirely different family of relationships between J and strain has been proposed for application to structures [15-17]. Structures usually have small cracks, which produce gross section yielding, whereas test specimens are deeply cracked, producing net section yielding.

Further Research

The most needed experimental extension of the present work is to geometries liable application to decisions about cracks in real structures. The most needed experimental extension of the present work is to geometries that produce yield elevation by constraint and materials with different stress-strain behaviors; such work is in progress. The two most important outstanding issues are: (1) the location of the transition from net section yielding to gross section yielding, and, (2) the dependence of J on remote strain and crack length for gross section yielding. Analytical methods, such as finite element analysis, also need to be developed to allow explanation and, eventually, prediction of the occurence of the various deformation patterns and their effects on applied J-integral values [18].

The question of the applicability of conventionally measured fracture toughness values to the variety of deformation patterns described here is an interesting problem that arises once net and gross section yielding behaviors have been clearly differentiated.

CONCLUSION

In this study, experimental data have been used to support the hypothesis that the variables controlling the applied J-integral in the linear elastic, contained yielding, net section yielding, and gross section yielding regions are, respectively stress, displacement, and remote strain. Experimental evidence in favor of the validity of existing relationships between J and stress, J and displacement, and J and strain in their appropriate regions has been presented, and a different relationship between J and displacement for the short crack net section yielding region has been derived from experimental results.

ACKNOWLEDGMENT

This report was prepared as part of the Fracture Control Technology Program under the sponsorship of Dr. H. H. Vanderveldt, Naval Sea Systems Command (SEA 05R15). The effort was directed by Mr. John P. Gudas, David Taylor Naval Ship R&D Center, Under Program Element 62761N, Task Area SF-61-544-504.

Technical assistance by J. D. McColskey and helpful discussions with H. I. McHenry, R. B. King, and R. H. Dodds, Jr. are gratefully acknowledged.

REFERENCES

1. J. R. Rice, "A Path Independent Integral and the Approximate Analysis of Strain Concentration by Notches and Cracks," Journal of Applied Mechanics 35, 1979, pp. 379-386.

2. J. A. Begley and J. D. Landes, "The J Integral as a Fracture Criterion," Fracture Toughness, ASTM STP 514, American Society for Testing and Materials, Philadelphia, 1972, pp. 1-20.

3. J. D. Landes and J. A. Begley, "The Effect of Specimen Geometry on J_{Ic}," Fracture Toughness, ASTM STP 514, American Society for Testing and Materials, 1972, pp. 24-39.

4. W. A. Logsdon, "Elastic Plastic (J_{Ic}) Fracture Toughness Values: Their Experimental Determination and Comparison with Conventional Linear Elastic (K_{Ic}) Fracture Toughness Values for Five Materials," Mechanics of Crack Growth, ASTM STP 560, American Society for Testing and Materials, 1976, pp. 43-60.

5. D. T. Read, "Direct Experimental Evaluation of the J Contour Integral," forthcoming.

6. V. Kumar, M. D. German, and C. F. Shih, "Estimation Technique for the Prediction of Elastic-Plastic Fracture of Structural Components of Nuclear Systems," General Electric Report SRD-80-094, 1980.

7. C. E. Turner, "Methods for Post-Yield Fracture Safety Assessment," Post-Yield Fracture Mechanics, edited by D. G. H. Latzko, Applied Science Publishers, London, 1979, pp. 23-210.

8. W. Soete and R. Denys, "Full and General Yield Behavior of Homogeneous Plates with Cracks," Proceedings of the International Institute of Welding Conference at Bratislava, Doc. X-921-79, 1979.

9. D. T. Read and H. I. McHenry, "Strain Dependence of the J-Contour Integral in Tensile Panels," Advances in Fracture Research, edited by D. Francois, Pergamon, New York, 1981, pp. 1715-1722.

10. R. J. Bucci, P. C. Paris, J. D. Landes, and J. R. Rice, "J Integral Estimation Procedures," Fracture Toughness, ASTM STP 514, American Society for Testing and Materials, Philadelphia, 1972, pp. 40-69.

11. H. Tada, P. C. Paris, and G. R. Irwin, "The Stress Analysis of Cracks Handbook," Del Research Corporation, Hellertown, Pennsylvania, 1973.

12. G. R. Irwin and P. C. Paris, "Fundamental Aspects of Crack Growth and Fracture," in Fracture, An Advanced Treatise, Vol. III, edited by H. Liebowitz, Academic Press, New York, 1971, pp. 1-46.

13. J. R. Rice, P. C. Paris, and J. G. Merkle, "Some Further Results of J-Integral Analysis and Estimates," Progress in Flaw Growth and Fracture Toughness Testing, ASTM STP 536, American Society for Testing and Materials, Philadelphia, 1973, pp. 231-245.

14. G. G. Chell, "Elastic-Plastic Fracture Mechanics," in Developments in Fracture Mechanics-1, edited by G. G. Chell, Applied Science, London, 1979, p. 76.

15. M. G. Dawes, "The COD Design Curve," in Advances in Elasto-Plastic Fracture Mechanics, edited by L. H. Larsson, Applied Science Publishers, 1980, London, pp. 279-300.

16. J. A. Begley, J. D. Landes, and W. K. Wilson, "An Estimation Model for the Application of the J-Integral," Fracture Analysis ASTM STP 560, American Society for Testing and Materials, Philadelphia, 1974, pp. 155-169.

17. C. E. Turner, "Elastic-Plastic aspects of fracture stress analysis: methods for other than standardized test conditions," in _Fracture Mechanics in Design and Service_, Phil. Trans. R. Soc. Lond. A299, 1981, pp. 73-92.

-18. R. H. Dodds, Jr., D. T. Read, and G. W. Wellman, "Finite Elment and Experimental Evaluation of the J-Integral for Short Cracks," to be published in Proceedings of the 14th National Symposium on Fracture, Los Angeles, 1981, sponsored by the American Society for Testing and Materials.

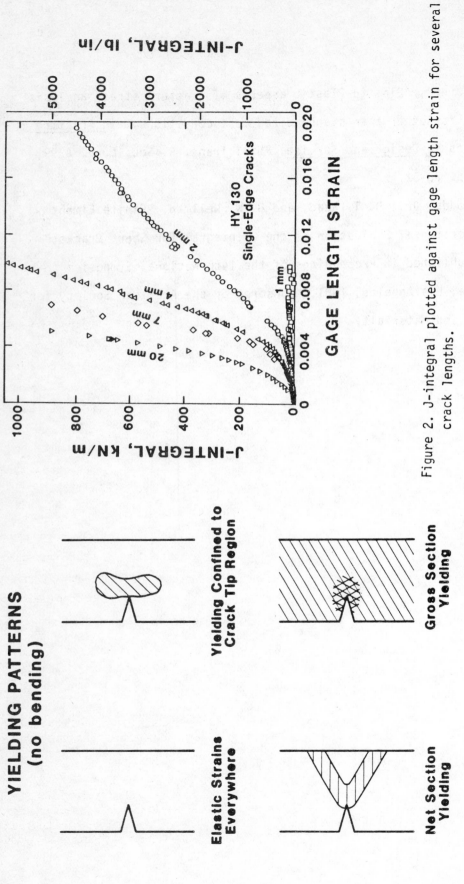

Figure 2. J-integral plotted against gage length strain for several crack lengths.

Figure 1. Deformation patterns.

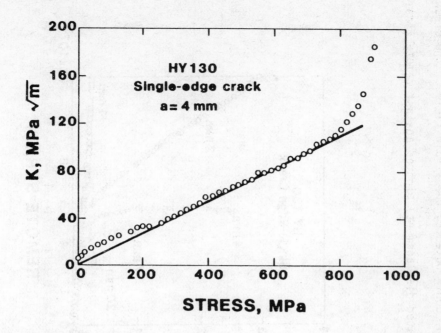

- Figure 3. Measured stress intensity factor plotted against remote stress for a single-edge-cracked specimen with a 4-mm crack.

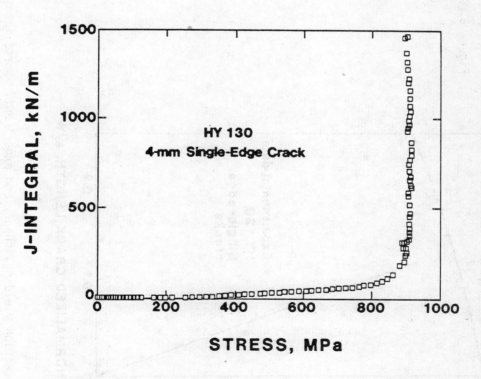

Figure 4. Measured J-integral plotted against remote stress for a single-edge-cracked specimen with a 4-mm crack.

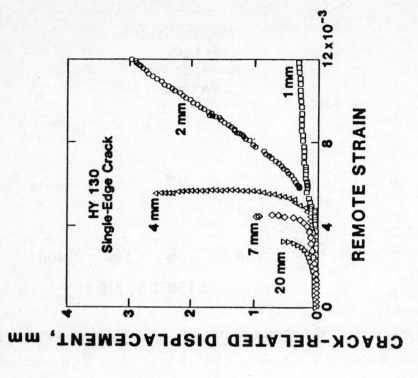

Figure 6. Displacement contributed by the strain field of the crack (crack-related displacement) plotted against remote strain.

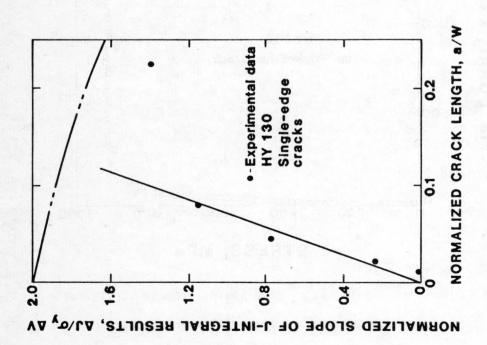

Figure 5. Net section yielding, theory and experiment. Normalized slopes of J-integral-vs.-displacement lines and experimental data.

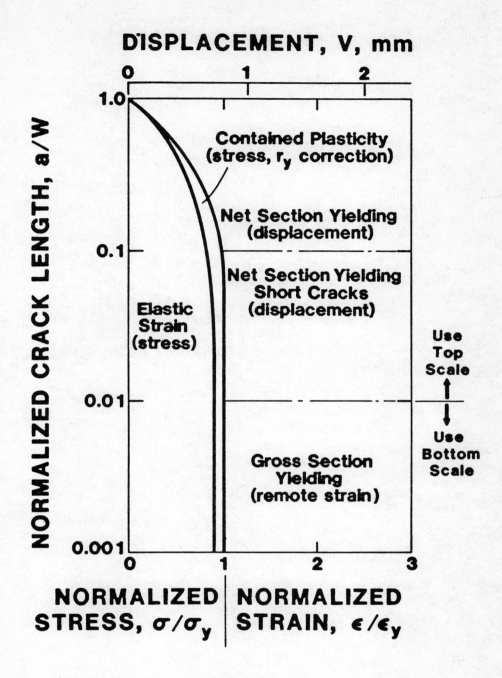

Figure 7. Deformation behavior map for tensile panels of HY-130 with cracks.

Session IV

TESTING OF STRUCTURAL ELEMENTS

Chairman - Président

L. CRESWELL

(United Kingdom)

Séance IV

ESSAIS SUR LES ELEMENTS DE STRUCTURE

SESSION NO. 4: TESTING OF STRUCTURAL ELEMENTS

Summary by the Chairman: L. Creswell

This fourth session of the Workshop dealt mainly with the application of ductile fracture methods to test pieces of direct structural relevance. Some interesting papers were also included which addressed new methods of fracture evaluation.

Perhaps the first point to emerge from this session's papers and discussions was the need for further development of fairly simple methods of analysis for ductile fracture for geometries other than those of the well-known standard test pieces. This is especially important when the testing of material from actual components is being considered and the amount of material is fairly limited.

A paper by Angelino and Reale, yielded some interesting conclusions which go some way to illustrate the above point. Their data showed J_I values for 100 mm diameter pipes falling within the scatterband obtained from small scale specimens. However, this same data indicated a far lower slope to the resistance curve for the large pipe than the small specimens, demonstrating apparent non-conservatism of the small-scale laboratory-type specimens. Although it was also made clear by the author, this observation as much as anything else, demonstrated the need for adequate methods of post-initiation analysis in situations such as the one described.

The results presented by MPA Stuttgart also indicate that initiation values (J_I) can remain invariant with a large change in specimen size, this conclusion was based on data from 14 mm to 500mm Compact Tension specimens.

Two speakers discussed the advantages of large wide plate tests over smaller laboratory specimens. During one of these presentations the proposition that the use of initiation-based integrity analysis is unnecessarily conservative was discussed. In addition, in a fitness-for purpose analysism all the available fracture resistance should be considered. However, these arguments did not distinguish between the need to establish a high level of assurance that a structure can withstand use, abuse, and re-use and the requirements to simply demonstrate survival under more severe but infrequent events. The former requires that no significant degradation of the structure should take place, implying an initiation assessment, whilst in the latter case, it may be appropriate to take account of more of the available fracture resistence, as simple survival is all that is required.

The three final speakers of this session dealt with the evaluation of fracture by methods other than those generally used in materials testing and evaluation. Papers on Fracture Kinematics, Fracture Propagation Energy (R_c) and Energy Dissipation Within Intense Crack Tip Strain Regions (W_p) were presented and discussed. Some common points between these and the more conventional approaches were identified.

Crack growth in flat plates

A. Quirk, SRD, UKAEA

Crack growth in a flat plate specimen can fall into three categories:-

a.   fast fracture,

b.   slow stable tearing.

c.   "pop-in".

In the fast fracture mode as the load is increased monotonically little or no crack extension occurs until sudden, and fast, propagation occurs.

Slow stable tearing is the propagation of a defect under rising load. This occurs in a continuous stable manner in that as the load is increased the crack length increases significantly but if the load is then held steady the crack growth stops (apart from creep effects). Eventually the crack starts to propagate under decreasing load.

With the "pop-in" failure mode as the load is increased at a particular load the crack will undergo a sudden significant increase in length and then stop. The load may then have to further increase to achieve further crack extension which may be by another "pop-in".

There are two categories of flat plate testing machines, the conventional type in which the load is applied to the specimen through stiff ends resulting in a non-uniform load application. The type used by SRD applies a uniform distribution of load to the specimen.

It is the author's contention that the loading method affects the fracture mode of the plate. Here discussion will centre on slow stable tearing and 'pop-in'. Consider a quadrant of a test plate as shown in Fig.1. The loading may be either uniformly distributed Fig.1b, or non-uniformly distributed Fig.1c. Schematic stress distributions along the quadrant boundaries are shown.

For the uniform distributed load case the positive and negative forces on the axis AB must balance because there are no transverse forces on FD or BC. The forces on AB create an anti-clockwise moment which must be reacted by a clockwise bending movement creating by bending stresses on FD. This means that the stress distribution on FD required to balance the applied loads on BC is reduced at F and increased at D. That is the existence of a bending moment on AB reduces the crack tip stresses, see Fig.2a. Moreover the greater the bending moment the greater the reduction at F. Thus if the applied load is such that crack extension occurs the bending moment on AB increases as the crack extends. (In our experiments an increase in crack size from 125 to 250mm caused a 5 five-fold increase in this bending moment). This means as the load increases the crack extends but this extension increases the bending stresses on AB which develop bending stresses on FD which reduce the crack tip stress therefore stabilising the crack. This process increases until the transverse stresses at A or B attain yield magnitude, at which point the bending moment ceases to increase sufficiently to maintain stability. The bending stresses are shown on Fig.3. and their effect on the crack tip stresses in Fig.4.

With a rigid end load machine the stressing situation is the opposite to that just discussed. Here the schematic stress distribution is as shown in Fig.1c., the transverse stresses arise because the plate has a Poisson contraction requirement and is restrained by the heavy ends. As a consequence the stresses on AB are not in equilibrium, Fig.5., they are however in equilibrium with the transverse stresses on BC. Hence a clockwise bending stress is created by these forces. This is balanced by an anti-clockwise moment on FD, which of course enhances the load equilibrium stress at F. When these forces cause the crack to propagate the non-uniformity of the applied stress will increase and the stresses on AB will decrease because the Poisson contraction requirement of the material above the crack will decrease. Hence there will be a reduction in the crack tip stress, this being the sum of the reaction due to the reduction of the applied load above the defect and the reaction of the bending moment on AB.

Thus theoretical treatments of flat plate test data should recognise the important effects that can be induced by the method of applying the load to the specimen.

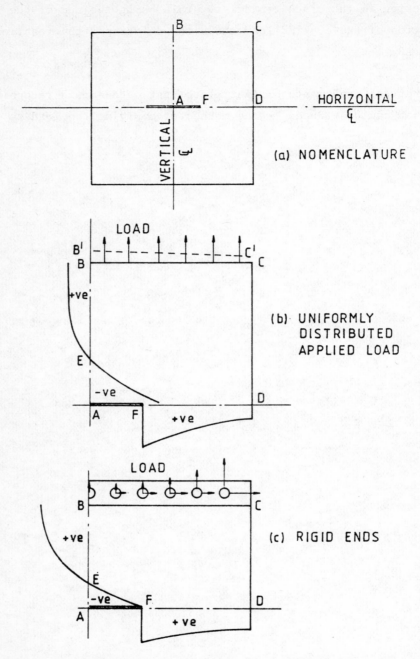

FIG. 1 SCHEMATIC STRESS DISTRIBUTIONS AROUND
THE QUADRANT OF A FLAT PLATE FOR
(b) UNIFORMLY DISTRIBUTED APPLIED LOAD,
AND (c) UNIFORM OVERALL EXTENSION

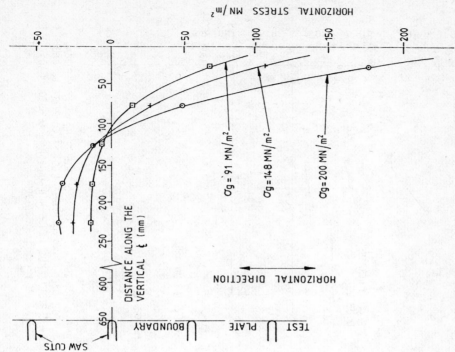

FIG. 3 HORIZONTAL STRESS DISTRIBUTION ALONG THE VERTICAL
ξ OF A THICK MILD STEEL PLATE WITH A TRANSVERSE WELD
AT AN APPLIED STRESS OF 200 MN/m². INSTABILITY OCCURRED
AT $\sigma_g = 203$ MN/m²

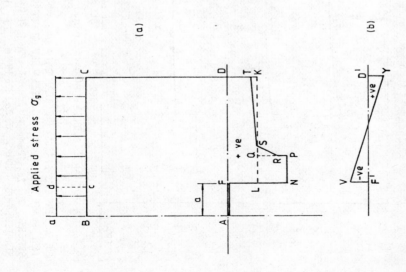

FIG. 2 (a) SCHEMATIC STRESS DISTRIBUTION ON
SECTION AD IN THE ABSENCE OF A BENDING
STRESS DISTRIBUTION ON AB AND (b) STRESS
DISTRIBUTION TO BE SUPERIMPOSED ON FD
STRESSES WITH MOMENT M APPLIED ON AB

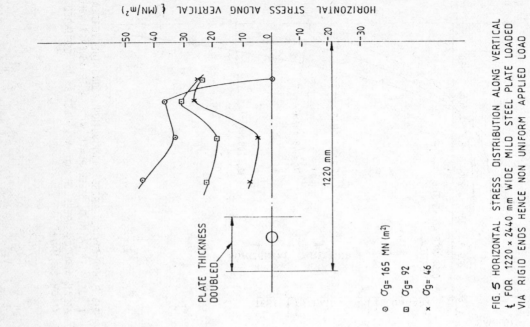

FIG. 5 HORIZONTAL STRESS DISTRIBUTION ALONG VERTICAL $\not{L}$ FOR 1220 x 2440 mm WIDE MILD STEEL PLATE LOADED VIA RIGID ENDS HENCE NON UNIFORM APPLIED LOAD DATA FROM REF 1. ($2a_0$ =305 mm)

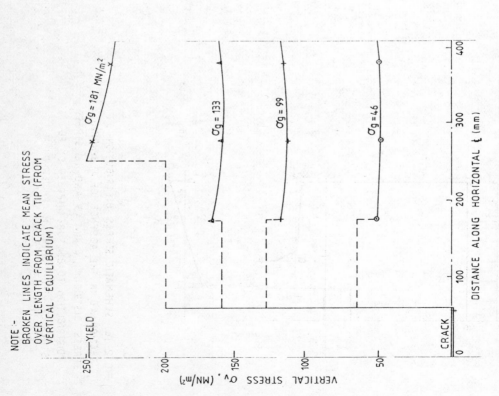

FIG. 4. VERTICAL STRESS DISTRIBUTION ALONG HORIZONTAL $\not{L}$ FOR TEST PLATE SSTP3 FOR VARIOUS VALUES OF UNIFORM APPLIED STRESS σ_g. $2a_0$ =125 mm NOTE THAT CRACK GROWTH HAS NOT STARTED, THIS OCCURS AT σ_g = 306 MN/m^2

NOTE –
BROKEN LINES INDICATE MEAN STRESS
OVER LENGTH FROM CRACK TIP (FROM
VERTICAL EQUILIBRIUM)

Fracture of stainless steel pipes with circumferential cracks in four-point bending

A. Brückner, R. Grunmach, B. Kneifel, D. Munz, G. Thun

Kernforschungszentrum Karlsruhe, Arbeitsgruppe Zuverlässigkeit
und Schadenskunde am Institut für Reaktorbauelemente, Karlsruhe,
FR Germany.

Experiments

Material: Austenitic stainless steel (18.1% Cr, 10.2% Ni,
1.54% Mn, 0.65% Si, 0.42% Ti, 0.051% C, 0.015% P,
0.005% S).

Pipe(Fig.1) Outer diameter: 33.7 mm
Wall thickness: 2mm
Length: 230 mm, welded on both ends to rods with a length
of 210 mm
Circumferential through wall cracks: fatigue crack starting
from an electro-erosive machined slot.

Loading: Four-point bending, outer span 500 mm,
inner span 300 mm (load points at the rods to avoid
indentation of the pipe).

Instrumentation of the pipe (Fig.2):
- load point displacement
- electrical potential drop
- two ovalization meter at 0° (vertical direction) and 90° (horizontal
direction)
- pipe angle meter

Aim of the investigation

Determination of criteria for the onset of stable and unstable crack
extension under load and displacement control

Methods: J-Integral approach
Plastic limit load analysis

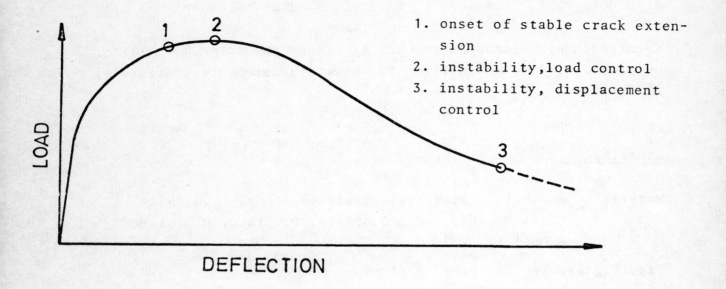

1. onset of stable crack extension
2. instability, load control
3. instability, displacement
   control

J-Integral evaluation

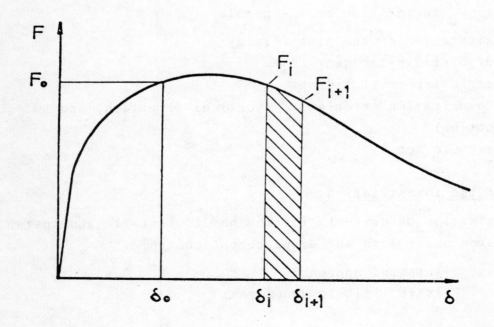

$$J_{plo} = \beta(\Theta_o) \cdot \int_o^{\delta_{plo}} F d\delta_{pl}$$

$$\beta = -\frac{1}{Rt} \frac{h'(\Theta)}{h(\Theta)} , \quad h(\Theta) = \cos\frac{\Theta}{2} - \frac{1}{2}\sin\Theta$$

Stable crack extension:

$$J_{pl,i+1} = J_{pl,i}\left[1 + \ln\frac{h'(\phi_{i+1})}{h'(\phi_i)}\right] - \frac{\beta(\phi_{i+1}) + \beta(\phi_i)}{2} U_{pl}^{i,i+1}$$

$$U_{pl}^{i,i+1} = \int_{\delta_{pl,i}}^{\delta_{pl,i+1}} F d\delta_{pl} = \int_{\delta_i}^{\delta_{i+1}} F d\delta + \frac{1}{2} F_{i+1}^2 C(\phi_{i+1}) - \frac{1}{2} F_i^2 C(\phi_i)$$

Necessary information: load, displacement, crack angle, compliance.

Plastic limit load analysis

Perfect plastic material with flow stress σ_f
Limit bending moment $M_L = 2\sigma_f t R^2 |2\cos\frac{\Theta}{2} - \sin\Theta|$
Necessary information: load, crack angle.

Stability analysis under displacement control

$$\delta_{tot} = \delta_{el,no\ crack} + \delta_{el,crack} + \delta_M + \delta_{pl}$$

$$= C_o \cdot F + C_r \cdot F + C_M F + \delta_{pl}$$

Instability criterion:

$$\frac{d\delta_{tot}}{d\theta} < 0$$

$$\frac{1}{C_o + C_M} < - \frac{\dfrac{dF}{d\theta}}{F\dfrac{dC_{cr}}{d\theta} + C_{cr} \cdot \dfrac{dF}{d\theta} + \dfrac{d\delta_{pl}}{d\theta}}$$

Necessary information: load, crack angle displacement, compliance.

Problems

1. Plastic deformation outside the crack region for $0 < F < F_{max}$
   Consequence: Error in J between F_i and F_{max}
   Correction: Determination of the deformation energy U_{crack} by subtracting for a given load F $U_{no\ crack}$ from the measured deformation energy U. $U_{no\ crack}$ is obtained from a test of a pipe without a crack.

2. Ovalization
   Diameter in vertical direction increases, diameter in horizontal direction decreases.
   Consequence: Correction necessary for
   - limit moment $M_L(\theta)$
   - compliance $C_{cr}(\theta)$
   - J-Integral evaluation

   Correction factor for M_L and for $h(\theta)$: $f = \dfrac{b(a+b)}{R^2}$

   a: semi-axis in horizontal direction
   b: semi-axis in vertical direction

Results

As an example results for an initial crack of $2\Theta_o = 62^o$ are shown in Figs. 3-5.

Fig. 3: Load and electrical potential drop versus displacement.
The onset of stable crack extension can be obtained from the deviation of the initial straight part of the potential drop-displacement curve.

Fig. 4: Ovalization
The vertical diameter increases, the horizontal diameter decreases.

Fig. 5: $J-\Delta\Theta$-curves
The effect of the ovalization correction and of the plasticity correction can be seen.

The complete results are presented at the ASME-Pressure Vessel and Piping Symposium in Portland, and shall be published in the conference volume.

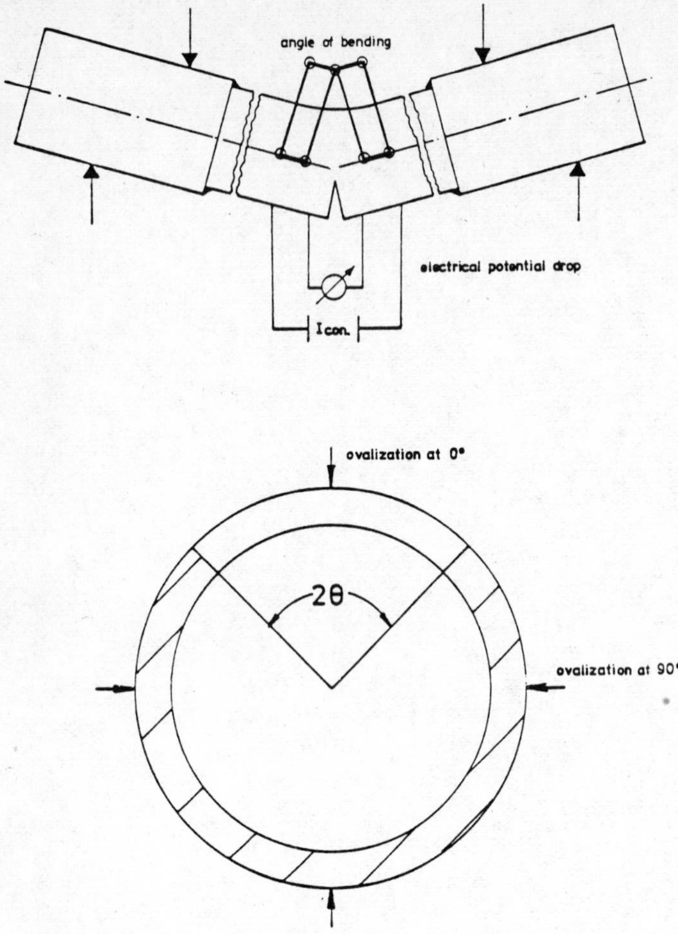

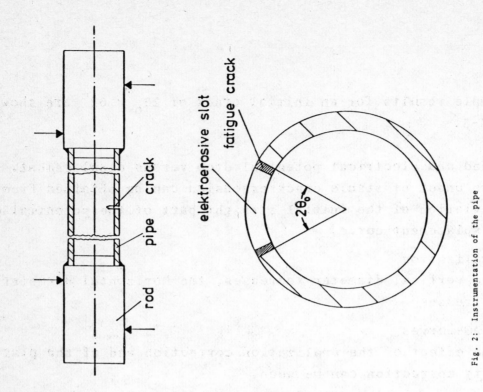

elektroerosive slot

fatigue crack

$2\theta_0$

pipe crack

rod

Fig. 2: Instrumentation of the pipe

electrical potential drop

I con.

angle of bending

ovalization at 90°

ovalization at 0°

2θ

Fig. 1: Pipe and loading arrangement

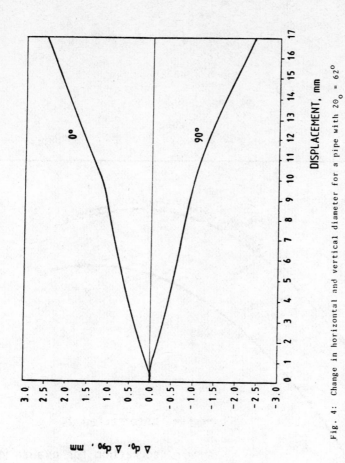

Fig. 4: Change in horizontal and vertical diameter for a pipe with $2\theta_0 = 62^\circ$

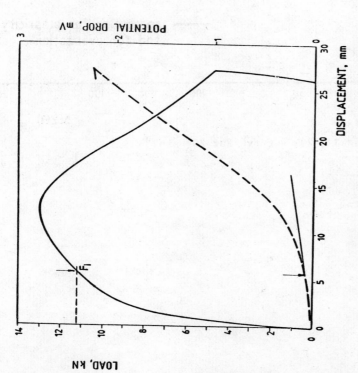

Fig. 3: Load and potential drop versus load point displacement for starting crack angle $2\theta_0 = 62^\circ$

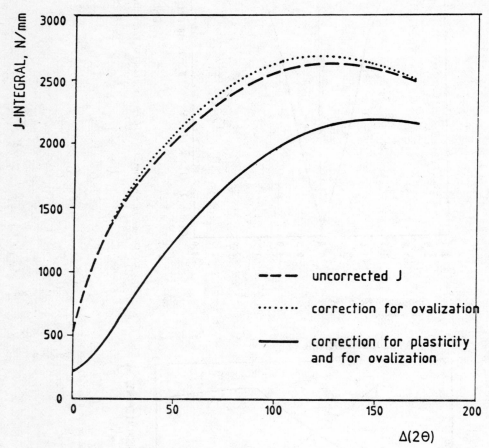

Fig. 5:   J-$\Delta(2\theta)$ - curve for $2\theta_o = 62^o$

THE DETERMINATION OF J_R-CURVES FROM IRRADIATED SPECIMENS FOR

CRITICAL CRACK LENGTH PREDICTION IN CANDU* REACTOR PRESSURE TUBES

by L.A. Simpson
Materials Science Branch
Whiteshell Nuclear Research Establishment
Atomic Energy of Canada Limited
Pinawa, Manitoba, Canada ROE 1LO

1.0 INTRODUCTION

The CANDU nuclear reactor system uses pressure tubes of cold-worked Zr-2.5% Nb as the primary coolant containment in the reactor core. (See Figures 1 and 2.) The safety criterion for these pressure tubes is based on the demonstration of leak-before-break. To satisfy this criterion, we must know the critical crack length for the tube for all operating conditions and anticipated material properties. To do this in a comprehensive manner requires many tests. Thus, even though we are able to test full-size sections of pressure tubing containing cracks (burst tests) (1), there is a need for a method of estimating the critical crack length using small specimens. Besides providing an economical method of obtaining a wide range of data, small-specimen methods are amenable to fracture-mechanism studies, which could lead to a predictive capability for untested conditions.

2.0 APPROACH

This presentation describes an approach to develop a relationship between the results of tests on compact specimens and the critical crack length in pressure tubes. In most situations, fracture is by ductile tearing and elastic-plastic analysis methods are appropriate. In particular, we have determined J-resistance (J_R) curves on compact specimens and related these to crack instability in pressure tubes by comparing them to curves of the crack driving force for axial cracks in the tubes.

---

*CANada Deuterium Uranium

3.0 J_R TESTING: SPECIAL CONSIDERATIONS PERTAINING TO THIS PROBLEM (2)

3.1 Specimen Preparation

Because it was important to test specimens having a microstructure identical to that of the pressure tube, specimens were cut directly from the pressure tube itself. The diameter (107 mm) and the curvature of the tube limited the maximum size of the specimen. With a minimum amount of flattening, we were able to produce compact specimens with a width of 34 mm (see Figure 3). After flattening, these specimens were annealed at 400°C to reduce any residual stresses imparted by the flattening process.

3.2 Test Temperatures

To reproduce reactor operating temperatures, tests had to be conducted in the range 25°C to 300°C. Difficulties associated with using displacement gauges at 300°C, plus space considerations in the testing furnace, led us to choose a photographic method for the displacement measurement (2). Hardness indentations, placed along both sides of the crack mouth, were photographed through a furnace window at intervals during the test (see Figure 4), and their displacement was used to determine load-point displacement, as shown in Figure 5.

3.3 Crack Length Measurement

The D-C potential drop method (3) was adapted to our test conditions for following crack growth.

4.0 CRACK DRIVING FORCE

The expression used for the crack driving force on an axial crack in a pressure tube is a linear, elastic, fracture-mechanics expression with a plastic-zone correction:

$$J_{CDF} = \frac{8a\bar{\sigma}^{2}}{\pi E} \, \ell n \, \sec \, \frac{\pi M \sigma_{H}}{2\bar{\sigma}}$$

where $\bar{\sigma}$ = flow stress

2a = crack length in the tube

M = stress enhancement factor arising from bending stresses
induced by tube curvature

σ_H = hoop stress

E = Young's modulus.

We are considering ways of improving this expression but, as yet, have not found one that is convenient. Errors arising from its application seem to be small.

5.0 TEST OF THE J_R-CURVE METHOD OF PREDICTING CRITICAL CRACK LENGTH

To test this approach, we carried out a series of burst tests on pressure-tube sections containing axial cracks of various sizes (4). Compact specimens were cut from undeformed parts of the burst tubes, and J_R curves were determined (see Figure 6). Then, a crack-driving-force curve, corresponding to the measured burst pressure in the parent tube, was used, with the J_R curve, to determine the "critical" crack length, which theoretically should have been equal to the actual crack length in the tube.

In all cases, the critical crack length was underestimated (see Figure 7). In a subsequent study (5), it was shown that a geometry dependence does exist, but that, using compact specimens, estimates of critical crack lengths will always be conservative.

6.0 APPLICATION TO RADIOACTIVE SPECIMENS

6.1 Source of Specimens

A most important variable in our industry is the effect of irradiation on critical crack length. We are studying two ways to assess the effect of irradiation:

(1) Preparing compact specimens prior to irradiation.

(2) Preparing small specimens from irradiated tubing.

The second approach is experimentally more difficult because irradiated tubing cannot readily be flattened at room temperature without cracking. Even if the specimen is flattened at an elevated temperature, the usual stress-relieving anneal may alter the radiation effects. We are, therefore, assessing some curved-specimen geometries.

We have, however, successfully applied the first approach to test radioactive specimens in a hot-cell facility using, essentially, the same procedures as described above for inactive testing. An Instron servo-hydraulic test frame is installed in the cell with the instrumentation and controls outside (see Figure 8). A split clam-shell furnace, which can be easily assembled using manipulators, is used to obtain temperatures up to $300^{\circ}C$. Crack length and displacement are obtained by modifying our standard procedures, as described below.

6.2 Crack-Length Measurement

(a) Prior to irradiation, the specimens are pre-tapped for fastening screws for the constant-current leads.

(b) Following irradiation, the current leads are attached in the hot cells using a special jig (see Figure 9) to align the screw holes, screws and screwdriver, using manipulators for assembly.

(c) Two electrically insulated parts of the jig are connected to a spot welder (outside the cell) and used to attach the potential drop leads at the crack mouth.

6.3 Displacement Measurement

(a) Prior to irradiation, hardness indentations are put on both sides of the crack mouth. Two additional indentations are used to determine magnification and their spacing carefully measured.

(b) The testing furnace has two windows, one for photography and one for illumination. The camera is set up outside the cell (about 4 metres from the test assembly). Using a high-power Questar telescope as a lens, the crack mouth and specimen ligament will fill a frame of 35-mm film and allow a reasonably accurate displacement measurement.

7.0 RESULTS

To date, we have tested forty specimens irradiated at 350°C for one year in our WR-1 reactor. All equipment has operated smoothly and reproducible results have been obtained (see Figure 10). The effect of irradiation on these specimens, as determined by hardness tests, was small (about 13% increase in strength). However, the resistance curves clearly reflected the change through a slight, but reproducible, reduction in slope. We are continuing this work on specimens irradiated at lower temperatures (to increase the hardening) and are planning to automate the test process. The one drawback with this approach is that data analysis is presently very time-consuming.

References

1. Langford, W.J. and L.E.J. Mooder. Int. J. of Pres. Ves. and Piping, 6, (1978), pp. 275-310.

2. Simpson, L.A. and C.F. Clarke. ASTM STP 668, American Society for Testing and Materials, Philadelphia, (1979), pp. 643-662.

3. Simpson, L.A. and C.F. Clarke. Atomic Energy of Canada Ltd. Report, AECL-5815 (1977).

4. Simpson, L.A. and B.J.S. Wilkins. Mechanical Behaviour of Materials, Vol. 3. Pergamon Press, Oxford, (1979), pp. 563-572.

5. Simpson, L.A. Advances in Fracture Research, Vol. 2, Pergamon Press, Oxford (1982), pp. 833-841.

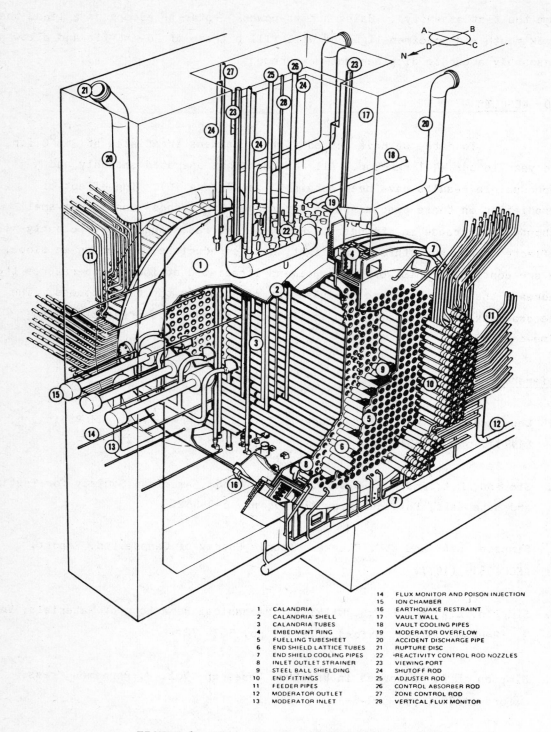

		14	FLUX MONITOR AND POISON INJECTION
		15	ION CHAMBER
1	CALANDRIA	16	EARTHQUAKE RESTRAINT
2	CALANDRIA SHELL	17	VAULT WALL
3	CALANDRIA TUBES	18	VAULT COOLING PIPES
4	EMBEDMENT RING	19	MODERATOR OVERFLOW
5	FUELLING TUBESHEET	20	ACCIDENT DISCHARGE PIPE
6	END SHIELD LATTICE TUBES	21	RUPTURE DISC
7	END SHIELD COOLING PIPES	22	REACTIVITY CONTROL ROD NOZZLES
8	INLET OUTLET STRAINER	23	VIEWING PORT
9	STEEL BALL SHIELDING	24	SHUTOFF ROD
10	END FITTINGS	25	ADJUSTER ROD
11	FEEDER PIPES	26	CONTROL ABSORBER ROD
12	MODERATOR OUTLET	27	ZONE CONTROL ROD
13	MODERATOR INLET	28	VERTICAL FLUX MONITOR

FIGURE 1: Cutaway of a CANDU Reactor Core

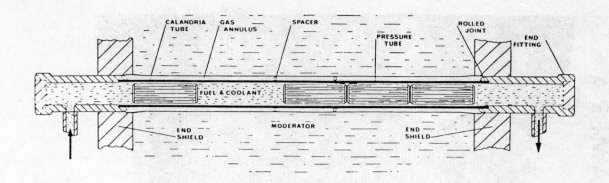

FIGURE 2: Schematic of a Fuel Channel for a CANDU Reactor
With Pressurized-Water Coolant

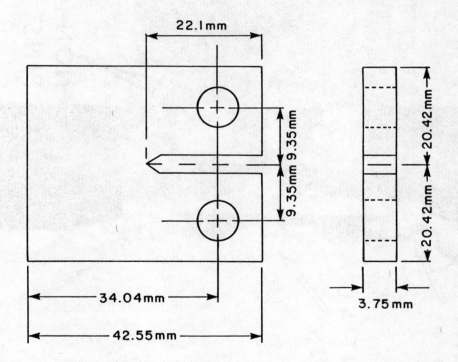

FIGURE 3: Compact Specimen Used for J_R-Curve Studies

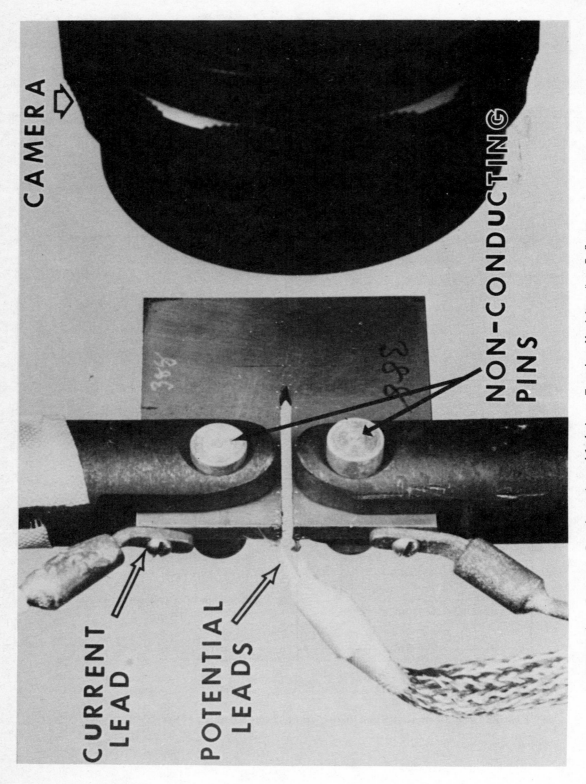

Text within the image: CAMERA, NON-CONDUCTING PINS, CURRENT LEAD, POTENTIAL LEADS

FIGURE 4: Compact Specimen Assembled in a Testing Machine for R-Curve Determination. Crack Length is Measured by a Change in Electrical Resistance and Displacements are Measured Photographically.

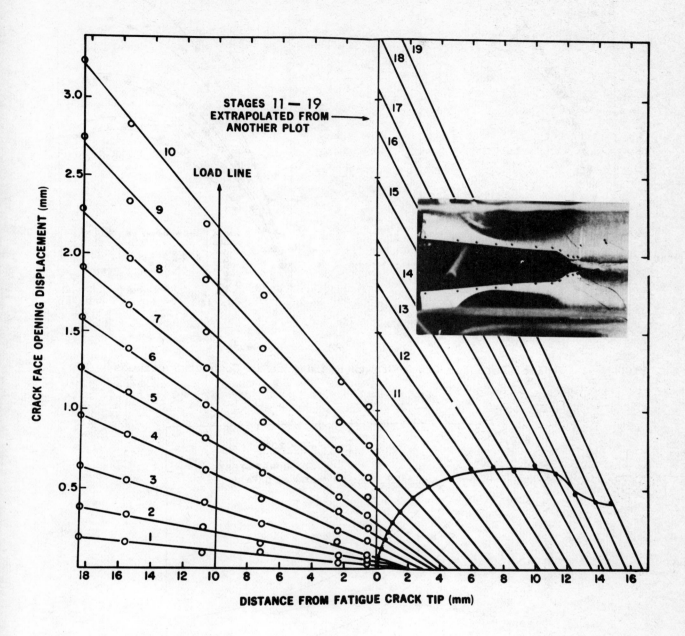

FIGURE 5: Method for Determination of Crack-Opening Displacement (COD)
From Displacement Measurements on Crack Face. Each Numbered
Straight Line Represents a Set of Displacements for a Given
Load Level. The Curved Line on the Right Indicates the
Magnitude and Position of the COD at the Actual Crack Tip.
The Insert Shows Microhardness Indentations Used for Crack-
Face Displacement Measurements.

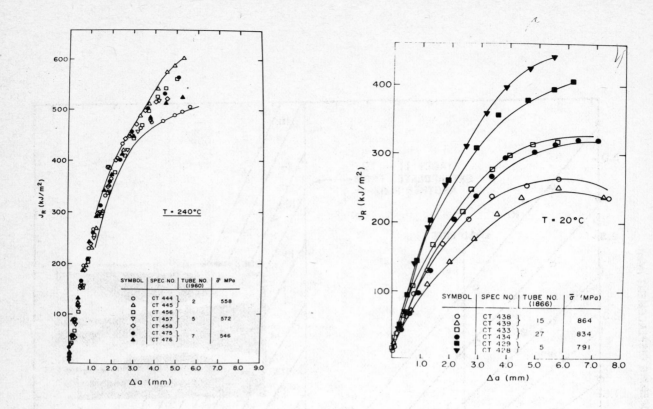

FIGURE 6: Typical Data for Crack-Growth Resistance in Compact Specimens

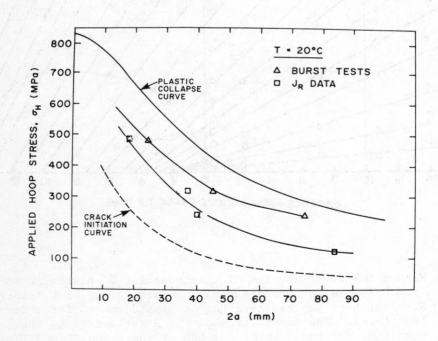

FIGURE 7: Comparison of Critical-Crack-Length Estimates Using
R-Curves, an Initiation Criterion, and a Plastic-
Collapse Criterion With Burst-Test Results

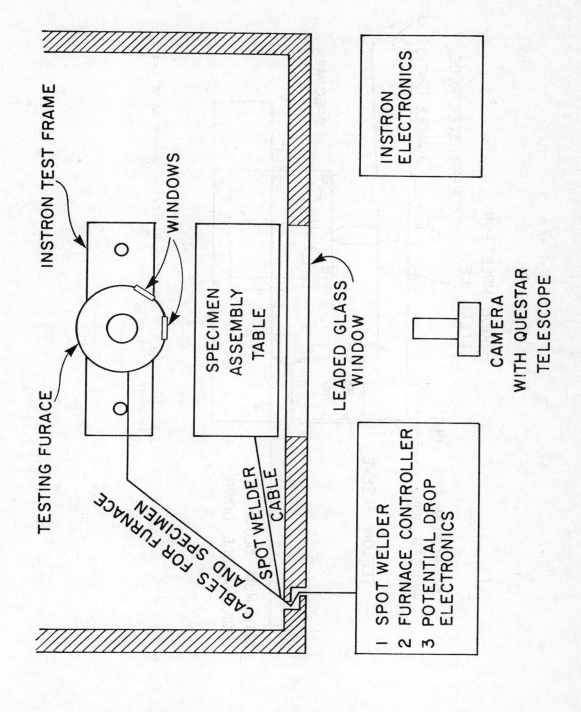

Figure 8. Arrangement for Testing Active Specimens in a Hot Cell.

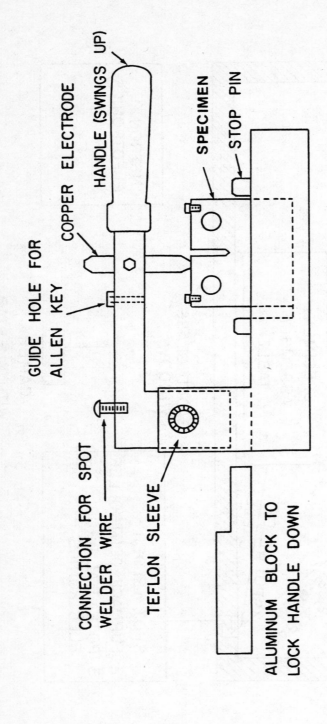

GUIDE HOLE FOR
ALLEN KEY

COPPER ELECTRODE

HANDLE (SWINGS UP)

SPECIMEN

STOP PIN

CONNECTION FOR SPOT
WELDER WIRE

TEFLON SLEEVE

ALUMINUM BLOCK TO
LOCK HANDLE DOWN

FIGURE 9:   Assembly Jig for Attaching Electrical Leads to Active
Compact Specimens.

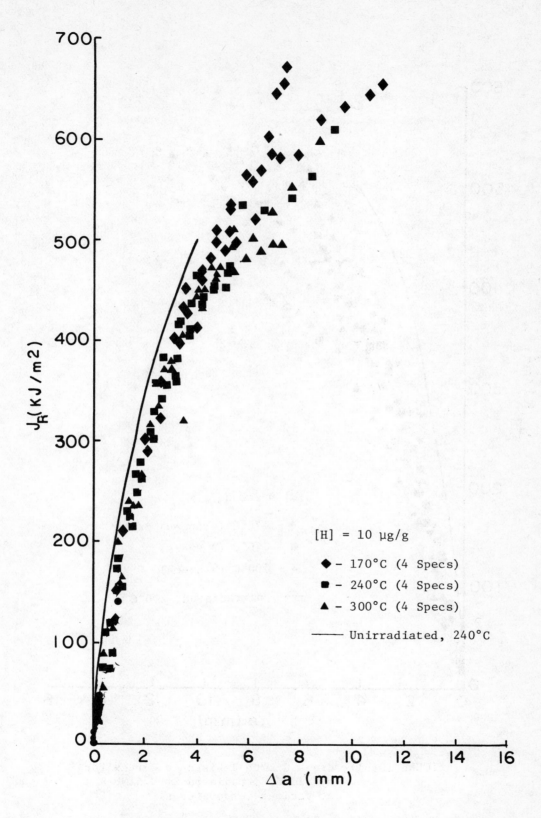

FIGURE 10(a): Crack Growth Resistance Curves for
unhydrided, irradiated Zr-2.5%Nb
at various test temperatures.

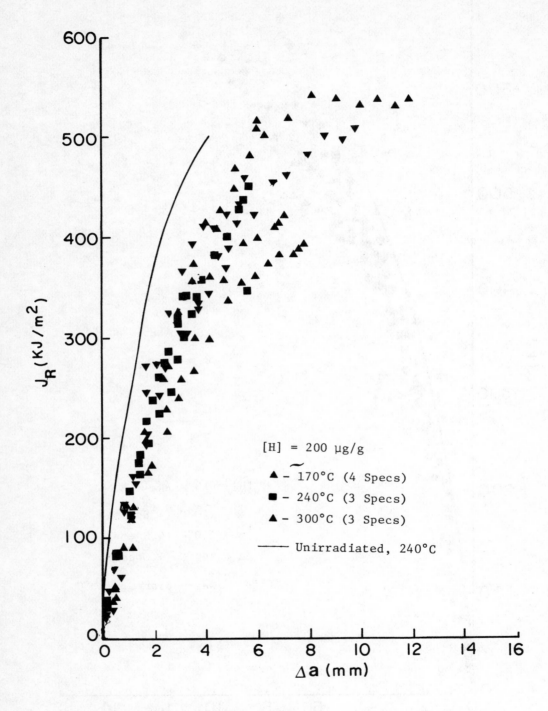

FIGURE 10(b):   Crack Growth Resistance Curves for
hydrided, irradiated Zr-2.5%Nb
at various temperatures.

EXPERIMENTAL APPLICATION OF EPFM
TO THROUGH THICKNESS CRACKED PIPES

G. Angelino, S. Squilloni, CISE Milan
S. Reale, University of Florence

INTRODUCTION

The ultimate goal of the research herein outlined is to provide some contributions to the general effort to reach experimental data or validated experimental
techniques to establish an improved basis for analysing, by means of Elasto-plastic
fracture mechanics, the effect of postulated or detected flaws on safety margins
or for making reasonable engineering evaluation of the integrity of potentially
flawed steel structures (Tab. 1).

In our investigation, developed jointly at University and CISE with the sponsorship of ENEA, attention has been focused on an experimental analysis of elasto-
plastic fracture behaviour of through thickness craked pipes.

The first statment of the research has been to evaluate and check potential criteria
(namely J and COD resistance curve) and experimental techniques able to provide
informations to decribe the crack onset under large scale plasticity conditions
of a through thickness cracked pipe. This step of the research has been performed
testing small scale tubular models.

The second statment, which is still going on, is to utilize the selected technique
for testing pipes in real scale. In this stage, material of the tested pipes is
deeply characterized through its standard mechanical properties and J - R curves
according with ASTM standards. The aim is to get comparison between J - R curves
of tested pipes and J - R material curves in order to verify whether the parameters
obtained from standard tests on specimens could be used in reliable way to a
quantitative description of fracture behaviour of pipes or, in general, of engineering structures. As development of the research, exsisting elasto-plastic
computational codes will be applied, tested by means the experimental results,
in order to have a validate tool to predict, realistically and reliably the fracture
behaviour of complex pipelines, when material fracture properties are known (Tab. 1)

TEST SETUP AND EXPERIMENTAL TECHNIQUE

The tests are carried out at room temperature on steel pipes. The dimension of
the pipes and mechanical properties of material are given in Tab. 2. The reported
values are obtained by standard tests on specimens taken from the pipes under
study.

Before tests pipes are longitudinally cracked. Cracks have been artificially made,
the crack tip is obtained by fatigue.

Burst tests under internal pressure are carried out: the experimental set up is
shown in Fig. I.

Pipes are loaded in quasi static and monotonic way. Pressure vs load application
point displacement are recorded.

Following Landes and Begley multi specimen calibration technique, monotonic pressure-
displacement curves, relative to models with different initial crack size, were
used to determine J calibration curve as energy release rate. J is computed as
$J = -\frac{1}{B} \left(\frac{dU}{da} \right)_P$ the derivative is performed at a constant value of P in order
to not take into account the energy related to the pressurization of testing setup.
J resistance curve is constructed testing a set of identical models . Each model
is loaded to different values of pressure and unloaded before unstable crack
growth. Ductile crack growth Δa is measured with the heat tint technique.
All tested models exibite slow-stable ductile fracture so that measurement of
fracture resistance could be made over a relatively large increment of crack
length.

The J -Δa curve is determined by a linear regression line. The critical value of
J is determined by the intersection of J curve with blunt line.
Standard fracture tests are also performed on standard specimens taken from the
pipes in order to obtain material related J_{Ic} and J resistance curve.

BACKGROUND AND VALIDATION OF THE TECHNIQUE.

The experimental technique described has been developed and validated in the frame
of the statment of the research focused on the methodological aspects of elasto-
plastic fracture test methods.
Tests are carried out at room temperature on thin tubolar models made by low carbon
annealed steel in three basic conditions (Fig. 1): burst tests under internal
pressure on longitudinally cracked models; four point bending tests on circumferentially cracked models; tensile test on pipe sectors.
In all conditions J - calibration curve and J - resistance curve is determined by
means the same above related technique.
J - controlled conditions are checked and, if verified, the crack initiation, growth
and stability are presented and discussed in terms of J. COD data are used for
monitoring the tests and for checking the unbiased results since the tests are
instrumented to provide in a straight forward way COD.
Good agreement in critical value of J is achieved in all test conditions. Experimentally determined J_c appears to be independent of geometry and loading (Fig. 2).
So as first statment it is possible to conclude that the related experimental
approach can give a reliable and validate technique to evaluate the critical value
of J and to detect some resistance correlated curve to forecast fracture behaviour
of pipes.

EXPERIMENTAL RESULTS AND CONCLUDING REMARKS.

As outlined, the second statment has been to carry out burst tests on pipes in a
real scale utilizing the described validated technique. J - Δa curves are determined for both A106B Heat 1 and Heat 2 pipes. The Fig. 3 reports a J - Δa curve of
burst test on a A106B Heat 1 pipe.
Materials have been characterized. J resistance curves have been obtained from
1/2 TCT specimens, orientation CL, according to ASTM. The Fig. 4 report, as example,
a J - R curve of A106B Heat 1.
As shown in Fig. 5 and 6, J_{Ic} on 1/2 TCT specimens well agrees, within the
scatterband, with the values measured on pipes.
Pipe bulging was large also at crack growth onset for the burst tests both on
small tubular models and on large, real scale pipes. In spite of that J_{Ic} from
specimens agrees with J_{Ic} from tubes and no effects of bulging has to be considered on J_{Ic} as a crack initiation parameter.
The R curve slope dJ/da from pressure test is markedly lower than dJ/da from specimens. This result does not agree with the Paris Tearing modulus approach for stable
crack growth. Indeed it states that during the stable propagation the crack is
"forced to follow" the material J - R curve independently of the structure's
geometry.
On the other hand it is to be pointed out that, following the Landes and Begly
approach, we have computed the J values through load-displacement curves related
to the initial crack size. This approach is correct up to the onset of crack growth
and than J_{Ic} value is unbiased. After, the computed J values are systematictly

over or understimated and than dJ/da is incorrect. Furthermore if we will focuse the interest on computing the actual slope of J – R curve, from a pressure tests the data reducting technique has to be improved holding in due consideration the testing conditions. This aspect has been displaied also within the tests on tubular models and, in the case of the bending test, an "ad hoc" data reduction has been successfully performed.

```
+--------------------------------------------------+
|            PROGRAM   OBJECTIVES                  |
|            =======   ==========                  |
|                                                  |
| - Develop a direct knowledge on the applicability|
|   of EPFM concepts & methods to assess the safety|
|   margins of a cracked pipe under different      |
|   loading conditions.                            |
|                                                  |
| - Develop an experimental methodology to         |
|   determine J resistance curves from pipes under |
|   different loading conditions.                  |
|                                                  |
|                 APPROACH                         |
|                 ========                         |
|                                                  |
| - Small diameter pipes, J – R curves:            |
|      i) four point bending                       |
|     ii) tension                                  |
|    iii) internal pressure (biaxial)              |
|                                                  |
| - large diameter pipes, J – R curves:            |
|     iv) internal pressure (uniaxial)             |
|      v) 1/2 TCT specimens                         |
|--------------------------------------------------|
|                  Tab. I                          |
+--------------------------------------------------+
```

```
+-----------------------------------------+
|              MATERIALS                  |
|              =========                  |
|                                         |
|  Fe – 35 steel pipes:                   |
|    O.D. 100 mm, thickness 4 mm          |
|    σ_o = 300 MPa                        |
|                                         |
|  A 106 – B pipes: two different heats   |
|    O.D. 406 mm, thickness 14 mm         |
|    1/2 TCT specimens, orientation CL    |
|    σ_o = 375 MPa                        |
|-----------------------------------------|
|               Tab. 2                    |
+-----------------------------------------+
```

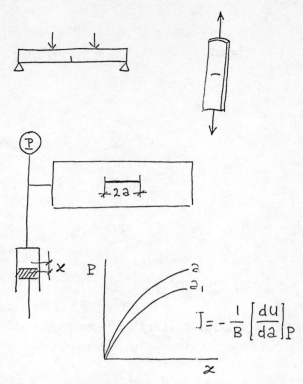

$$J = -\frac{1}{B}\left[\frac{dU}{da}\right]_P$$

Fig. 1 Experimental arrangements

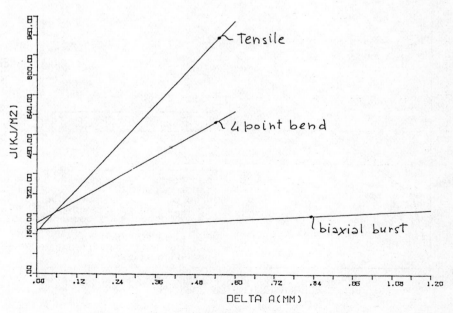

Fig. 2 Fe-35 steel 100 mm pipe
J-R curves for different loading

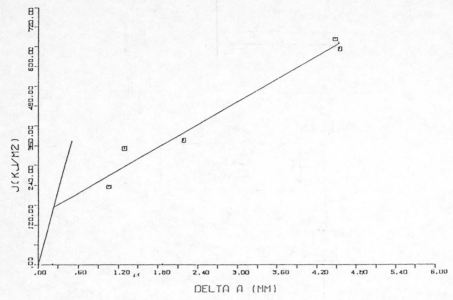

Fig. 3   A 106 B steel   406 mm pipe, uniaxial burst
J-R curve

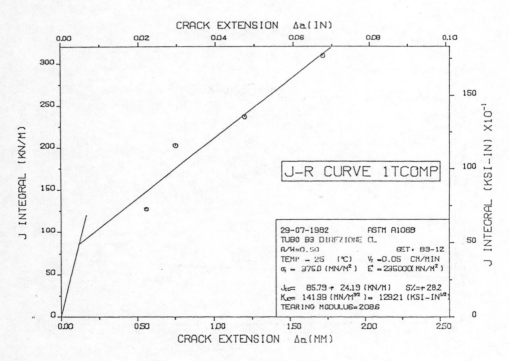

Fig. 4   I/2 TCT   A 106 B steel
J-R curve

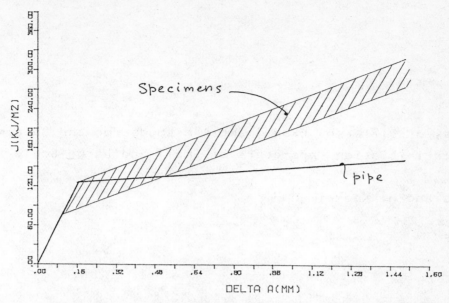

Fig. 5   A 106 B, Heat 1   406 pipe, – Comparison between
1/2 TCT specimens (5 J–R curves envelop) and pipe
results

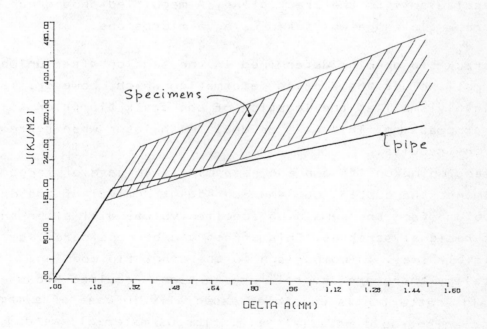

Fig. 6   A 106 B, Heat 2   406 pipe – Comparison between
1/2 TCT specimens (5 J–R curves envelop) and pipe
results

A Survey of Different Experimental Methods for the Determination of Crack Initiation Parameters of Small and Large Scale Specimens

E. Roos[*] and H. Kockelmann[*]

1.1. Mechanical Sectioning

The procedure in mechanical sectioning to determine the crack tip opening displacement is shown by means of an example of a CT-25 specimen in Fig. 1. The deformed specimen was separated perpendicularly to the crack plane. A magnified photograph of the crack tip range was taken using a microscope.

The crack tip opening determined in the section after unloading does not include the elastic deformation which, however, is negligibly small at the location of the crack tip prior to further cracking. This will be dealt with later when representing the results.

The sectioning on its own and, especially in case of larger specimens, the cutting out and subsequent removal of the crack tip volume from the remaining specimen-volume will alter the macro residual stresses. This effect can be considered as negligibly small in comparison to the crack tip opening. Generally, there are several more sections required because a certain scattering is possible, especially in case of a macroscopic anisotropic material (e.g. fibrous material, welds). The

[*] Staatliche Materialprüfungsanstalt Universität Stuttgart
Director: Professor Dr.-Ing. K. Kussmaul

microscopic anisotropy (grain structure and crystallinity) may well be of little influence at least in fine-grained steels where the crack tip opening displacement is large as against the grain dimensions. A reliable statement is only possible as long as the specimen is not completely broken. In this case, crack tip blunting and propagation are mostly very distinct. The completely broken specimens can only be evaluated with restrictions and utmost care; especially in tough fractures, the complementary fracture surface profiles cannot be joined exactly anymore; in addition to that, it may be possible that the sections in the two fracture pieces will not exactly be in alignment.

1.2. Replica Technique

A replica of the shape of the crack is made in a certain deformation state. Borings, which have to be located as near as possible to the crack tip, are required for infiltration, however, they shall not considerably change the stress state there. In Figs. 2 and 3 the borings are sketched in a CT 50 specimen and a plate seam containing a natural weld crack along the HAZ. The infiltrate is a two-component silicone rubber, which does not need any humidity for solidification. If there are multiple infiltrations each of the replica will be characterized by a certain colour.

The silicone rubber is pressed through the boring into the opened crack. The displacement is kept constant until complete solidification is reached (approx. 10 min), prior to continuing the test. After fracturing the replica is peeled off from the fracture surface. Because of its little mechanical strength it is not possible to make direct cross sections, but the replica has to be vapor-deposited with silver, electroplated with copper, and moulded in casting resin. This will also guarantee the required sharpness of the edge of the microsections. Various sections of the replicas are shown in Figs. 2 and 3.

The practical temperature application range is confined to a maximum of 100 $^{\circ}$C, because solidification will be very fast at higher temperature and a complete penetration into the crack is prevented.

1.3. Moiré Method

The Moiré method is in principle a method of measuring displacement. The Moiré fringes appear at locations of the same displacements with reference to the original position. The principle of crack and notch opening measurement according to the Moiré method is represented in Fig. 4. The number of fringes, crossing a connection line around the crack tip between two opposite points of the notch or crack border, is determined. This number indicates the opening at this point in grid line distances of the used Moiré grid, Fig. 4. The sensitivity of the method depends on the pitch of the Moiré grid and lies between 5 and 25 μm for grids with 10 to 50 lines per mm. The measuring range is not limited. With the help of high temperature Moiré grids it is possible to carry out measurements also at extremely high temperatures. By using the Moiré method, the notch and crack shape on the side face can be completely determined up to nearly the crack tip or the notch tip.

It has to be considered that this surface measurement in the plastified area at the crack tip or the notch tip cannot directly be transferred to the decisive centre of the specimen. According to experience there are only inconsiderable differences between surface and interior measurement in elastic deformed areas.

1.4. Endoscopy

The crack opening inside the specimen can be determined with the help of endoscopy. Borings in the precracked structure are required to insert the endoscope. On the one hand, they have to be situated as near as possible to the crack tip, and on the other hand, they should not influence measurably the deformation condition. Therefore, measurements are only possible in greater distances (minimum approx. 10 mm) from the crack tip. In Fig. 5, can be seen a schematic representation of the apparatus. The depicted set-up can be used without cooling up to +80 $^{\circ}$C. Because of condensation, it can hardly be used at temperatures below 10 $^{\circ}$C.

Fine cracks can definitely be seen. However, a quantitative measurement can only be made from an opening of approx. 0.03 mm.

The apparatus enables measurements to be made down to 250 mm
boring depth. In case of cracks in inhomogeneous materials
(e.g. welds) of complicated crack structure (natural cracks),
it is problematic to determine the position of the borings.
Measurements are hardly possible in the critical testing phase,
because at fracture of the specimen, the inserted endoscope
may be destroyed.

1.5 Potential Drop Method

The potential drop method is employed for measuring both the
crack start (initiation) as well as the medium crack depth and
the crack growth within the scope of fracture mechanics investigation under steady and fatigue load. The principle of the method
is that potential distribution is changed in a characteristical
manner in a metallic conductor through which a current flows
if there are geometrical changes (e.g. crack formation and
crack extension). This change in the potential is recorded with
an appropriate measuring apparatus, Fig. 6 and can be translated
with the help of geometry - specific calibrating curves in crack
depths or changes in crack length. The potential drop characteristics for different instrumentations are shown in Fig. 7, where
the crack has been simulated by a saw cut. In Fig. 8 can be seen
the calibrating curve for large scale SEN and DEN specimens.

The ac-method works with alternating current of a certain
frequency. The so called skin effect occurs: depending on frequency and material, the current is pushed back to the surface
and the center areas remain approximately without current; this
is called a limited penetration depth of the current.

In case of large scale specimens, the ac-method has some advantages over the dc-method:

- The ac-method is much more sensitive because of the mentioned
  frequency-dependent skin effect. The required current intensity
  is about one magnitude smaller than that of the dc method.

- Using an alternating current, the specimen needs not to be
  insulated against the testing machine because of the skin
  effect. Thus it can be applied to large scale specimens.

- There will not be any disturbances due to thermoelectric
  currents as they occur at the connections between copper
  conductors and specimen.

It is uncertain whether at all the crack initiation moment in
tough materials can be determined by this method, since prior
to cracking considerable plastic deformation including crack
tip blunting takes place. Furthermore, the actual cracking is
combined with continous deformation around the crack tip, especially in quasi-static fracture mechanics tests. This problem
increasingly occurs in the ac-method, because apart from the
electric properties the magnetic ones are also included and
change considerably with increasing deformation.

1.6. Stretch Zone Determination in SEM

In tough material, the crack tip will be very strongly plastic
deformed prior to cracking. Like the COD, the stretch zone
at the crack tip is a material depending characteristic deformation value. It is determined in the SEM as an average value
over the specimen.

2. Results
Comparison of different methods applied to
small and large scale specimens

2.1. Crack (Tip) Opening Displacement C(T)OD

The COD of a CT-200 specimen with a fatigue crack and that of a
CT-50 specimen without fatigue crack on the side surface of the
specimens and in the centre plane of the specimens are compared
in Figs. 9 and 10. The COD is somewhat larger in the interior
than on the side surfaces. This may be attributed to the curvature
of the fatigue crack front inside the specimen.

The changes within the notch shape of large and small scale DENT-
specimens are insofar different, as in large scale specimens
the notch edges are curved while they remain straight in small
scale specimens, apart from the immediate notch tip area, Fig. 11.

The crack tip and notch tip opening displacement at the instance of crack initiation in differently loaded small and large scale specimens (DENT, DECT, SECB, CCT, and CT specimens with and without fatigue crack) made of 20 MnMoNi 55 at test temperatures of -20 and +80 ^{o}C can be found in the diagrams of Figs. 12 and 13, in which a large scattering, especially of the surface values can be found. The latter can be determined with much more difficulties and uncertainties caused by a local reduction of area. As expected, the initiation values of notched specimens are clearly above those of specimens with fatigue cracks. There is no difference within the scatter band between large and small scale specimens (cf. Fig. 14). The CTOD at initiation in the specimen centre are independent of the specimen thickness within the range of 10 to 250 mm (Fig. 13).

The influence of unloading on the COD-measurement by means of mechanical sectioning is shown in Figs. 15 and 16. An elastic spring-back resilience, which is fairly strong, can be found at greater distances from the crack tip or notch tip, especially under bending conditions. This effect becomes smaller with increasing approximation to the crack tip and is, even at the crack tip, so small that it is unmeasurable, cf. Fig. 16. In there, the COD of a CT 46-specimen subjected to load is shown. A wedge has been inserted at the load-line of the loaded specimen to avoid elastic spring-back resilience. Having reached this state after unloading and removing the wedge, sections without providing any measurable differences have been made.

The COD of bending specimens is determined according to BS 5762:1979. These values are compared in Fig. 17 with experimental results of the Moiré-method and the mechanical sectioning on the example of the austenitic steel X 10 CrNiNb 18 9. The results according to BSI are a little below the Moiré-extrapolation but distinctly above the results of the mechanical sectioning, especially in the centre plane.

The rotational factors of CT-25 and CT-50 specimens with and without fatigue crack are plotted in Fig. 18. They have been experimentally determined with the help of the Moiré-method and clip gauges. The rotational factor is, first of all, increasing with the deformation. Finally, a value of 0.35 to 0.5, depending on the specimen type, will be achieved at δ_g/W of approx. 0.03.

2.2. Stretch Zone

The stretch zone has been determined using a SEM photograph. The centre plane of two CT-25 specimens, which were tested in the toughness upper-shelf region (C_v = 160 J, T = 80 $^{\circ}$C) and at NDT-temperature (C_v = 40 J, T = -20 $^{\circ}$C) can be found in Fig. 19. In this very region, the significant stretch zone is accompanied by stable crack extension, while there is a considerably smaller stretch zone and no stable crack growth at NDT-temperature. CT-50 specimens of the same material with notches but without a fatigue crack have been prepared in order to compare their stretch-zone with those having a fatigue crack. The testing temperature has been at 80 $^{\circ}$C, cf. Fig. 20. In comparison to the precracked specimen, the stretch zone is, in this case, more than twice as big. The same tendency can be seen in the stretch zones of large scale specimens, Fig. 21. The GS5 specimen in there is provided with a fatigue crack and has a somewhat larger stable crack growth than the notched specimen G5. This is due to the fact that a high bending stress has been superimposed on specimen GS5. The measuring values of all investigated specimens are compiled in Fig. 22. With the help of this, it can be proven that the values in the centre of the speci- men are higher than at the edge of the specimen. It has to be mentioned that the values for the same notch and crack configur- ation and the same test temperature are placed in a narrow scatter band.

The results of the double-edge notched small scale specimens, which were also tested at T = 80 $^{\circ}$C, are much more scattered. The reason is that with this type of specimen geometry, the de- formation is completely concentrated in the notch tip, Fig. 23.

2.3. Determination of the J_R-Curves

The way of ascertaining the J-Integral value and determining the crack initiation values from the J_R-curve is described in ASTM E 813-81. Consequently, if the points on the left side of

the 0.15 mm offset-line are omitted, the J_R-curves in Fig. 24
will be the result of the present testing points (multi-specimen technique). As expected, the increase in the J_R-curve of
the CT-50 specimens is flatter than of the CT-25 specimens,
both having a fatigue crack, whereby the crack initiation value
of the J_R-curve obtained from the CT-25 specimens is smaller
than those of the CT-50. The J_R-curve determined with the CT-50
specimens without fatigue crack is situated higher and runs steeper
than those with fatigue crack. This behaviour was to be expected considering the higher values of the stretch zone and the
COD. Using all testing points, and representing the J_R-curve
also as regression line, the distributions will be according to
Fig. 25. The slope of the blunting line is flatter according to
ASTM than the one determined experimentally over the stretch
zone, Fig. 25. This will yield lower crack initiation values
for the J-integral. However, the J_i-value for the not precracked
CT-50 specimen is below that for the precracked CT-50 specimen.
This, however, cannot be possible from the physical point of
view, because of the stretch zone and COD values.

Therefore, the J_R-curves were represented[*] fitted as smoothed
curves according to the least squares method using all test
values. Employing the experimentally determined blunting line to
ascertain the J_i-value yielded the lowest initiation values
according to which the precracked CT-50 specimen provided the
lowest, but the not precracked specimen the highest J_i-value.
The precracked CT-25 specimen yielded a little higher J_i-value
than the CT-50 specimen with fatigue crack, Fig. 26. From the
physical point of view, this way of representing seems to be
the most reasonable.

Using those J_R-curves, the best results were obtained when
calculating large scale specimens with the help of fracture
mechanics concepts (R-curve method according to EPRI, Two-
Criteria-Approach according to CEGB (R6-curve)). The J_R-curve
and the blunting line according to both methods is shown for
the precracked CT-25 specimen in Fig. 27.

---

[*]Loss, F.J.: Structural Integrity of Water Reactor Pressure
           Boundary Components NRL Memorandum Report 4400 (1981)

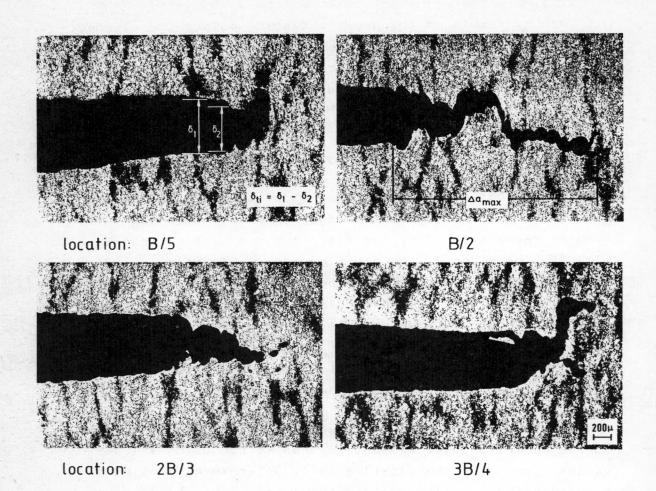

location: B/5   B/2

$\delta_{ti} = \delta_1 - \delta_2$

Δa_{max}

200μ

location: 2B/3   3B/4

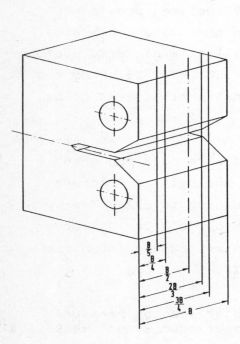

Fig. 1:

Experimental determination of crack
tip opening by means of mechanical
sectioning (CT 25; 20 MnMoNi 55; 80°C)

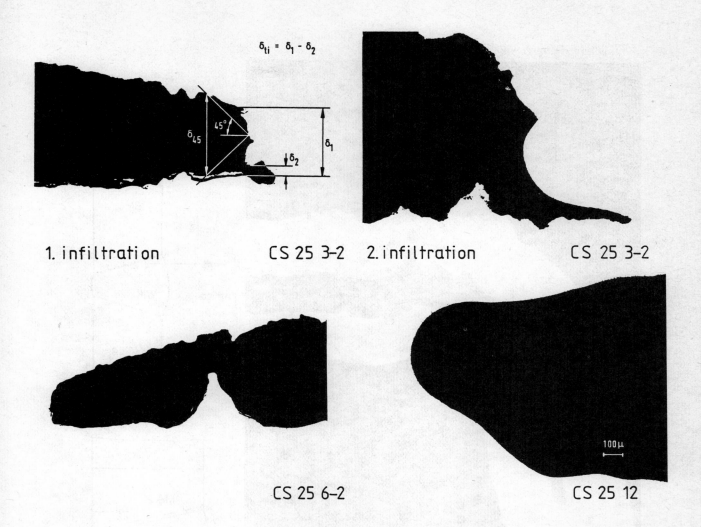

$$\delta_{ti} = \delta_1 - \delta_2$$

45°

δ_{45}

δ_1

δ_2

1. infiltration   CS 25 3–2   2. infiltration   CS 25 3–2

CS 25 6–2   CS 25 12

100μ

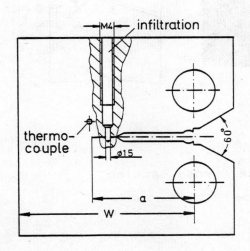

M4   infiltration

thermo-
couple

$\varnothing 15$

60°

a

W

Fig. 2:

Experimental determination of crack
tip opening by means of replicas
(CT 25; 20 MnMoNi 55; 80°C)

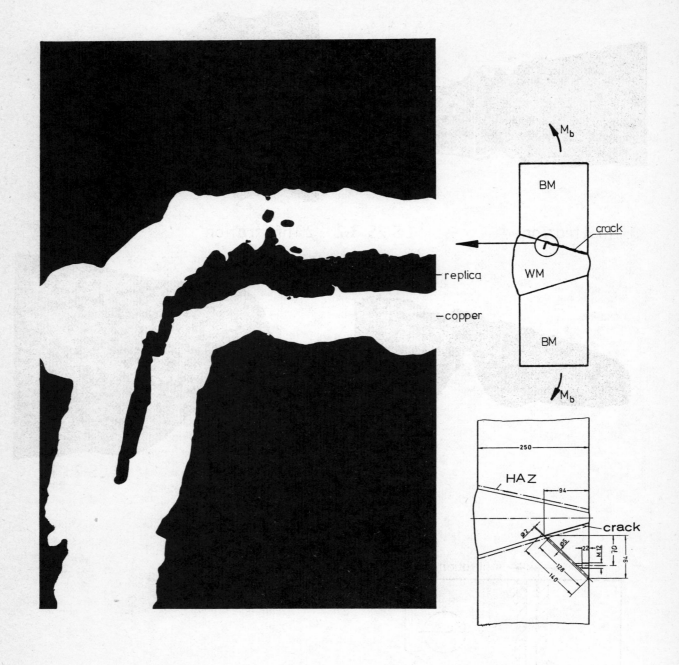

Fig. 3:   Replica of a welding crack (cross section:
             20 MnMoNi 55; RT)

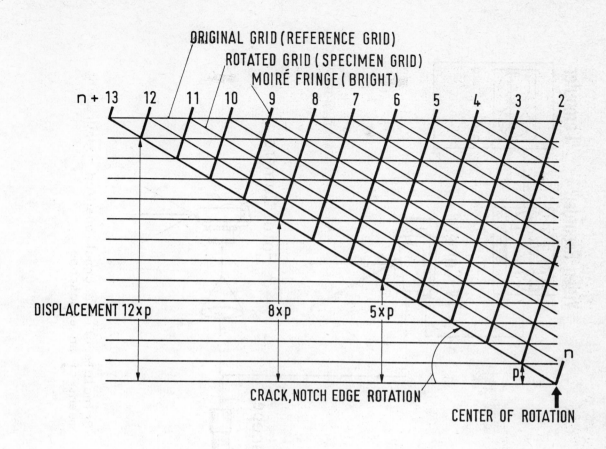

ORIGINAL GRID (REFERENCE GRID)
ROTATED GRID (SPECIMEN GRID)
MOIRÉ FRINGE (BRIGHT)

n + 13   12   11   10   9   8   7   6   5   4   3   2

DISPLACEMENT 12×p   8×p   5×p

CRACK, NOTCH EDGE ROTATION

CENTER OF ROTATION

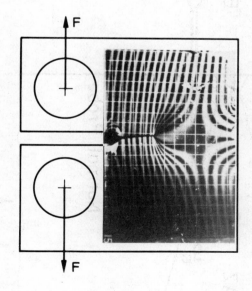

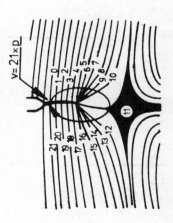

$v = 21 \times p$

<u>Fig. 4:</u>   Crack opening measurement by moiré method

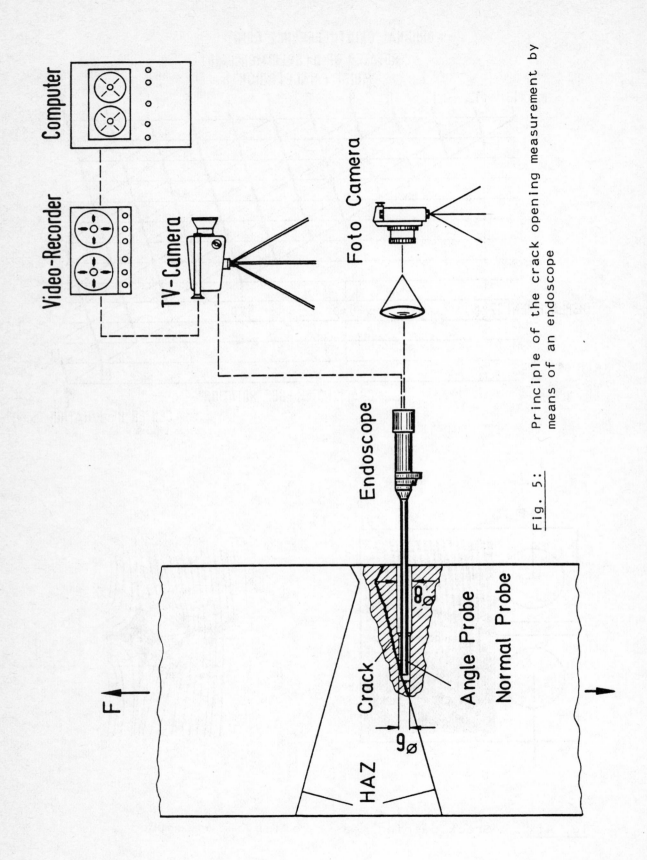

Fig. 5: Principle of the crack opening measurement by means of an endoscope

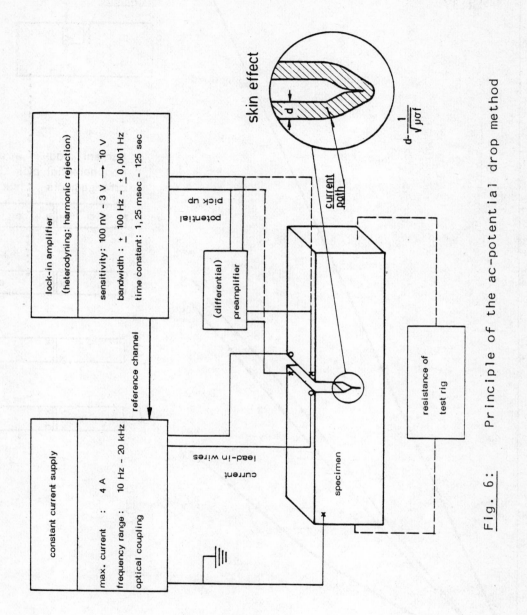

Fig. 6: Principle of the ac-potential drop method

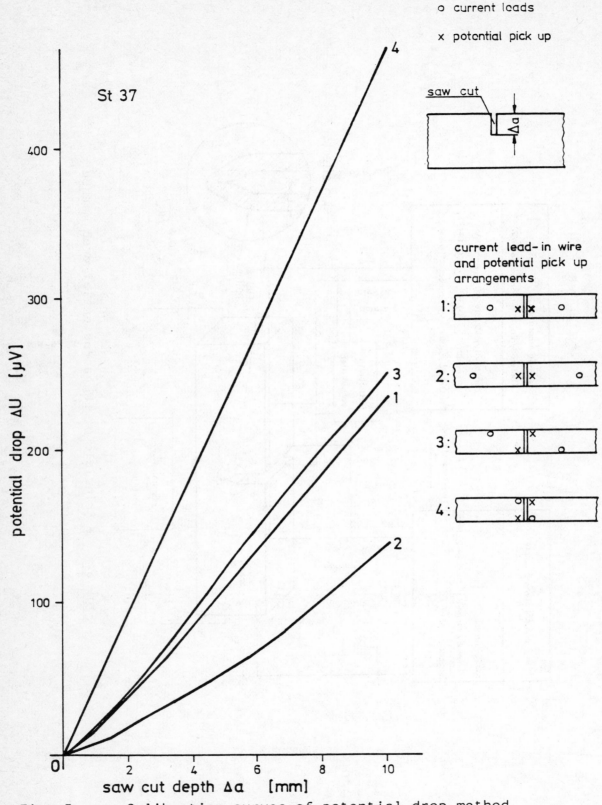

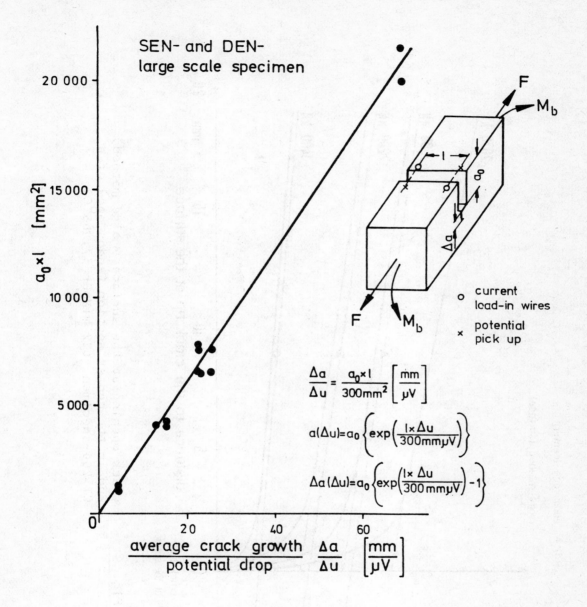

Fig. 8:   Calibration curve for crack growth measurement
           by potential drop method
           (1 A ac, 333 Hz, 22 NiMoCr 37)

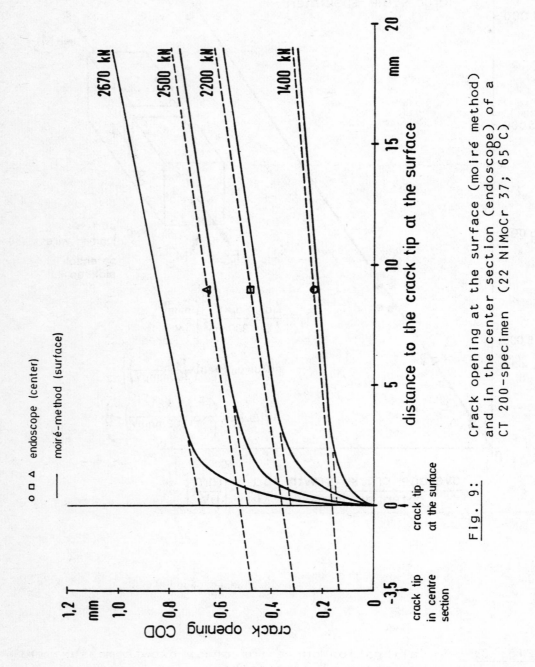

Fig. 9: Crack opening at the surface (moiré method) and in the center section (endoscope) of a CT 200-specimen (22 NiMoCr 37; 65°C)

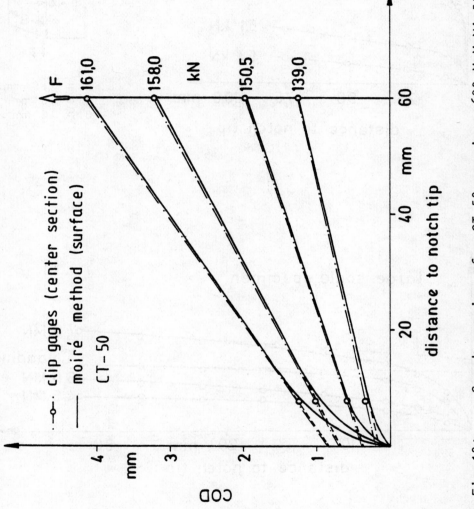

Fig.10: Crack opening of a CT 50-specimen (20 MnMoNi 55; 80°C) without fatigue crack

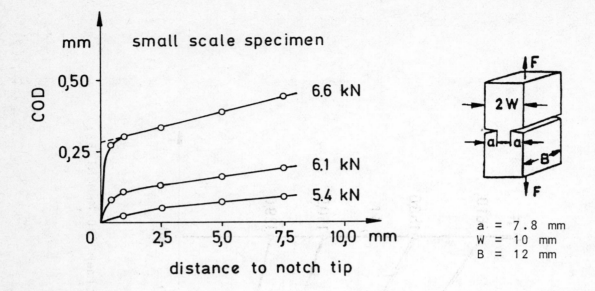

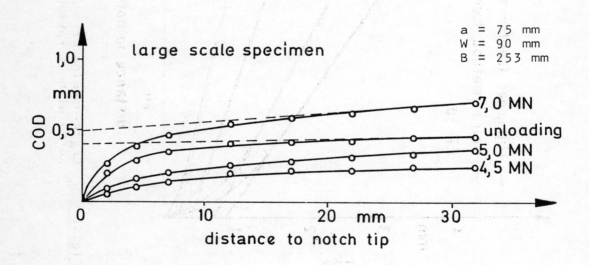

Fig.11:   Notch opening of small and large scale specimens
          (20 MnMoNi 55; + 80°C)

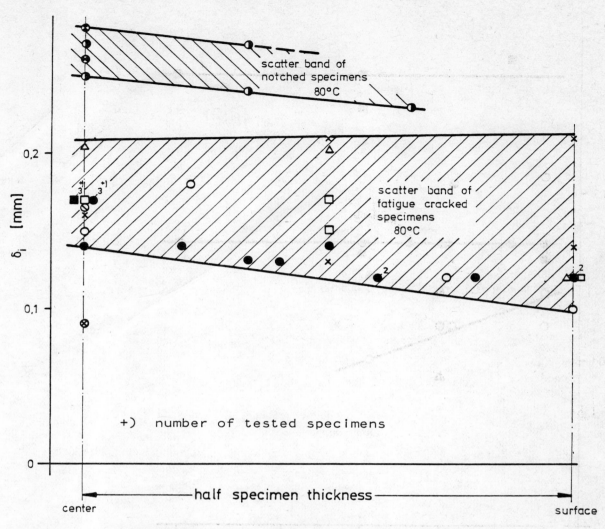

specimen	symbol	flaw	thickness (mm)	testing temperature (°C)
SECB	△		22,5	
DECT	☐		10	
DECT	■		12/31,6/252	
CCT	✕	fatigue	10	+ 80
CT 25	●	crack	25	
CT 46	○		46	
CT 50	⊘		50	
CT 25	⊗		25	− 20
CT 50	◐		50	
DENT	☻	notch	253	+ 80
DENT	⊛		12	

Fig.12: Crack and notch tip opening at crack initiation measured by mechanical sectioning (20 MnMoNi 55)

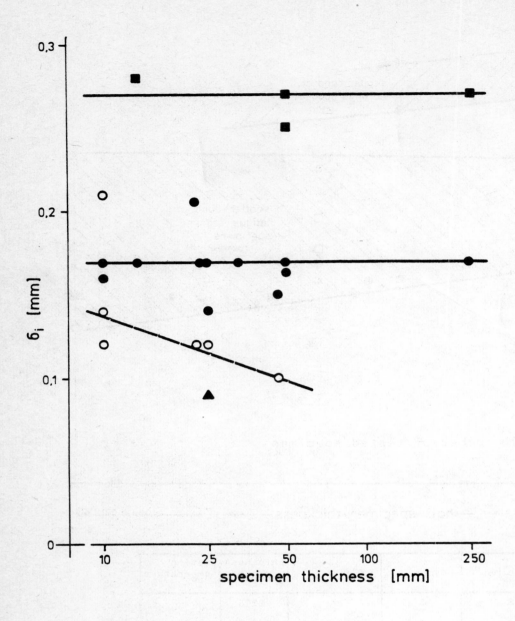

| section | flaw | testing temperature | |
		− 20°C	+ 80°C
surface	fatigue crack		○
center	fatigue crack	▲	●
	notch		■

<u>Fig.13:</u>   Crack and notch tip opening at crack initiation measured by mechanical sectioning vs specimen thickness (20 MnMoNi 55)

small scale specimen (center section)

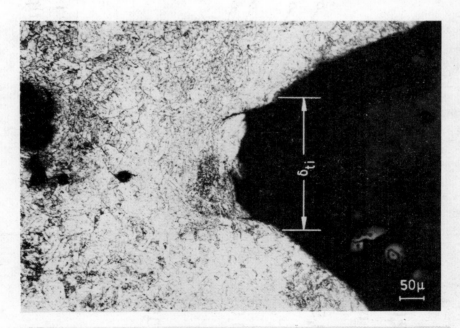

large scale specimen (center section)

Fig. 14: Notch tip opening of small and large scale specimens (20 MnMoNi 55; +80°C)

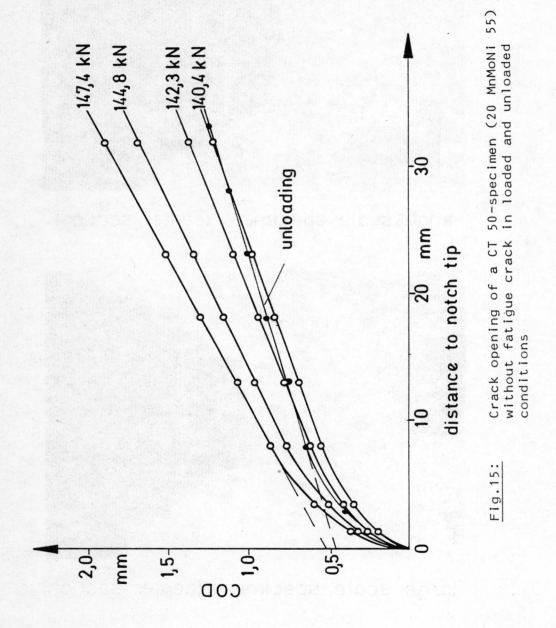

Fig.15: Crack opening of a CT 50-specimen (20 MnMoNi 55)
without fatigue crack in loaded and unloaded
conditions

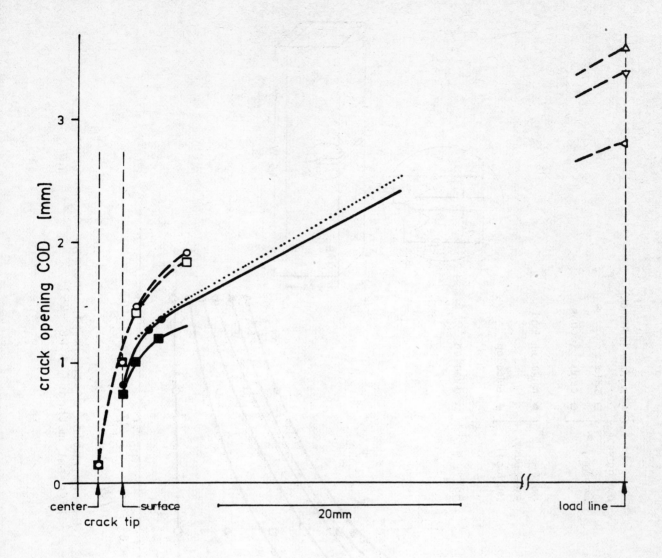

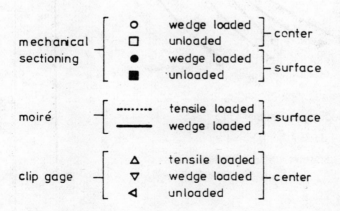

Fig.16: Crack opening in the tensile loaded, wedge loaded and unloaded condition (CT 46; 20 MnMoNi 55; + 80°C)

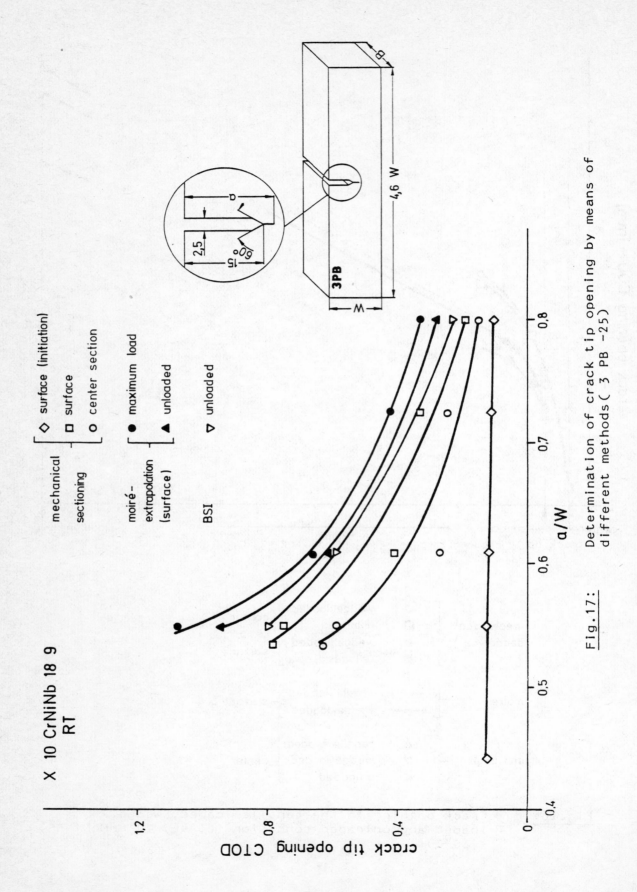

Fig.17: Determination of crack tip opening by means of different methods (3 PB –25)

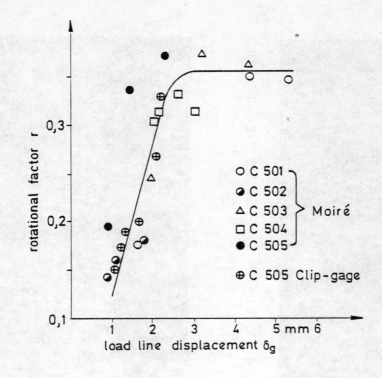

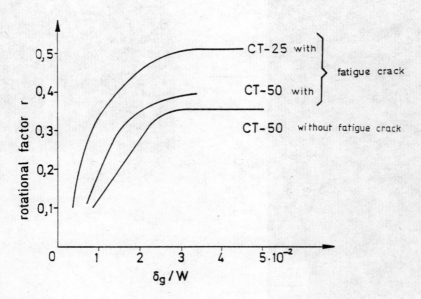

Fig.18:   Rotational factors of CT 25- and CT 50-specimens
with and without fatigue cracks (20 MnMoNi 55;
+ 80°C) vs. load line displacement

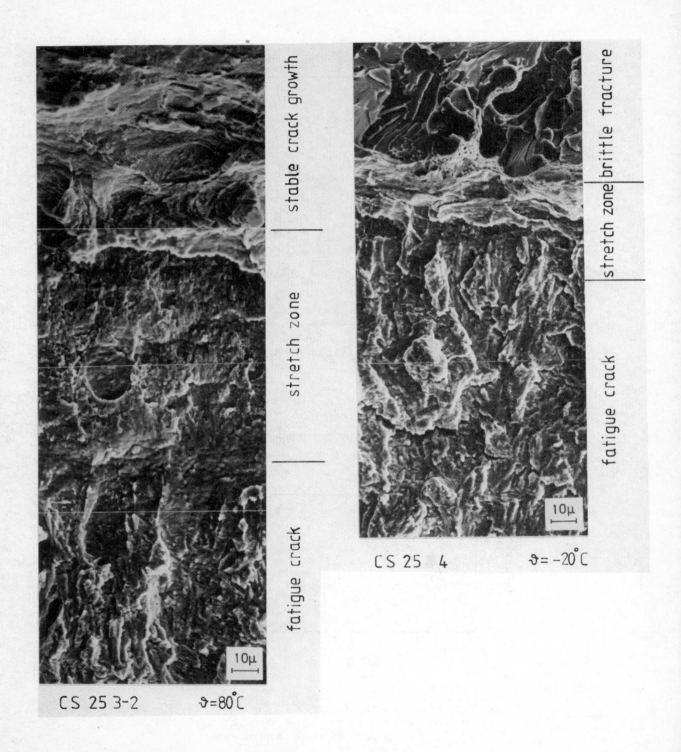

Fig.19:   Fracture surfaces of CT 25-specimens (center section)
            SEM photography (20 MnMoNi 55)

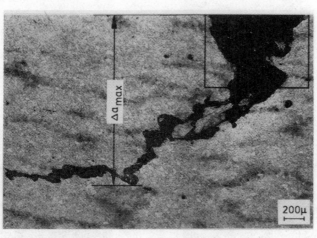

C 50 6 center section

SEM (center)

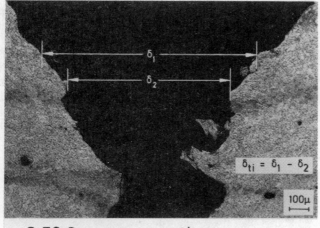

C 50 6 center section

Fig. 20: Mechanical section and fracture surface of a CT 50-
specimen without fatigue crack (80°C; 20 MnMoNi 55)

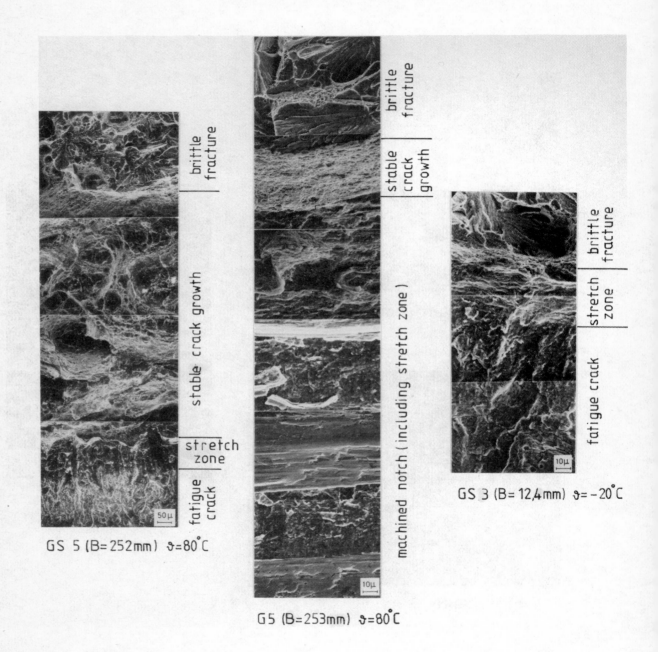

brittle fracture

stable crack growth

stretch zone

fatigue crack

GS 5 (B=252mm) ϑ=80°C

brittle fracture

stable crack growth

machined notch (including stretch zone)

G5 (B=253mm) ϑ=80°C

brittle fracture

stretch zone

fatigue crack

GS 3 (B=12,4mm) ϑ=-20°C

Fig.21: Fracture surfaces of large scale specimens
(DENT; 20 MnMoNi 55)

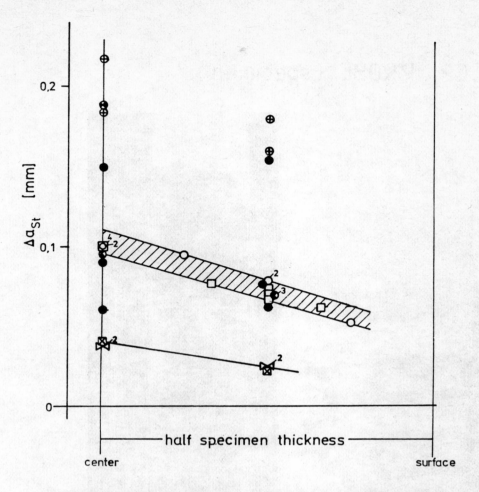

specimen	symbol	flaw	specimen thickness (mm)	testing temperature (°C)
DECT	⊕	fatigue crack	12,2 / 31,6 / 252	+ 80
CT 25	□		25	
CT 50	○		50	
DECT	⋈		12,4	− 20
CT 25	⊠		25	− 20
DENT	●	notch	4 / 12 / 14	+ 80
	◕		253	
CT 50	⊕		50	

Fig. 22: Stretch-zone width of different specimens
(20 MnMoNi 55)

DECP PROBE (specimen)

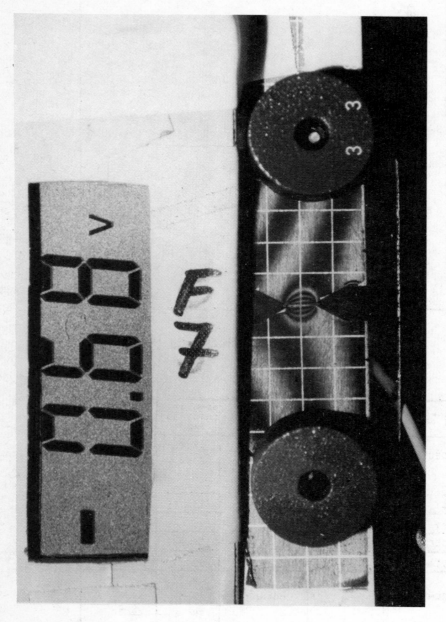

F 7   (B = 2mm) $\vartheta = 80°C$

Fig.23:   DENT small scale specimen (20 MnMoNi 55; T = 80°C)

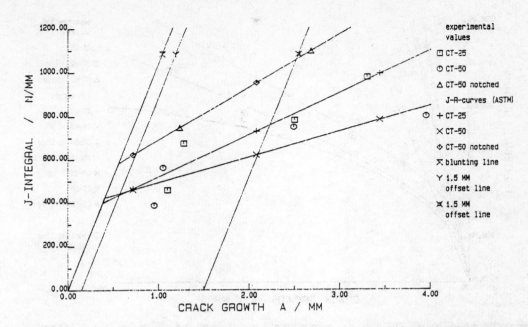

Fig. 24: J-R-curves of CT 25- and CT 50-specimens with and
without fatigue crack according to ASTM-standard
(20 MnMoNi 55; T = 80°C)

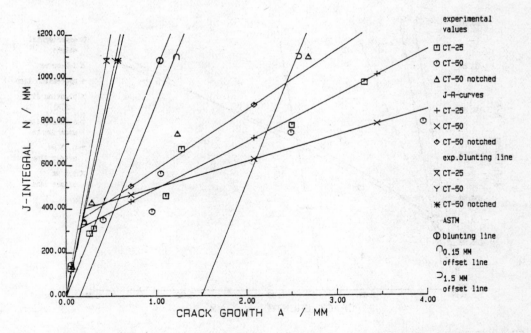

Fig. 25: J-R-curves of CT 25- and CT 50-specimens with and
without fatigue crack including all experimental
results (20 MnMoNi 55; T = 80°C)

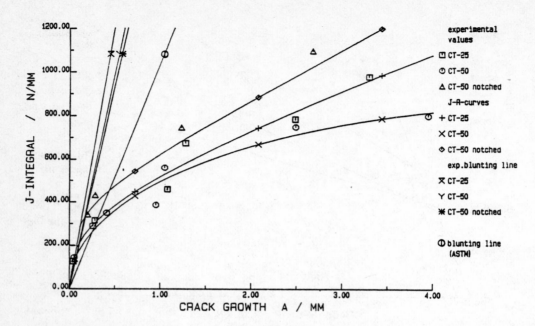

Fig. 26: J-R-curves of CT 25- and CT 50-specimens with and without fatigue crack fitted with smooth curves (20 MnMoNi 55; T = 80°C)

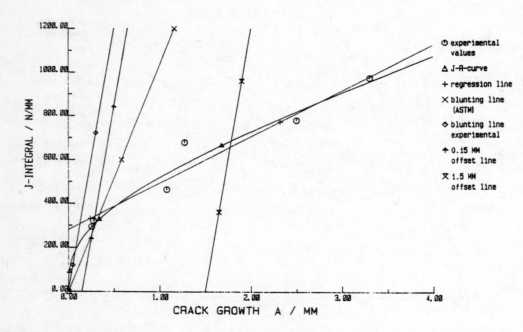

Fig. 27: J-R-curves of CT 25-specimen with fatigue crack according to ASTM-standard and fitted with smooth curve (20 MnMoNi 55; T = 80°C)

Dormagen, D.; W. Hesse, H.-J. Rosezin

Application of J-Resistance Curves for the Prediction of Failure Loads of Wide Plates

Communication of the Institute for Ferrous Metallurgy, RWTH Aachen

The Shih and Kumar engineering approach has been used to predict failure loads of compact specimens and wide plate tests. For two different steels, a pressure vessel steel 20 MnMoNi 5 5 and a structural steel St 52-3, the predicted instability loads are compared with the maximum loads.

According to the Shih and Kumar methodology the material's strain hardening behaviour and the state of stress have to be taken into account to calculate the applied J.

Fig. 1 shows the comparison between predicted and measured instability loads of 1 CT, 2 CT-28, 2 CT-28/20 % SG and 3 CT-25 CT-specimens. Only for the side-grooved specimen failure occurs in plain strain condition, whereas the failure state for the 1 CT-, 2 CT-28 and 3 CT-25 specimen is between plain stress and plain strain.

Fig. 2 and 3 shows the different kinds of approximation of the stress strain-curve with the Ramberg-Osgood equation. Because of the extended Lüders strain curve 1 is describing the material's work hardening behaviour and curve 4 is describing the elastic part of the stress-strain graph best of all. The variation in the α- and N-values is very big and must have of course an influence for calculating the $J_{applied}$-values.

Fig. 4 and 5 shows the ratio of F_{inst}/F_{max} depending on α- and N-values, where failure loads of wide plates are predicted using R-curves of CT-specimens. One can see that for steel 20 MnMoNi 5 5 the predicted failure loads are always conservative, when J-resistance curves with side-grooved specimens were used only, independent of α- and N-values. For steel St 52-3 conservative instability loads are predicted for the 18 mm flaw, when method 1 and 2

is used to describe the strain hardening behaviour.
For the 30 mm flaw no conservative values are predicted.
There is a general tendency that with decreasing N-values
and corresponding increasing α-values lower instability
loads are predicted.

Conclusions:

The constants of the Ramberg-Osgood equation severely
influences the results of the Shih and Kumar engineering
approach. The aim of future research must be to develop
a physically significant description of the stress strain
approximating as well the elastic as plastic deformation
behaviour.

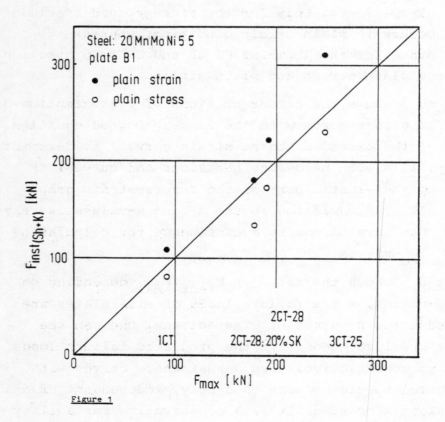

Figure 1

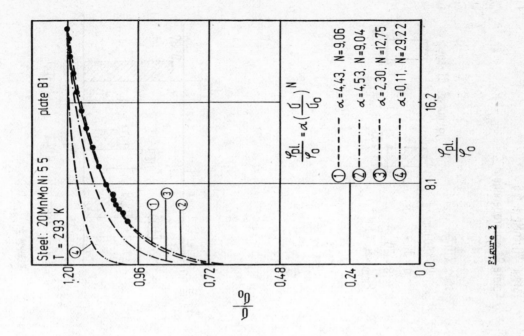

Figure 3

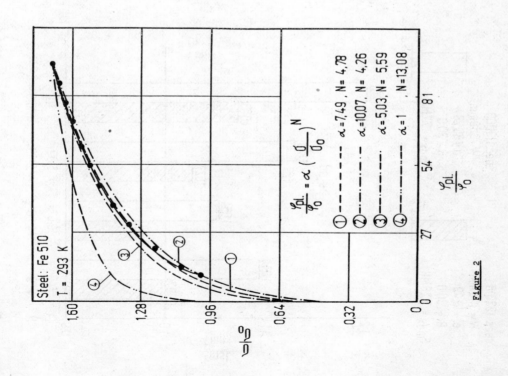

Figure 2

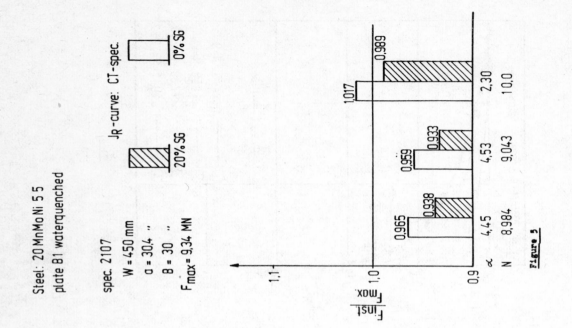

Steel: 20 MnMoNi 5 5
plate B1 waterquenched

spec. 2107
W = 450 mm
a = 30,4 ..
B = 30 ..
F_{max} = 9,34 MN

J_R-curve: CT-spec.

20% SG   0% SG

Figure 5

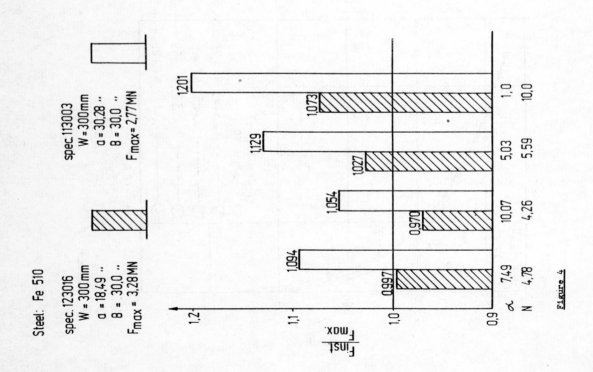

Steel: Fe 510

spec. 123016
W = 300 mm
a = 18,49 ..
B = 30,0 ..
F_{max} = 3,28 MN

spec. 113003
W = 300 mm
a = 30,28 ..
B = 30,0 ..
F_{max} = 2,77 MN

Figure 4

THE APPLICATION OF GROSS SECTION YIELDING FOR SAFETY ASSESSMENT IN STRAIN HARDENING MATERIALS.

Dr. ir. Rudi DENYS
Laboratorium Soete Gent University
St Pietersnieuwstraat, 41,
B 9000 Gent - BELGIUM

INTRODUCTION

The engineering prediction of service performance of structures can be realized either by intermediate scale or by large scale testing. With the former method, a crack tip fracture toughness parameter is obtained from which an indirect assessment of defects can be performed (see e.g. the CTOD design curve approach). However, it is evident that the preference goes to the latter method since tests on large specimens simulate service or full scale behaviour in a more realistic way.

BEHAVIOUR OF WIDE PLATE TESTS

The attempt of simulating service behaviour has led us to utilize and test large specimens with various types of defects (through thickness -, surface-and buried defects) at the minimum OPERATING temperature.

The behaviour of approximately 400 wide plate tests on plate material and weldments (plate width $\geq$ 350 - 1000 mm and plate thickness $\geq$ 15-100 mm) that have been carried out in Gent, indicate that in general:

- The macro mechanism of yielding occurs by slip along lines starting from the ends of the initial butterfly shaped plastic zones at the crack tips.
Depending on the material properties (yield strength, Lüders elongation, strain hardening exponent ...) and the specimen geometry (crack dimensions, plate thickness, ...) there are two quite distinct ways in which incipient net section yielding can occur, namely by in plane slip at 45° to the outer edges of the specimen or by cross slip at 45° to the plate surface (Fig. 1)

- The strain of all slip lines converging at the end of the notch (sawn as well as fatigued) gives extremely large values. A strain of the order of 1 is quite normal, while the material in the same section, some distance away, is still in the elastic state.

- The net section stress at fracture is above the material's yield stress (0.2% proof)and that the flawed wide plate specimens deform either by net section or gross section yielding. Moiré pictures of plastic displacement patterns at the plate surface are used to provide confirmatory evidence on this point (Fig. 2 to 3)

- Gross section yielding occurs soon after the net section passes into the strain hardening range, i.e. when the overall stress-strain curve intersects the materials yield stress before fracture initiation.
On the other hand, gross section yielding is accompanied by considerable plastic deformation. This is e.g. typified by the gross stress-overall strain record, as obtained from an actual test, on weldments, (Fig. 4).

- For situations where the weld metal yield strength overmatches that of the plate, the load extension behaviour of
the whole plate is controlled by the plate material properties and that gross section yielding is then easily obtained
when the defect is located within the width of the weld.

- Although high local (crack tip) strains occur, very few fai-
lures involve 100% slant-ductile fracture . From a macroscopic viewpoint, fracture occurs mostly with raising load i.e.
the fracture surface is flat and perpendicular to the plate
surface and herringbone, or chevron, patterns with or without
shear lips are visible. In other words, failure occurs predo-
minantly by cleavage.(Fig. 5)

These and many other observations led to the consideration of a
direct approach of defect tolerance, based upon wide plate testing solely. The use of such an engineering assessment is characterised by a direct experimental measurement of the overall
behaviour of constructional elements.

THE APPLICATION OF WP DATA TO DEFECT SIGNIFICANCE.

As indicated, a wide plate with a defect loaded in tension de-
forms either by net section or by gross section yielding. The
shift from net to gross section is caused by the strain hardening properties of the crack tip material.
Considered as such, the demarcation between net to gross section yielding permits an estimation of a critical defect
size.(Fig. 6 and 7)

Thus, a defect smaller than the critical defect induces gross section yielding guaranteeing a good overall strain and a strength of at least yield magnitude, from which by agreeing on both properties, it can be construed that subcritical defects are acceptable.

Experimental evidence for the previously defined approach of defect assessment is given in table 1. This table summarizes a selection of results obtained from wide plate ten- sile tests on normalized (St E 355) and quenched and tempered (St E 690) plate material and their weldments (see also Fig. 5).

Fracture tests were carried out on 30 mm thick plates at normal operating temperature (T = −20°C).

It can be seen that the critical defect sizes are surprisingly large, although gross section yielding was imposed.

ADVANTAGES AND USE OF LARGE SCALE TESTING

The preceding observations emphasize that we must utilize the actual material properties, which differ from the idealized elastic perfectly plastic characteristics, for safety assessments in ductile strain hardening materials. Following this line of thought, it is evident that quite some difficulties and ambiguities associated with the crack tip fracture toughness experiments in the elastic-plastic range can be solved by testing large specimens. The advantages of large scale testing are indeed numerous :

- The relevance of the elastic-plastic fracture mechanics procedures can be assessed. This is especially needed in the field of fracture mechanics testing of structural steel and weldments with yield strengths less than 500 N/mm^2.

- The actual constraint conditions at the crack tip can easily be simulated.

- A variety of specimen geometries and specimen sizes, which mirror the structural component, can be tested.

- It is possible to examine directly:
    * Various defect geometries.
    * Defect location.
    * Defect orientation.

- The different strength ratios in weldments i.e. for situation where the weld metal overmatches or undermatches the plate metal, are assessed directly.

- Complex stress patterns existing in structures are easily assessed e.g.:
    * Angular distortion
    * Misalignment
    * Geometrical strain concentrations
    * Residual stresses

- It is possible to study strain ageing and thus embrittlement effects by introducing defects before welding.

- The necessity of a stress relief treatment can be assessed.

- It provides a quantitative evaluation of ductile materials i.e. the behaviour in the near upper shelf and upper shelf conditions.

- With wide plate testing, the gross section yielding approach can be utilized when the crack tip fracture toughness parameter approaches becomes invalid.

- Finally, a large scale test gives data of direct relevance to evaluate fitness for purpose.

FINAL OBSERVATIONS

The following conclusions are in the context of the preceding statements and experimental observations :

1. It is common knowledge that a lot of structures fabricated with C—Mn structural steel are still in service, without satisfying the present requirements based on fracture mechanics toughness concepts. This may be explained by and is due to the fact that the effect of many important, beneficial and relevant factors can not be accounted for by standard testing procedures.

2. The ultimate use of elastic plastic methods of fracture mechanics is in the engineering prediction of the behaviour of stuctures. The proposed methods of inelastic fracture analysis for service components vary widely in nature, depending greatly on the technical background of the analysts and the type of experimental data analyzed. However, it is to recommend that the development of any single crack tip fracture toughness parameter should be considered in the context of the behaviour of a large structure or a large scale testing component.

3. For structural and thus real materials, which exhibit substantial plastic deformation accompanied with strain hardening, the ultimate failure behaviour in the ductile and indeed cleavage regime cannot always be characterised by a single deformation parameter. Instead, an overall approach is recommended to avoid safety factors which are unnecessarily generous.

4. It must be recognized that unstable crack growth in small specimens does not mean that the same situation will occur in large structures with different constraint conditions. This implies that the requirement to match the constraint condition in the fracture mechanics specimen to that in service or large scale test components militates against handsome test procedures and that the criterion of fracture appearance is liable to be misinterpreted.

5. The different areas of engineering utilize different toughness levels and may therefore require different fracture characterizing parameters; e.g. most structures that serve an important safety function are deliberately designed not to fail in the elastic range of stress, but to fail only under gross overload in the plastic range. This implies that the single deformation parameters such as δ and J are not necessarily to be rejected; however, the extent to which these and other fracture mechanics parameters can be used must be clearly defined.

6. Looking through the enormous amount of literature on fracture mechanics one finds few data of service experience, to be referred to, in order to translate practically sized laboratory test specimen results into unconservative fitness for purpose information.

7. Not every laboratory is equipped with the necessary large scale testing machines. Nevertheless, wide plate or large scale testing should be promoted, in particular for huge and expensive structures for which conservative crack tolerance levels may cost unnecessarily a lot of money. In the authors'experience, despite the requirement of gross section yielding, larger defects may be allowed than those normally predicted by the traditional fitness for purpose approaches.

8. Both theoretical and experimental fracture mechanics ap-
    proaches are required to assess the response of structures.
    However, scientists and researchers should be aware that
    there is a difference between an academical approach and
    real life evaluations of flawed structures. To this extent,
    literature is sometimes misleading since one almost exclu-
    sively finds fracture toughness data of specimens with lar-
    ge defects, i.e. to simulate a plane strain situation.
    Whilst "small" defects, of the type used in wide plate
    tests giving rise to gross section yielding, are suffi-
    ciently large to be reliably detected using the available
    NDT methods.

9. With regard to nomenclature, it would be better in future
    to refer to net or gross section deformation, rather than
    plastic collapse. The adjective collapse is unfortunate and
    only applies, without ambiguities, to perfectly elastic-
    plastic material but not always real materials and
    structures. In this context, it has been shown that the
    defect dimensions play a major role.

CONCLUSIONS

The gross section yielding concept must be considered as a direct fitness for purpose approach. In any event, this approach,
which gives an adequate margin of safety in terms of strain,
should be considered as an alternative, a complementary and a
necessary method that should be used together with the single
crack tip frac- ture toughness parameter methods to ensure
safety.

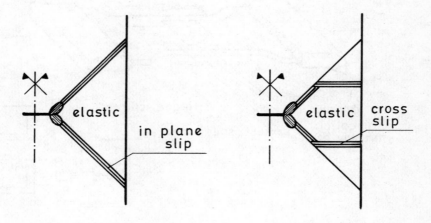

Fig. 1. Possible modes of INCIPIENT net section yielding.

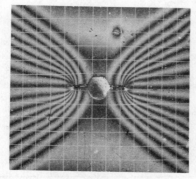

Fig. 2. Moiré picture of NET section yielding.

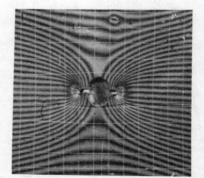

Fig. 3. Moiré picture of GROSS section yielding.

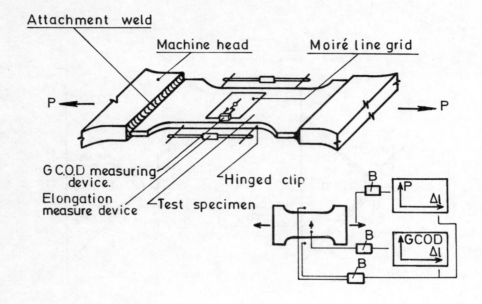

Test set - up

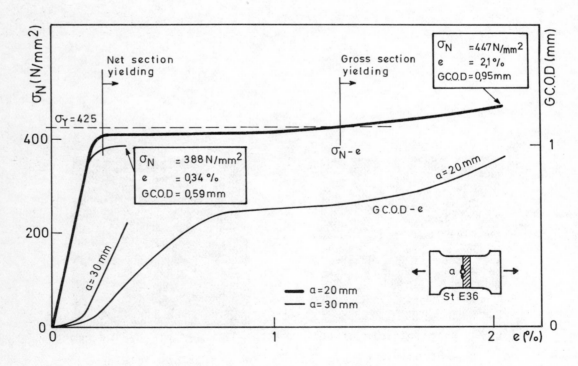

Fig. 4. Typical gross stress (Gent COD) - gross strain
records (Gauge length = plate width = 350 mm)

Wide plate specimen

StE 355 − a = 20 mm − Plate material − Deformation mode:
Gross section yielding − Ductile tear: 4 mm − Cleavage
fracture − e = 4.28 % −− Plate thickness 30 mm

StE 355 − a = 15 mm − Weldment − Deformation mode : Gross
section yielding − No ductile tear − Cleavage fracture −
e = 2.18% − Notch position HAZ − PLate thickness 30 mm

Fig. 5. Ste 690 − a = 8 mm − Weldment − Deformation mode:
Gross section yielding − No ductile tear − Cleavage
fracture − e = 1.26% − Notch position HAZ − Plate
thickness 30 mm

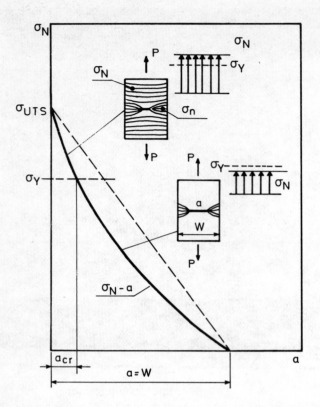

Fig. 6. Gross stress dependence of crack length.

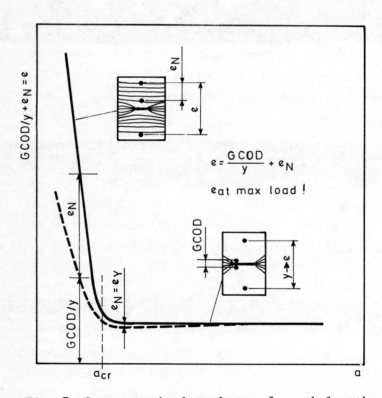

$$e = \frac{GCOD}{y} + e_N$$

$e_{\text{at max load}}$!

Fig. 7. Gross strain dependence of crack length.

YIELD STRENGTH OF UNNOTCHED WIDE PLATES* (N/mm2).

Plate material		Welded joint	
StE 355	StE 690	StE 355	StE690
425	751	410	757

* Material properties of steels tested at
test temperature (-20°C)

CRITICAL NOTCH DIMENSIONS AND RELATED NOTCH TOUGHNESS VALUES.

Notch Geometry	Material	Critical length or depth (mm)	GCOD* at a_{cr} or c_{cr} (mm)	e** at a_{cr} or c_{cr} (%)
Plate	StE 690	a_{cr}=12	3,3	1,2
	StE 355	a_{cr}=28	2,8	1,5
Weldment	StE 690	a_{cr}= 8	0,3 - 1,0	0,4 - 1,3
	StE 355	a_{cr}=21	0,9	1,5
Plate	StE 690	c_{cr}=3,5	-	3
	StE 355	c_{cr} = 9	-	2,5
Weldment	StE 690	c_{cr} = 2	0,3	0,5
	StE 355	c_{cr} = 6	-	1,7

* Gent COD - measured at the crack tip on a gauge length of 8 mm.
** e - overall strain, gauge length equalled plate width : 350 mm

FRACTURE CHARACTERISATION OF LINEPIPE STEELS

A.H. Priest, British Steel Corporation, Swinden
House, Moorgate, Rotherham

Synopsis

In Laboratory tests on fully ductile steel the energy absorbed per unit area in shear fracture is $U/A = R_c + S_c (W-a)$. Where R_c is the fracture propagation energy, S_c is the mean energy density in the necked zone adjacent to the fracture and $(W-a)$ is the ligament size.

A similar expression has been used to determine the energy absorbed in the fracture of high pressure gas pipelines using $2/3\ C_v$ and Battelle DWTT energies to estimate the value of the two constants.

By relating this value to the total energy of the compressed gas, a high proportion of the results of full scale burst tests have been predicted. A third term incorporating yield stress and work hardening rate has enabled ambiguous data from highly work hardened steels to be successfully included in the analysis. It is also suggested that R_c values are related to G_{IC} and offer an alternative to J_{IC} tests as a method of measuring static and dynamic toughness from fully ductile materials.

Fracture Propagation

Work on pipeline steels has indicated that two parameters are required to describe the fracture process in the fully shear and post yield situation. (1, 2 and 3).

The first parameter, R_c, is the energy per unit area for fracture per se. It is the energy required to create a given area of fracture surface. If there was no other energy involved in fracture, then the total energy required to fracture a test piece or a structure would be

$$U_f = R_c.B.(W-a) \qquad (1)$$

where $(W-a)$ is the length of the fracture surface, B is the thickness. However, in these steels, fracture is accompanied by plastic deformation and energy is absorbed in this process simultaneously with crack propagation. The energy absorbed in plastic deformation is equal to the second constant, S_c, multiplied by the volume of material deformed. S_c is the energy required to deform a given volume of material adjacent to the crack.

The volume of plastically deformed material is, by virtue of geometrical considerations, proportional to the square of the fracture length.

Hence, the plastic deformation energy

$$U_p = S_c.B.(W-a)^2 \qquad\qquad (2)$$

The total energy to fracture test pieces of a range of size is therefore

$$U = U_f + U_p = R_c.B.(W-a) + S_c.B.(W-a)^2 \qquad (3)$$

Tests can be carried out on compact tension specimens broken in tension, or on bend specimens broken in three point bending at either slow strain rate or under impact loading.

Values of R_c and S_c are determined experimentally by plotting the fracture energy per unit area values $\dfrac{U}{B(W-a)}$, against $(W-a)$ for a series of test piece sizes, Fig. 1. The positive intercept of $\dfrac{U}{B(W-a)}$ at zero ligament length is equal to R_c and the slope of the straight line is equal to S_c.

R_c and S_c are found to vary with test temperature and strain rate and both variables need to be defined when test results are reported. Whilst R_c values are found to be the same in bend and compact tension tests, the values are approximately doubled in pure tension tests. Values of S_c are the same when determined by all three test methods. This variation in R_c value has been put down to the compressive zone towards the back face of the former tests causing material damage before the crack propagates through it.

So long as fully shear fracture occurs and there is no inhomogeneity in the material, the values of R_c and S_c appear to be independent of thickness. Fig. 2. The difference in R_c values obtained from machined notched or fatigue cracked test pieces is small; S_c is unaffected.

Before this can be applied to fracture occurring in a pipeline, it is necessary to know two things:

(1) The measurement of a pipeline equivalent to $(W-a)$.

(2) The energy absorbed per unit area of fracture surface $\dfrac{U}{A}$ as opposed to the total fracture energy from Equation (3).

It is known that the value of $(W-a)$ is related to the distance, normal to the crack path, over which necking of the test piece or structure occurs. This distance, L, is approximately equal to $\frac{2}{3}(W-a)$ (1) so that instead of $(W-a)$ we can write 1.5 L in expression (3). Little information is currently available regarding the value of L in actual pipeline fracture. It has been stated (2) that 200 to 400 mm is a common size but an understanding of the factors controlling this aspect must be the objective of future research. The current opinion is that it is likely to increase with the diameter of the pipeline because this is the dimension that governs the size of the region of plastic deformation.

The energy absorbed per unit fracture area is obtained by substitution. In test pieces:

$$\frac{U}{A} = \frac{U}{(W-a)B} = R_c + S_c(W-a) \qquad (4)$$

where A is the crack area.

In pipelines:

$$\frac{U}{A} = R_c + 1.5\, S_c.L \qquad\qquad (5)$$

However since L is probably limited by the diameter in a linepipe and since the stress at the crack tip in a linepipe approximates to pure tension we can write

$$\frac{U}{A} = 2\,R_c + Z\,S_c D \qquad\qquad (6)$$

where D is the pipe diameter,
Z is a constant which compensates for determining R_c and S_c values from Charpy and Battelle tests which are known to be in error for reasons given below.

Values for the required fracture energy from expression (6) have been plotted against the energy available for fracture (4) i.e.

$$\text{Potential Energy} = \frac{\sigma D}{2}\left(\ln\frac{P_1}{P_2} + \frac{5}{4}\frac{\sigma}{E}\right) \qquad (7)$$

where P_1 is pressure in the pipe

P_2 is ambient pressure

σ is applied hoop stress

E is Elastic Modulus

Values of both R_c and S_c vary with increasing strain rate and therefore values from high rate tests will be more representative of rapidly propagating cracks in pipelines. The current proposal is that impact tests should be adopted because of this. However, it may be possible to establish a simple correlation between impact and low rate values and this approach is also being pursued, because of the ease and relative accuracy of low rate tests.

To establish S_c and R_c values, it is necessary to establish the slope and intercept of the straight line relationship above. In theory, this is possible with just two tests, having two different ligament sizes. However, recent work has indicated that, below a certain ligament size in impact tests, the U/A values tend to fall below the straight line of the theoretical relationship. To be certain of avoiding this, a minimum ligament size of 30mm is recommended, precluding the Charpy test which has a ligament size of 8mm. The Charpy test, because it gives results below the straight line relationship, provides results which are conservative by an unknown amount which may vary with metallurgical factors. Another difficulty is that in bend tests, if the span:width ratio is small, the fracture load increases and indentation and friction can cause unacceptably high errors in measurement of fracture energy. At present it is thought that a span:width ratio of 4:1 or more is desirable. Certain Dynamic Tear Tests would therefore be suitable provided test pieces were of the 4:1 span to width ratio type. The Battelle Drop Weight Tear Test piece is considered to be unsuitable because the span to width ratio is only 3.3 to 1 which results in an increase in load by a factor of 1.2 relative to a 4:1 span.

Furthermore, experience has shown that notch depth to width ratios (a/W) of at least 0.2 are required to avoid spuriously high energy values due to plastic deformation around the notch. This would rule out the standard 3 inch wide Drop Weight Tear Test. The pressed notch of such tests is also known to cause errors.

Some of the above observations indicate why only poor correlations have been obtained between laboratory tests and service performance. This is particularly the case with materials exhibiting separations or splitting where the departure from linearity of the $\frac{U}{B(W-a)}$ values is particularly marked at low ligament sizes. Any relationships between Charpy values and R_c would therefore be expected to show differences for different classes of steel and heat treatments.

Results were obtained from AISI and EPRG burst test programmes with pipe diameters varying from 406 to 1219mm and thicknesses from 7.92 to 25.44mm, (5,6). The results of these full scale burst tests are available as propagate or arrest data points.

In order to obtain the value for Z, an optimisation technique was employed for the ratio:-

$$R = \frac{(\text{Shear fracture propagation energy})}{(\text{Potential energy available})} \qquad (8)$$

The minimum ratio for the arrest points divided by the maximum ratio for the propagate points gave a series of gradient ratios for different values of Z. The optimum value of Z which gave the highest gradient ratio was equal to 0.12. A gradient ratio of 1 would have indicated that the possibility of fracture is governed totally by expression (8).

The shear fracture propagation energy as a function of the available potential energy is presented for various thicknesses of pipe in Fig. 3; the lines shown indicate the boundary positions for both arrest and propagate data points. It can be clearly seen that there is no evidence of any effect due to pipe wall thickness. As stated above the degree of overlap between the two boundaries, representative of the gradient ratio, is probably largely accounted for by the necessity of using existing $2/3C_v$ and Battelle test data in order to calculate R_c and S_c.

The gradient ratio for the above data was 0.81. However, certain AISI steels having high levels of tensile properties were excluded from this analysis because these were known to show longitudinal fissures in the fracture surface.

Such separations were suspected of causing a reduction in the fracture energy of running cracks. If these latter steels were included in the analysis the gradient ratio was reduced to 0.58. However, inclusion of a third term in the expression related to the tensile properties raised the gradient ratio to 0.76. At this stage the true significance of the third term is not understood. It may be a real effect; it may be related to the influence of the geometry of Battelle and Charpy tests or it may be related to strain rate effects. The three term expression is:-

Fracture Energy $= 2R_c + 0.12\ S_c D + 0.08\ \sigma_y n\ D.$ \qquad (9)

where σ_y is the yield stress

n is a measure of the strain hardening rate (3)

Fracture Initiation

In order to establish the original relationships it was necessary to conduct slow rate tests on a range of test pieces having different geometries but similar thicknesses. During these tests the electrical potential technique was used to determine the point of fracture initiation. This method had previously been shown to give almost identical J_R curves to those derived by the multi-test piece method. Figure 4.

For the linepipe steels J values were calculated according to the procedure of Rice et al (7) and J_R curves were derived using the modified expression of Garwood, Robinson and Turner (8). The J_i values are in agreement with $\delta_i \sigma_y$ values from the same tests[1], Figure 5 δ_i is the COD value at initiation of ductile fracture, calculated according to BS 5762, with the exception that compact tension test pieces were used. The values of δ_i are therefore probably in error but close to the correct value, in relation to the wide variation in values due to the difference in test piece size. Both the J_i and δ_i values varied with test piece size, and were not related to the toughness of the material as determined by Charpy and Battelle tests, or in terms of R_c and S_c. The relationship between the load at fracture initiation P_i and the expression in Figure 6 indicates that fracture initiation was characterised by the nominal stress, probably related to the tensile properties of the material.

This supports the view that shear fracture initiation is governed by the tensile properties (9).

A similar variation in J_R curves with test piece dimension is indicated in Figure 7.

A comparison of the J_i values for the three steels investigated indicates that for none of the test pieces were these values as high as the R_c values illustrated in Figure 1. For a given geometry, R_c is independent of test piece size, provided a minimum ligament size of 30mm is exceeded. Conceptually it is the fracture propagation energy for a ligament size tending to zero which is a plane strain configuration. It is therefore probably realistic to suggest that R_c is what J_{IC} would be if large enough test pieces were available to measure it. Certainly, in comparison with the difficulties and apparent errors inherent in J_R tests the relative simplicity and lack of ambiguity of the two parameter method has much to recommend it.

Similarly, the slope of the fracture resistance curve, as defined by S_c is also probably related to the value dJ/da in test pieces of adequate size. In the authors experience, however, it has not been possible to define the J_R curve, using standard methods, with sufficient confidence to make a credible comparison of the two methods. However Venzi (10) has provided theoretical evidence to support the equivalence of $\frac{dJ}{da}$ and S_c; the relationship given being:-

$$\frac{dJ}{da} = -S_c \qquad (10)$$

which is in accordance with the negative slope of the U/A versus (W - a) curve with respect to Δa (Fig. 1).

ACKNOWLEDGEMENT

The author wishes to thank Dr. K.J. Irvine, Director, Research and Development BSC, for permission to publish this paper and B. Holmes for much of the effort in preparing it.

Part of the work was sponsored by the ECSC.

References

(1) A.H. Priest and B. Holmes, "A Multi-Testpiece Approach to the Fracture Characterisation of Linepipe Steels". International Journal of Fracture, Vol. 17, No. 3, June 1981.

(2) A.H. Priest and B. Holmes "The Influence of Test Conditions on the Fracture Resistance of Linepipe Steels" Advances in Fracture Research, 1980, Pergamon Press.

(3) B. Holmes, A.H. Priest and E.F. Walker "Prediction of Linepipe Fracture Behaviour from Laboratory Tests" International Journal of Pressure Vessels and Piping 11, 1983.

(4) H.C. van Elst et al "Fracture Behaviour of Linepipe Steels", Progress Report No. 3 (EPRG Phase IV), Part 3, Appendix 1.

(5) Sub-committee of Large Diameter Linepipe Producers, "Running Shear Fracture in Linepipes", AISI, Technical Report, Sept. 1974.

(6) "Fracture Behaviour of Gas Transmission Pipelines Full Scale Fracture Tests", EPRG Reports, ECSC Agreement No. 7210 KE 8/804 May, 1980.

(7) J.R. Rice, P.C. Paris and J.G. Merkle "Progress in Flow Growth and Fracture Toughness Testing", ASTM STP 536, American Society for Testing and Materials, 1973, pp 231-245.

(8) S.J. Garwood, J.N. Robinson and C.E. Turner "The Measurement of Crack Growth Resistance Curves (R Curves) Using the 'J' Integral", Int. Journal of Fracture, Vol. 11, No. 3, June 1975.

(9) S.E. Webster, T.M. Banks and E.F. Walker "The Influence of Yield and Tensile Strength on Fracture Toughness Testing and Defect Tolerance Calculations", BSC Report SH/PT/8239/-/78/B.

(10) S. Venzi, G. Re and D. Sinigaglia "Part 2 - COD Calibration - Crack Growth Measurement by TWO Clip Gauges - Plastic Instability Determination". This workshop proceedings.

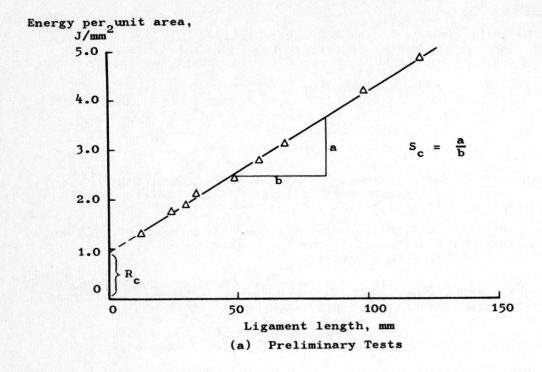

(a) Preliminary Tests

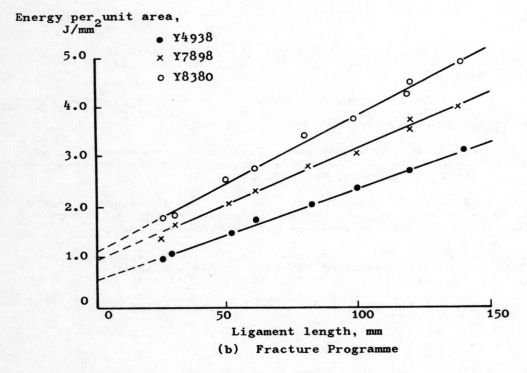

(b) Fracture Programme

ENERGY PER UNIT AREA AS A FUNCTION OF THE LIGAMENT LENGTH   FIG.1.

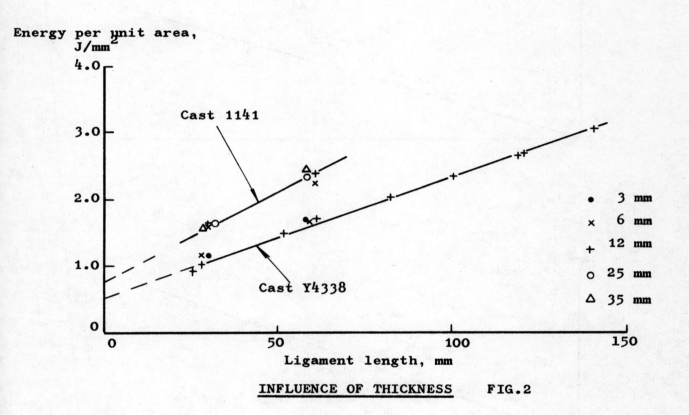

INFLUENCE OF THICKNESS   FIG.2

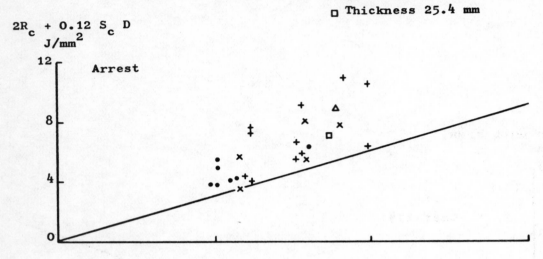

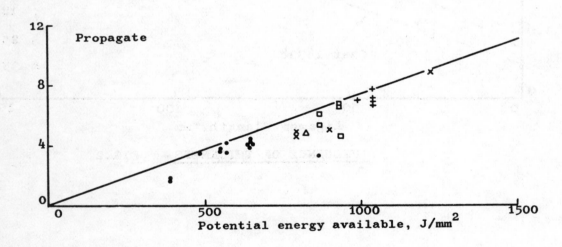

PROPAGATION ENERGY AS A FUNCTION OF AVAILABLE POTENTIAL ENERGY

FOR BASE LINE AISI AND COMBINED EPRG DATA   FIG.3

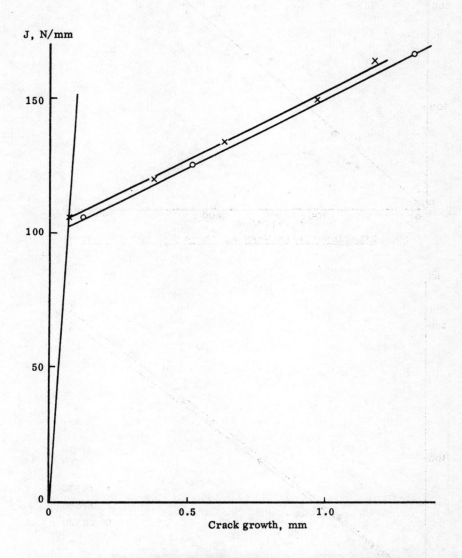

○ Multi-test data

× Electrical Potential Method

J, N/mm

Crack growth, mm

J RESISTANCE CURVES FOR 25 mm THICK TEST PIECES   FIG.4

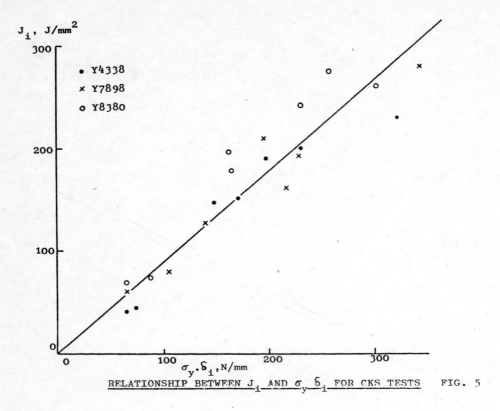

RELATIONSHIP BETWEEN J_i AND $\sigma_y \delta_i$ FOR CKS TESTS   FIG. 5

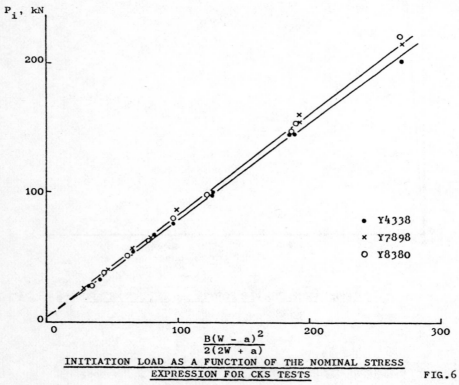

INITIATION LOAD AS A FUNCTION OF THE NOMINAL STRESS
EXPRESSION FOR CKS TESTS   FIG.6

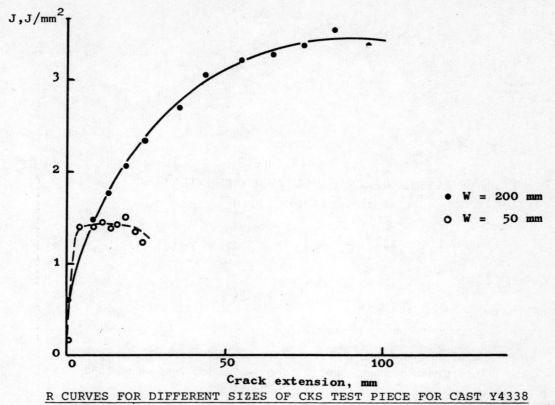

R CURVES FOR DIFFERENT SIZES OF CKS TEST PIECE FOR CAST Y4338

Fig. 7

ANALYSIS OF FRACTURE TOUGHNESS TESTING
AND CRACK GROWTH

S. Venzi** - G. Re** - D. Sinigaglia***

*   Paper presented at: "Fracture Toughness Testing-Methods,
    Interpretation and Application" - London 9-10 June 1982

**   SNAM S.p.A.

***   MILAN UNIVERSITY

Snam is a Company in the ENI Group (Ente Nazionale Idrocarburi) which
has charge of the supply, transport and sale to the Italian Market of
natural gas produced both in Italy and imported from abroad.
SNAM owns and operates the Italian natural gas distribution network
and is engaged, abroad, with the transport of imported natural gas.
Snam also has its own tanker fleet for the transport of crude oil and
other petroleum products as well as a system of pipelines which feeds
several oil refineries in Italy and other European countries.

Part 1

FRACTURE TOUGHNESS TESTING BY MEANS OF FRACTURE KINEMATICS

SUMMARY

In this paper the fundamental aspects of fracture kinematics are explained both in fracture initiation and propagation. Assuming a suitable load displacement curve the energy absorbed during a test is calculated and this model is compared with the results obtained by the researchers of British Steel. An instability criterion is defined for fracture tests and it is applied to the evaluation of the critical conditions for SENT specimen and for ductile fracture propagation in pipeline.

NOMENCLATURE

a_o = Initial Crack Length

a = Crack Length

δ = Crack Tip Opening Displacement = CTOD

V = Clip Gauge Displacement = Notch Opening

δ_L = Limit CTOD

σ_Y = Tensile Yield Strenght

E = Young Modulus

W = Specimen Width

$r_i^{\,*}(w-a_o)$, $r^{\,*}(w-a)$ = Hinge Point Distances

$\gamma_p = \frac{1}{2}$ CTOA (Crack Tip Opening Angle)

γ = Deformation Angle

S = Span Length

SENT = Single Edge Notch Tension

e = Load Point Displacement

SENB3 = Three-point Loaded Single Edge Notch Bend

x_F, y_F = Fracture Profile Coordinates

B = Specimen Thickness

q = Displacement

P = Load

Z = Non-dimensional Elastic Compliance Function

P_L = Limit Load

U   = Energy

M_L = Limit Bending Moment

R_c = Energy (associated with the initiation and propagation of the shear fracture)

S_c = Energy (absorbed as a consequence of associated plastic deformation)

α_1, α_2 = Constraint Factors

$P_S; P_{LS}; P_a; P_m$ = Specimen; Load System; Specimen Arms, Machine Loads

$q_S; q_{LS}; q_a; q_m$ = Specimen, Load System; Specimen Arms, Machine Displacements

$C_S; C_{LS}, C_a; C_m$ = Specimen, Load System, Specimen Arms, Machine Compliances

L   = SENT Length

T_{mat} = Material Tearing Modulus

J   = J-integral

m   = Constant = 1÷2

v_d = Crack Tip Velocity

v_g = Gas Decompression Velocity

σ_n = Nominal Hoop Stress

α_o, β = Venzi   Calibration Constant Parameters

C   = α/α_o

x_C, y_C = Centroid Coordinates

1. INTRODUCTION

Broken test specimens contain a great deal of information that until now has not been sufficiently utilized in the study of ductile fracture propagation.

After a fracture has advanced on a test specimen, if a/W is large enough, no further plastic deformations occur behind the front of the fracture. Consequently, rigid body kinematics can be applied to the two "non-deformable" parts of the test specimen that is cracking, at the same time that these same parts are linked by a plastic zone that is becoming deformed. Fig. 1 shows how it is possible to reconstruct the entire fracture process by examining the two broken parts of the test specimen.

Fracture kinematics (F.K.) allows us to determine fracture initiation under large scale yielding conditions when suitable hypotehses are assumed /1-3/. During fracture propagation, it is also possible to correlate the displacement of the applied load points to crack length. The resulting equations can be applied to evaluate the condition of plastic instability and the energy balance during fracture propagation.
Finally, F.K. allows us to calculate the fracture profile, assuming only that the crack tip opening angle (CTOA) is known during propagation. F.K. will be applied in an analysis of two typical test specimens, i.e. SENT and SENB3.

2. KINEMATIC DEFINITIONS

In this paper, F.K. will only be applied to plane motion, which necessarily predicts a fracture in plane strain (3). Plane strain occurs in the midplane of specimens in almost all of the usual tests. In addition, the analysis will be restricted to fracture in Mode I which, in F.K., is defined in such a way that the bisector of the CTOA passes through the center of rotation.

It therefore follows that the fracture surfaces:

a) are symmetrical;
b) have the same extension.

The relative motion of two bodies can be calculated when the following are known:

I)   the centroids of the motion (the locus of the instantaneous centers of rotation);
II)   the relative displacement of a point (at all times) and the instantaneous center of rotation (at all times);
III) the relative displacement of two points.

Case III) is applicable to a specimen which is deforming in the area around a deep notch, if we know the calibration or relationship between the clip gauge displacement V and the crack tip opening displacement δ both before and after crack growth initiation. Presently available calibrations enable us to obtain a reasonable description of the specimen motion at least until the onset of the fracture /4,5/ (see also Appendix 1).

3. KINEMATICS OF FRACTURE INITIATION

By tracing the centers of rotation on to a half specimen, we can obtain the centroid of the motion, which is the envelope of the symmetry axis during deformation (Fig. 2). If we analytically derive these centroids from the calibrations presently used, at $a=a_o$ (i.e., prior to fracture initiation), the centroid will necessarily have a cusp if the δ -V relationship becomes linear when V is large enough (see Appendix 2).

The centroid cusp is physically associated with the complete yielding of the ligament, which means that the slip lines have reached the back side of the specimen. We assume that at this point (cusp) the ductile fracture starts if other fracture mechanisms, controlled by toughness, have not yet occurred.

The value of δ calculated at the cusp using the Venzi calibration /4/ is given by the following formula as a function of a_o :

$$\delta_L = \text{const. } \Pi \cdot \frac{\sigma_Y}{E} \cdot a_o \cdot (1 - a_o/W)^2 \div \alpha_o/\beta \qquad (1)$$

where α_o and β are given in Appendix 1. From this relationship, it follows that the critical condition for the initiation of a ductile fracture depends upon both the ligament and the crack length , and on the ratio between yield strength and the Young modulus, rather than the toughness.

The cusp point is located at a distance from the crack tip given by $r_i^{\text{R}} \cdot (W-a_o)$, and by again using the Venzi calibration, we can obtain :

$$r_i^{\text{R}} = r_i^{\text{R}} (a_o/W) \tag{2}$$

The numerical values of this function are given in Table 1.

From this analysis of fracture initiation, we can conclude that δ_{meas}, i.e. the value of δ obtained in a test conducted according to a well-defined procedure (i.e., BS 5762-79 and $\delta_{meas} = \delta_c$ or δ_u or δ_m or δ_i), verifies one of the following conditions :

$$\delta_{meas} < \delta_L \quad \text{or} \quad \delta_{meas} > \delta_L$$

where δ_L is the value of δ at the cusp calculated, as shown in Appendix 2, using the same calibration for δ_{meas}. In the first case, fracture initiation is controlled by toughness; δ_{meas} is independent of the ligament and the load at instability depends on δ_{meas} and a_o. In the second case, fracture initiation is controlled by the yield strength and the ligament, thus F.M. is no longer valid. After crack initiation, the toughness (i.e. the resistance to crack growth) is no longer defined by a value of δ, but rather by its derivative versus "a", i.e. $d\delta/da$.

4. KINEMATICS OF FRACTURE PROPAGATION

Every plane motion can be considered as a rotation around an instantaneous center. Rectilinear movement falls into this category since it places the center of rotation at infinity. In order to study the motion of the two specimen parts during fracture propagation in Mode I, these two basic hypotheses must be made:

a) the instantaneous center of rotation is found on the symmetry axis at a distance from the crack tip equal to $r^{\text{R}} (W-a)$;

b) the fracture surfaces and the plane of symmetry form an angle equal to

γ_p (CTOA = $2 \cdot \gamma_p$) /2,6,7/.

In propagation, r^{R} can be selected according to two criteria. The first is that of the complete development of the plasticization of the ligament. According to the plasticity theory, r^{R} is equal to 0.45 for SENB3. For SENT specimens, r^{R} is ∞, provided that the specimens are held with hydraulic clamps so as to prevent them from rotating.

Propagation can also be considered as a continuous series of reinitiations and therefore r^{R} can be related to crack length "a" according to eq.(2) and Table 1, if $r^{\text{R}}(a/W) = r_i^{\text{R}}(a/W)$. Such an assumption is very similar to $r^{\text{R}} = $ const. since when a/W falls within the range of 0.3 to 0.95, the r_i^{R} values are between 0.52 and 0.45; i.e. the hypothesis of continuous re-initiation leads to results which more or less coincide with the plasticity theory.

Appendix 2 shows the complete analytical development of the geometric relationships between the parameters of deformed specimens (V, δ, a, γ). For easier physical interpretation of the following equations, we will hereafter use an approximation that is valid when γ is small and $\gamma \ll \gamma_p$. The fundamental geometric formulas are (Fig. 3):

$$d\delta/da \simeq 2 \; ; \; tg \; \gamma_p \tag{3}$$

$$d\delta/d\gamma = 2 \cdot r^{\bigstar} \cdot (W - a) \tag{4}$$

Therefore:

$$\frac{d\gamma}{da} \simeq \frac{tg \; \gamma_p}{r^{\bigstar} \cdot (W-a)} \tag{5}$$

By integrating eq. (5) with constant $\gamma_p (=\gamma_{po})$ and $r^{\bigstar}$, we obtain :

$$\gamma \overset{\sim}{=} \gamma_i + \frac{tg \; \gamma_{po}}{r^{\bigstar}} \cdot \ln \frac{W - a_o}{W - a} \tag{6}$$

This equation is applicable to all types of specimens. By means of the fundamental eqs. (3) and (4) of F.K., it is possible to obtain the relationships between "a" and the deformation parameters which can be measured in usual tests (e,V). In fact:

a) For SENT specimens ($r^{\bigstar} = \infty$), the relationship is

$$\gamma \overset{\sim}{=} \gamma_i$$

$$\frac{de}{da} \simeq \frac{dV}{da} = \frac{d\delta}{da} \simeq 2 \; tg \; \gamma_p \tag{7}$$

in which V represents the notch opening, i.e., the displacement of the test specimen in the area of the notch, measured by an extensiometer; it does not coincide with the stroke ("e") which is affected by the elastic deformation of the load machine and of the two elastic ends of the test specimen. By integrating eq. (7), we obtain:

$$V = V_i + 2 \cdot (a - a_o) \cdot tg \; \gamma_{po} \tag{8}$$

b) For SENB 3 specimens, the relationship is :

$$e = (S/2 - W. \; tg \; \gamma/2) \cdot tg\gamma \simeq S/2 \cdot tg \; \gamma \tag{9}$$

where "S" is the span length. From eq. (6) we obtain :

$$e \simeq S/2 \; tg \; (\gamma_i + \frac{tg \; \gamma_{po}}{r^{\bigstar}} \; \ln \frac{W - a_o}{W-a}) \tag{10}$$

The V-a relationship is given in Appendix 3.

Furthermore :

$$\frac{dy_F}{dx_F} = \frac{dy_F}{da} = tg \; (\gamma_p - \gamma) \tag{11}$$

By integrating eq. (11), taking into account eq. (6), the fracture profile can be determined (see Appendix 3).

5. LOAD-DISPLACEMENT CURVE AND ENERGY CALCULATION

5.1 Load-displacement curve

For the purposes of our analysis, the load-displacement curve for a SENT specimen can be schematized as shown in Fig. 4. The first section of this curve is the purely elastic one, and during loading the crack length remains constant, thus $d\delta/da$ cannot be defined at this stage. This stage ends when $\delta = \delta_L$ (see eq. (1)).

In the second stage (the transient stage), the following phenomena occur:

a) the crack advances and $d\delta/da$ can be defined;

b) the ligament area is deformed over the yielding point showing the necking through the thickness, and the material in this area becomes strain-hardened.

These two factors act in opposite ways on the load bearing capacity of the specimen, and the load is the result of these contrasting factors.

Finally, in the third stage, steady state crack propagation occurs. Further deformation of the material implies necking of the ligament and the formation of the CTOA. The fracture advances in a region strain-hardened during stage II, and $d\delta/da = 2 \ tg\gamma_{po}$ is assumed to be constant in this stage, as stated in Appendix 3.

5.1-a) Transient Stage

A simple model for the transient stage can be obtained by assuming that $d\delta/da$ varies in a suitable way. The function $d\delta/da$ vs. a has been assumed in this paper, taking into account that $d\delta/da$ decreases from a high value $(= 2 \ tg\gamma_{pi})$ for $a = a_o$, to a limit value equal to $2 \ tg \ \gamma_{po}$ for $a-a_o=B/2$, and that it remains constant during further crack propagation.

A tentative function could be :

$$\delta - \delta_i = K_1 (a-a_o)^2 + K_2 (a-a_o) \tag{12}$$

and :

$$d\delta/da = 2 \ K_1 (a-a_o) + K_2 \tag{13}$$

Since:

$$d\delta/da\Big|_{\Delta a=B/2} = 2 \ tg \ \gamma_{po} \tag{14}$$

it is :

$$K_1 B + K_2 = 2 \ tg \ \gamma_{po} \tag{15}$$

and:

$$\frac{d\delta}{da}\bigg|_{a=a_o} = K_2 = 2 \ tg \ \gamma_{pi} \qquad (16)$$

Therefore :

$$K_1 = \frac{2}{B} \ (tg \ \gamma_{po} - tg \ \gamma_{pi}) \qquad (17)$$

$$(\delta - \delta_i)_{\Delta a = B/2} = \delta_T - \delta_i = K_1 \frac{B^2}{4} + K_2 \frac{B}{2} = \frac{B}{2}(tg\gamma_{po} + tg\gamma_{pi}) \qquad (18)$$

5.1-b) Steady State Stage

Eq. (7) is valid in the steady state crack propagation for SENT specimens. Thus the function is :

$$\delta = \delta_T + 2 \ (a - a_o - \frac{B}{2}) \ . \ tg \ \gamma_{po} \qquad (19)$$

and

$$\delta_F = \delta_T + 2(W - a_o - \frac{B}{2}) \ tg \ \gamma_{po} \qquad (20)$$

5.2 Energy Calculation

In the purely elastic range, the relationship between load and displacement is linear, and thus is :

$$q = \frac{P \ Z}{B \ E} \qquad (21)$$

where Z is a suitable non-dimensional elastic compliance function /8/.

Assuming that the crack starts to propagate when the ligament has completely yielded, the maximum load for the elastic section for SENT specimens /9, 10/ is:

$$P_L = \alpha_1 \ . \ \sigma_Y \ . \ B \ . \ (\ W - a_o) \qquad (22)$$

In the transient range, the load remains quite constant for a certain a-mount of crack extension, due to the opposite effects of crack growth and strain-hardening.
The objective of the present analysis is to estimate the total energy absorbed during a fracture test, and for this purpose, the load-displacement curve has to be conceived in such a way as to simplify the energy calculation. The model proposed here (schematized in Fig. 4), assumes that the load does not remain constant for the entire transient range, but only in the range $\delta_i \leqslant \delta \leqslant \delta_1$, since :

$$\delta_1 = \delta_T - B \ tg\gamma_{po} = \frac{B}{2} \ . \ (tg \ \gamma_{pi} - tg\gamma_{po}) + \delta_i \qquad (23)$$

Since the energy absorbed in the elastic range is negligible, it is :

$$U_{tot} = P_L \cdot (\delta_1 - \delta_i) + \frac{1}{2} \cdot P_L \cdot (\delta_F - \delta_1) \tag{24}$$

$$U_{tot} = \alpha_1 \cdot \sigma_Y \cdot B \cdot (W - a_o) \cdot \left[\frac{B}{2}(tg\gamma_{pi} - tg\gamma_{po}) + \right.$$
$$\left. + (W-a_o) \, tg\gamma_{po} \right] \tag{25}$$

In this way, the total energy is obtained as a function of two parameters: $tg\gamma_{pi}$ and $tg \gamma_{po}$. These results refer to SENT specimens. For SENB 3 specimens, the energy can be calculated from the equation:

$$\frac{dU}{B \, da} = 2 \cdot \frac{M_L}{B} \frac{d\gamma}{da} = \frac{\alpha_2}{4r^{\stackrel{*}{R}}} \cdot (W-a) \cdot \cos\gamma \cdot \frac{d\delta}{da} \tag{26}$$

If γ is small enough, this equation differs only in a constant $(\alpha_2/4r^{\stackrel{*}{R}}$ rather than α_1) from the one valid for SENT specimens..If we also consider that eq. (A3-1) is valid for both specimens, total energy for SENB 3 can again be expressed by eq. (25) by substituting $\alpha_2/(4 r^{\stackrel{*}{R}})$ for α_1.

According to some experimental results /11/, eq. (25) can also be formulated for SENB 3 specimens in the following way :

$$\frac{U_{tot}}{B \cdot (W-a_o)} = R_c + S_c \, (W-a_o) \tag{27}$$

where R_c and S_c are constant in a fairly wide range of ligaments and thicknesses. By comparing eqs. (25) and (27), we can obtain:

$$R_c = \frac{\alpha_2}{4r^{\stackrel{*}{R}}} \cdot \sigma_Y \cdot \frac{B}{2} (tg \, \gamma_{pi} - tg \, \gamma_{po}) \tag{28}$$

$$S_c = \frac{\alpha_2}{4r^{\stackrel{*}{R}}} \cdot \sigma_Y \cdot tg \, \gamma_{po}$$

Since R_c, S_c and $tg \, \gamma_{po}$ are constant, $tg \, \gamma_{pi}$ must depend on B and is:

$$tg \, \gamma_{po} = \frac{4 \, r^{\stackrel{*}{R}}}{\alpha_2 \sigma_Y} \, S_c$$

$$tg \, \gamma_{pi} = \frac{4r^{\stackrel{*}{R}}}{\alpha_2 \, \sigma_Y} \, (\frac{2 \, R_c}{B} + S_c) \tag{29}$$

6. ENERGY ABSORPTION DURING CRACK PROPAGATION

The energy absorbed per unit crack length $\frac{dU}{B \, da}$ can be evaluated :

a) by directly measuring the load-displacement curve during a test:

$$\frac{dU}{Bda} = \frac{P}{B}\frac{dq}{da} = \frac{P}{B} \cdot \frac{d\delta}{da} \tag{30}$$

where the relationship between δ and a can be obtained by F.K. if no direct measurement is made of the crack propagation .

b) by the analysis developed in par. 5; the derivative of energy vs. crack length must be calculated separately for the transient and the steady state stages.

The model proposed here for the transient stage gives an acceptable approximation for calculating total energy, but it is not accurate enough to evaluate dU/(Bda) (nor obviously the second derivative either). In fact, we have assumed here that the reduction of load due to crack extension is counterbalanced by strain-hardening and thus dU/(Bda) is constant and the second derivative is zero.

During steady crack propagation, the following are valid:

$$\frac{dU}{Bda} = \frac{P}{B} \cdot \frac{d\delta}{da} = 2 \, \alpha_1 \cdot \sigma_Y \cdot (W-a) \cdot tg \, \gamma_{po} \tag{31}$$

and

$$\frac{d^2U}{B\,da^2} = -2 \cdot \alpha_1 \cdot \sigma_Y \cdot tg \, \gamma_{po} \tag{32}$$

If we compare these results with those obtained by means of the J-integral theory, we have the formulations reported in Table 2. The difference between the different formulations derives from the fact that while the J-integral theory is valid at the stages of crack initiation and slow crack growth under increasing load, our formulas in Table 2 take into account the so-called"unstable crack growth", i.e., crack propagation under decreasing load.

7. INSTABILITY AND PROPAGATION CRITERIA

The analysis developed in the preceeding paragraphs implies a knowledge of the quantities P_S and q_S, i.e. the load and the displacement of the specimen. These quantities are known in some mechanical tests, while in others, we know only the quantities P_{LS} and q_{LS}, i.e., the load on the loading system and the total displacement of two fixed points. If no instability occurs /12/,we can assume the following :

$$P_S = P_{LS} \tag{33}$$

$$q_{tot} = q_S + q_{LS} \tag{34}$$

where:

$$q_S = C_S \, P_S \; ; \qquad q_{LS} = C_{LS} \, P_{LS}$$

and q_{tot} is the displacement due to oil in an hydraulic machine or to head cross in a screw drived one.

If the loading system is elastically loaded (the machine and arms of the test specimen are loaded in the elastic range), two possibilities exist. If the machine and arms charge the specimen according to a series disposition, we obtain:

$$P_{LS} = P_a = P_m \quad \text{and} \quad q_{LS} = q_a + q_m \tag{35}$$

$$C_{LS} = C_a + C_m = L/(EBW) + C_m = \text{const.} \tag{36}$$

where L is the arm length of a SENT specimen.

If the specimen is charged according to a parallel disposition :

$$q_{LS} = q_a = q_m \quad \text{and} \quad P_{LS} = P_a + P_m \tag{37}$$

$$\frac{1}{C_{LS}} = \frac{1}{C_a} + \frac{1}{C_m} = \frac{EWB}{L} + \frac{1}{C_m} = \text{const} \tag{38}$$

If instability occurs, we can assume the following :

$$P_S \lessgtr P_{LS} \quad \text{and} \quad P_{LS} = P_s + P_i \qquad \tag{39}$$

i.e., the equilibrium of the loads is obtained by means of the inertial load. Since the response frequency of the machine engine is usually very low, due to the limited power of the engine itself, we obtain the following:

$$\frac{dq_{tot}}{dt} \ll \frac{dq_s}{dt} \quad \text{and} \quad \frac{dq_{LS}}{dt} \; ; \; \text{so} \quad dq_{tot} = 0 \quad \text{and}$$

$$dq_S = - dq_{LS} \tag{40}$$

$$\frac{dP_S}{dq_S} \leq \frac{dP_{LS}}{dq_S} = - \frac{dP_{LS}}{dq_{LS}} = - \frac{1}{C_{LS}} \tag{41}$$

The entire $P_S - q_S$ curve can be obtained from the $P_{LS} - q_T$ recording only if no instability occurs during the test, i.e., if eq. (41) is never verified during crack propagation. Furthermore, ductile instability can occur at or after maximum load if C_{LS} is infinite or finite respectively.

If we apply this condition to the so-called "unstable crack growth" in a SENT specimen, from the eq. (28), we can obtain :

$$\frac{dP_S}{dq_S} = \frac{dP_S}{d\delta} = \frac{dP_S}{da} \cdot \frac{da}{d\delta} = - \alpha_1 \cdot \sigma_Y B \cdot \frac{1}{2 tg \, \gamma_{po}} \tag{42}$$

If the specimen is loaded by a series system, as is the case in normal tests we can obtain:

$$\alpha_1 \cdot \sigma_Y \cdot B \cdot \frac{1}{2 \, tg \, \gamma_{po}} \geq \frac{EWB}{L + C_m \, E \, W \, B}$$

Hence:

$$\frac{L}{EW} \geq \frac{2 \, tg \, \gamma_{po}}{\alpha_1 \sigma_Y} - C_m B \tag{43}$$

and for parallel systems, it is :

$$\alpha_1 \sigma_Y B \frac{1}{2 tg\gamma_{po}} \geq \frac{EWB \, C_m + L}{L \, C_m}$$

Hence:

$$\frac{L}{EW} \geq 2 \, tg \, \gamma_{po} \, / \, (\alpha_1 \sigma_Y - \frac{2 \, tg\gamma_{po}}{B \, C_m}) \tag{44}$$

If we rewrite eq. (43) in the following way :

$$L/W \geq T_{mat} - C_m BE$$

being $T_{mat} = \dfrac{E}{\sigma_Y^2} \cdot \dfrac{dJ}{da} = \dfrac{mE}{\sigma_Y} \dfrac{d\delta}{da} = \dfrac{2mE}{\sigma_Y} \, tg \, \gamma_{po}$

it agrees with the equation given by Paris /13;14/, assumed m=1, except for the formal difference, that eq. (43) also takes into account the compliance of the machine.

A major difference between these two equations is that the above analysis is applied to ductile instability which only occurs at or after maximum load, whereas the Paris tearing modulus is measured at the very first stage of crack propagation.

8. DUCTILE FRACTURE PROPAGATION IN A PIPELINE

The condition for ductile instability in a pipeline is given by eq. (41) in which dP_S/dq_S can be obtained by considering the section of the pipeline around the crack tip as a SENT specimen; thus, during steady state propagation, we have the following condition :

$$\frac{dP_S}{dq_S} = - \alpha_1 \sigma_Y \, t \, \frac{1}{2 tg\gamma_{po}} \tag{45}$$

where t is the thickness of the pipe wall.
The compliance of the loading system can be calculated by observing that the gas and the elastically loaded section of the pipe apply the load to the crack tip (specimen) according to a parallel disposition, i.e. $q_{LS} = q_{pipe} = q_{gas}$. We thus obtain :

$$C_{LS} = (1/C_{pipe} + 1/C_{gas})^{-1} = C_{pipe} = \frac{\Pi D}{EWt} \tag{46}$$

with $C_{gas} \simeq \infty$, D = the pipe diameter, and W = the suitable width of the pipe section considered here.

On this section, the nominal stress applied behind the crack tip during ductile fracture propagation, as related to the nominal hoop stress, can be expressed by the following equation /15/:

$$\sigma_{CF} = \sigma_n \left(\frac{v_d}{6v_g} + \frac{5}{6}\right)^7 \tag{47}$$

in which v_d is the constant crack tip velocity and v_g is the gas decompression velocity.

If (W-a) is the length of the equivalent ligament of the section of pipe whose width is W, σ_{CF} is related to σ_Y in such a way :

$$\sigma_{CF} \, W = \alpha_1 \cdot \sigma_Y \cdot (W-a) \tag{48}$$

Finally, eq. (41) can be rewritten in the following manner:

$$\sigma_n = \frac{pD}{2t} \geqslant \frac{2E \, (W-a) \, tg \, \gamma_{po}}{\Pi D \left[v_d/(6v_g) + \frac{5}{6} \right]^7} \tag{49}$$

in which p is the gas pressure.

This equation has been obtained from the expression of dP_s/dq_s in steady crack propagation since the entire fracture process has been divided into three stages.

In the first stage, a high degree of plastic deformation occurs in a limited area of the crack tip /16/; this plastic enclave moves, in fact, at the same velocity as the crack and can travel only a very limited distance from the crack tip, due to the fact that in a metallic material, a very high plastic deformation has a low propagation speed. The equivalent ligament length must be on the order of this plastic enclave, and this agrees with the assumption made in the British Steel Co.'s theory /11/.

In the theory presented here, the equivalent ligament must be calculated by means of experimental data on full-scale burst tests. By optimizing the theory prediction with respect to the data given in /11/, (W-a) is on the order of a few centimeters and seems to be proportionated to D. It thus follows from eq. (49) that :

$$\sigma_n \geqslant \text{const. } tg \, \gamma_{po}$$

In the second stage, the fracture travels in strain-hardened material which has already absorbed energy, due to the transient range. The material resistance to crack propagation is due only to the steady state range. In the third stage, the gas dissipates the last of its energy by deforming the body of the pipe and removing the backfill; this stage does not influence the critical conditions for ductile fracture propagation.

9. CONCLUSIONS

F.K. is a mathematical tool that allows a complete analysis of the geometry of a cracked specimen during deformation and fracture. By using F.K. and a simple model of a load-displacement curve, the following information can be obtained:

a) The initiation of ductile fracture does not depend on the metal toughness, but rather on geometry and dimensions of the specimen and on the
   ratio between yield strength and the Young modulus.

b) Fracture energy is associated with the transient and steady propagation
   stages, the former being essentially a plastic deformation energy controlled by strain-hardening, and the latter a real fracture energy con -
   trolled by ligament length. The energy of both stages depends on the
   CTOA.

c) The CTOA is the toughness parameter in steady state crack propagation.

At last F.K. has been used to analyze the conditions for instability, and
the relationships obtained have been applied to ductile crack propagation
in pipelines.

ACKNOWLEDGMENTS

We would like to thank CSM and Italsider for their help in critically discussing the ideas presented here.

TABLE 1 - Relationship between a_o/W, r_i^* and r_i^* obtained from the cusp condition using the Venzi calibration. In the last column is reported the ratio between the δ at the cusp (δ_L) and α_o/β multiplied by the constant of eq. (A1-4) ($=16/\pi \simeq 5$).

a_o/W	r_i	r_i^*	const.
.10	.2326	.8856	8.80733
.15	.2647	.6551	8.80711
.20	.2843.	.5796	8.80682
.25	.2975	.5422	8.80649
.30.	.3071	.5198	8.80616
.35	.3142	.5049	8.80585
.40	.3199	.4943	8.80560
.45	.3244	.4863	8.80542
.50	.3281	.4801	8.80533
.55	.3312	.4752	8.80533
.60	.3338	.4711	8.80542
.65	.3361	.4678	8.80560
.70	.3380	.4649	8.80585
.75	.3397	.4625	8.80615
.80	.3412	.4604	8.80648
.85	.3426	.4585	8.80681
.90	.3438	.4569	8.80711
.95	.3449	.4554	8.80732

TABLE 2 - Expression of the first two derivatives of U vs a according J-integral theory and F.K.

Quantity	J-integral Theory	This Analysis in Steady State
$\dfrac{dU}{Bda}$	$=J=m\,\sigma_Y\,\delta$	$= \alpha_1\sigma_Y(W-a)\dfrac{d\delta}{da} = 2\alpha_1\sigma_Y(W-a)\,tg\,\gamma_{po}$
$\dfrac{d^2U}{Bda^2}$	$= \dfrac{dJ}{da} = m.\sigma_Y.\dfrac{d\delta}{da}$	$= -\alpha_1\sigma_Y.\dfrac{d\delta}{da} = -2\alpha_1\,\sigma_Y\,tg\,\gamma_{po}$

APPENDIX 1 - δ-V CALIBRATION

The Venzi calibration was used in the analytical development of the theory presented in this report. It is valid for both SENB 3 and CT geometries and explicitly expresses the dependency of the apparent rotational constant "r" upon V. In fact, it is :

$$r = \frac{a_o \, \alpha_o \, C}{(W-a_o)(1-\alpha_o C)} \tag{A1-1}$$

where :

$$\alpha_o = \left[1 + \frac{a_o}{r_o(W-a_o)}\right]^{-1} \tag{A1-2}$$

$$r_o = 0.40$$

$$C = 1 - \exp(-\beta V) \tag{A1-3}$$

$$\beta = \alpha_o \left[\text{const.} \; \pi \frac{\sigma_Y}{E} \cdot a_o \cdot \left(1 - \frac{a_o}{W}\right)^2\right]^{-1} \tag{A1-4}$$

The constant appearing in the expression of β has a value of 5, based upon the original calibration data. However, the latest data seem to suggest that the best adaptation is obtained around 1 or 2. The original expression of β was modified, keeping in mind the fact that δ_L must tend toward 0 when a_o or $(W-a_o)$ tend toward 0.

The expression of the CTOD therefore differs according to the meaning which is attributed to this quantity (see for example Fig. 2), where:

$$\delta_{phis} = \alpha_o(V-C/\beta) \quad \text{and} \quad \delta_{geom} = \alpha V$$

α and C are respectively :

$$\alpha = \left[1 + \frac{a_o}{r(W-a_o)}\right]^{-1} \quad ; \quad C = \alpha/\alpha_o$$

δ_{phis} indicates the physical opening at the tip of fatigue crack; δ_{geom} on the other hand, indicates a crack tip opening when the sides of the crack are extended, and it therefore has only a geometric meaning. δ used in the F.K. equations is always δ_{geom}. Really, different techniques allow us to measure different quantities; for example δ_{phis} is the quantity measured by means of optical or infiltration techniques, while δ_{geom} could be measured by means of double clip gauge.

APPENDIX 2 - CRACK INITIATION

With reference to Fig. 2, the test specimen centroids prior to fracture initiation have the following coordinates:

$$x_C = a_o + r(W-a_o) + \frac{dr}{d\gamma}(W-a_o)\frac{\sin 2\gamma}{2} \tag{A2-1}$$

$$y_C = -\frac{dr}{d\gamma}(W-a_o)\sin^2\gamma$$

where γ, r and $\dfrac{dr}{d\gamma}$ can be derived from a calibration in the following manner:

$$\sin \gamma = \frac{V - \delta}{2\,a_o} \tag{A2-2}$$

$$r = \frac{V - 2a_o \sin\gamma}{2(W-a_o)\sin\gamma} \tag{A2-3}$$

$\dfrac{dr}{d\gamma}$ can be obtained numerically or, by using the Venzi calibration (4); its expression is

$$\frac{dr}{d\gamma} = \frac{2a_o\alpha_o\,\beta\exp(-\beta V)\,\left[a_o+r(W-a_o)\right]\cos\gamma}{(W-a_o)\,\{(1-\alpha)^2-2a_o\alpha_o\exp(-\beta V)\sin\gamma\}} \tag{A2-4}$$

APPENDIX 3 - CRACK PROPAGATION

With reference to Fig. 3 it is :

$$\frac{d\delta}{da} = \frac{2\sin \gamma_p}{\cos(\gamma_p-\gamma)} \tag{A3-1}$$

$$\frac{d\gamma}{da} = \frac{\sin \gamma_p \cos\gamma}{\cos(\gamma_p-\gamma)\,r^{\bar{\lambda}}(W-a)} = \frac{\cos\gamma}{2r^{\bar{\lambda}}(W-a)} \cdot \frac{d\delta}{da} \tag{A3-2}$$

The following hypotheses were used in defining the geometry of the test specimen during fracture propagation :

a) γ_{po} = const.;   b) $r^{\bar{\lambda}}$ = const. = 0.45.

By integrating eq. (A3-2), with these hypotheses, we get :

$$a = W - (W-a_i)(\frac{\cos\gamma}{\cos\gamma_i})r^{\bar{\lambda}} \exp\left[-\frac{r^{\bar{\lambda}}}{tg\gamma_{po}}(\gamma-\gamma_i)\right] \tag{A3-3}$$

"i" represents the conditions of initiation, and

$$a_i = a_o + \frac{\delta_L}{2} \sin \gamma_i \tag{A3-4}$$

For SENB 3 specimen, by using eq. (9) in the approximate form, we obtain:

$$\frac{de}{da} \simeq \frac{S \sin \gamma_{po}}{2 \cos\gamma \cdot \cos(\gamma_{po}-\gamma) \cdot r^{\bar{\lambda}}(W-a)} \tag{A3-5}$$

It is also :

$$V = 2\{\left[a+r^{\bar{\lambda}}(W-a)\right]\sin\gamma + y_C \cos\gamma\} \tag{A3-6}$$

$$\frac{dV}{da} = 2\{\left[a+r^{\bar{\lambda}}(W-a)\right]\cos\gamma - y_C \sin\gamma\} \cdot \frac{\sin \gamma_{po}\cos\gamma}{\cos(\gamma_{po}-\gamma)r^{\bar{\lambda}}(W-a)} \tag{A3-7}$$

The coordinates of the centroid are:

$$x_C = x_{Ci} + \int_{a_i}^{a} (1-r^{\bullet}) \, da + \int_{r_i^{\bullet}}^{r^{\bullet}} (W-a) \, dr^{\bullet}$$

$$y_C = y_{Ci} - \int_{a_i}^{a} (1-r^{\bullet}) \, tg\gamma \, da + \int_{r_i^{\bullet}}^{r^{\bullet}} (W-a) \, tg\gamma \, dr^{\bullet}$$

(A3-8)

Obviously, for hypothesis b), $dr^{\bullet} = 0$ and therefore the last two integrals are equal to zero.

The coordinates of the fracture profiles are:

$$x_F = a$$

$$y_F = y_{Fi} + \int_{a_i}^{a} tg(\gamma_p - \gamma) \, da$$

(A3-9)

in which y_{Fi} is equal to $(\delta_L/2) \cos\gamma_i$.

The following formulas are therefore valid :

$$x_C = x_F + r^{\bullet}(W-a)$$

$$y_C = y_F - r^{\bullet}(W-a) \, tg\gamma$$

(A3-10)

Part 2

COD CALIBRATION – CRACK GROWTH MEASUREMENT BY TWO CLIP GAUGES –
PLASTIC INSTABILITY DETERMINATION

1. DETERMINATION OF DEFORMATION PARAMETERS ON A BEND SPECIMEN
    UNDER STATIC LOAD

Scope of this investigation is to provide analytical relationships
which permit determination of certain significant parameters during
fracture toughness testing using suitable clip gauge instrumentation.

In the case of symmetrical deformation and fracture of the specimen
(fig. 3) load (P) and clip gauge displacements (V_1, V_2) were measured
during the test.

The two clip gauges were positioned approximately symmetrical to the
mid point of the initial ligament for a better revealing of the defor
mation phenomena.

1.1. DEFORMATION BEFORE CRACK INITIATION

The following relationships are deduced from fig. 2   which defines
the reference system and the geometric significance of the symbols used:

$$\sin(\gamma + \gamma_0) = \frac{V_1 - V_2}{2\left[(z_2 - z_1)^2 + (h_1 - h_2)^2\right]^{1/2}} \; ; \; \gamma_0 = \text{arctg}\, \frac{h_1 - h_2}{z_1 - z_2} \quad (1)$$

$$z = \frac{V_2(a_0 - z_1 - h_1 \cot g \gamma) - V_1(a_0 - z_2 - h_2 \cot g \gamma)}{(W - a_0)(V_1 - V_2)} \quad (2)$$

$$z^* = z + \frac{dz}{d\gamma} \sin \gamma \cos \gamma \quad (3)$$

$$\delta_G = 2 z (W - a_0) \sin \gamma \quad (4)$$

$$V = 2\left[z(W - a_0) + a_0\right] \sin \gamma \quad (5)$$

The difference between the apparent centre and the real centre of rotation, and therefore the difference between r and r* has been expained in previous works (2,3), and in part 1 of this paper.

In this phase, first of all the specimen is deformed elastically, then plastically in the zone of the ligament with propagation of slip lines from the crack tip proceeding towards the opposite end of the specimen.

1.2. CRACK INITIATION

Deformation of the crack can continue without crack propagation until the crack tip reaches a certain critical condition whereby:

a) either the limit of toughness

b) or the limit of ductility is reached.

The first condition is defined by a critical value of a fracture toughness parameter K_{IC} or δ_C, respectively depending on whether the plastic region satisfies the condition laid down in ASTM E-399 or it can be greater but always small compared with ligament. Under these conditions, K_{IC} or δ_C are characteristic parameters of the material.

In the second case instead, it is reasonable to expect the critical condition to depend on the tensile properties of the material (σ_Y or $\bar{\sigma}$ or $\bar{\sigma}_u$), on the geometry and the dimensions of the specimen.

Equations (1 to 3) give the geometric relationships between the parameters γ, r and r* and the experimental quantities V_1 and V_2. A sufficiently accurate COD calibration relationship, expressed in purely geometric terms, must permit, instead, deduction of the three above indicated parameters where just one of the experimental data is known, that is ΔV ($\simeq \Delta V_1$ for small deformations, $Z_1 = 0$ and $h_1 \neq 0$).

For example, by using calibration (17) we have:

$$r = \frac{\partial_o \alpha_o C}{(W - \partial_o)(1 - \alpha_o C)}$$

with $\quad \alpha_o = \left[1 + \frac{\partial_o}{.4(W - \partial_o)} \right]^{-1} \quad$ and $\quad C = 1 - exp\left(-\beta \Delta V\right)$

β being a parameter of dimensions $\left[L^{-1} \right]$ to be determined experimentally.

In calibration (17) $\quad \beta = \frac{2\alpha_o E}{\pi m \sigma_y (W - \partial_o)} \quad$ with m being from 1 to 3.

We then have

$$\delta_c = \delta = \alpha \Delta V \quad \text{as} \quad \alpha = \left[1 + \frac{\partial_o}{2(W - \partial_o)} \right]^{-1}$$

$$\sin \gamma = \frac{\Delta V - \delta}{2 \partial_o}$$

Lastly r* can also be calculated from equation (3) or else from

$$d\delta = \alpha^* dV \text{ as } \alpha^* = \left[1 + a_o/r^*(W-a_o)\right]^{-1}, \text{ hence}$$

$$r^* = \frac{a_o}{(W-a_o)\left(\frac{d\Delta V}{d\delta} - 1\right)} = \frac{a_o}{(W-a_o)\left\{\frac{1}{\alpha_o\left[1-(1-\beta\Delta V)\exp(-\beta\Delta V)\right]} - 1\right\}}$$

It has been found that by using this calibration, the centroid of the relative motion of the two specimen halves has a cusp when r* reaches a maximum and the curve δ -V has an inflexion. Thus it is assumed that this point corresponds to the full plasticity of the ligament and that it constitutes the upper crack initiation limiting condition through reaching the limit of toughness or the lower limiting condition through reaching the limit of ductility.

This limiting condition (dr*/dV=0) occurs when $\beta\Delta V=2$, and is:

$$\delta_L = 2\left(1 - e^{-2}\right)\frac{\alpha_o}{\beta} \qquad (6)$$

Bearing in mind the physical meaning first given to this limiting condition, it can be seen how δ_L can be determined by another method (18), that is, by combining (for each type of specimen or structure) the expressions giving yielding due to plastic collapse and failure in function of fracture toughness.

For example, the two critical load conditions (19) for a SENB 3 specimen can be expressed thus:

$$P_{cz} = \frac{2}{3} \sigma_y \frac{W^2 B}{S} \left(1 - \frac{a}{W}\right)^2 \qquad \text{plastic collapse}$$

$$P_{cz} = K_{1c} \frac{W^{3/2} B}{S} f\left(a/w\right)^{-1} \qquad \text{brittle fracture}$$

Since

$$\delta = \frac{K^2}{\sigma_y E} \qquad \text{being m=1}$$

follows

$$\delta_L = \frac{4}{9} \varepsilon_y W \left(1 - \frac{a}{w}\right)^4 f\left(\frac{a}{w}\right)^2 \qquad (7)$$

Lastly by combining the two equations (6) and (7) we have:

$$\beta = \frac{9}{2}\left(1 - e^{-2}\right)\alpha_0 \left[\varepsilon_y\, w\left(1 - \frac{a}{w}\right)^4 f^2\left(\frac{a}{w}\right)\right]^{-1}$$

Analogous relationships can be obtained for the different types of standard specimens.

After determining parameter δ_L, it is also possible to determine the corresponding parameters γ_L, r_L and r_L^* :

$$\sin\gamma_L = \frac{\Delta V_L - \delta_L}{2 a_0} = \frac{1 - \left(1 - e^{-2}\right)\alpha_0}{a_0\, \beta}$$

$$r_L = \frac{a_0\, \alpha_0\left(1 - e^{-2}\right)}{(w - a_0)\left[1 - \alpha_0\left(1 - e^{-2}\right)\right]}$$

$$r_L^* = \frac{a_0}{(w - a_0)\left\{\dfrac{1}{\alpha_0\left(1 + e^{-2}\right)} - 1\right\}} = \frac{a_0\, \alpha_0\left(1 + e^{-2}\right)}{(w - a_0)\left[1 - \alpha_0\left(1 + e^{-2}\right)\right]}$$

1.3. CRACK PROPAGATION

The following relationships, which are valid after crack initiation, can be deduced from fig. 3 :

$$\sin(\gamma + \gamma_0) = \frac{V_1 - V_2}{2\left[(z_2 - z_1)^2 + (h_2 - h_1)^2\right]^{1/2}} \quad ; \quad \gamma_0 = \operatorname{arctg}\frac{h_1 - h_2}{z_1 - z_2} \tag{8}$$

$$r = \frac{V_2\left(a - z_1 - h_1 \cot g\, \gamma\right) - V_1\left(a - z_2 - h_2 \cot g\, \gamma\right)}{(w - a)(V_1 - V_2)} \tag{9}$$

$$r^* = \frac{a - z_1 + h_1\, tg\, \gamma - r(w - a)\, tg^2 \gamma - \dfrac{dV_1}{dV_2}\left[a - z_2 + h_2\, tg\, \gamma - r(w - a)\, tg^2 \gamma\right]}{\left(\dfrac{dV_1}{dV_2} - 1\right)(w - a)\left(1 + tg^2 \gamma\right)} \tag{10}$$

$$tg\,\gamma_p = \frac{r^*(W-a)\frac{d\gamma}{da}}{1-r^*(W-a)\frac{d\gamma}{da}tg\,\gamma}$$

(11)

In order to solve the above system, one of the 5 parameters (γ , a, r, r*, $\dot{\gamma}_p$) requires determination through a not purely geometric method. The best choice appears to be that of fixing r* according to a suitable criterion.

On the basis of its kinematic significance, it should be assumed that r*, as it defines the point at zero instantaneous "deformation also identifies, from the tensile point of view, the neutral axis of stress on the ligament section.Hence r* can be deduced through the theory of slip lines which gives, for example, r*=0.45 for a SENB3 specimen. Or else r* can be obtained analytically in relation to "a" by using the equation giving r*.
L

An experimental aspect which complicates the analysis of the tests conducted on the basis of equations (8 to 11) is the crack front cur vature. This can make it difficult to define the value of "a" (mean) to be introduced in the equation and to identify the neutral axis.

It should be observed that the angle 2 γ_p calculated according to equation (11), is assessed under load; this angle is formed by the tangents to the fracture surfaces in the region of the crack tip during the test (neglecting obiously the crack tip rounded by the metal lurgical fracture processes – coalescence of microcavities). Thus this parameter incorporates two components: a plastic component which then remains "frozen" in the fracture surface, and an elastic component which depends on the applied load and on the instantaneous compliance of the specimen.

2. PLASTIC INSTABILITY DETERMINATION

2.1. J_R and J_{Pr} approaches

With reference to fig. 5, if we consider load-displacement curve of a tensile specimen having initial crack length a_o, at the point 1 (initiation point) the crack starts to propagate in a ductile way (tearing). At the point 2 the crack has reached a length a at which we unload the specimen to point 3.

If we consider now a second specimen having an initial crack length equal to $a_o + \Delta a$ ($\Delta a \ll a - a_o$) and we load this up to a crack length $a + \Delta a$ and then we unload the specimen.

We can define two hatched areas (energies) in the load-displacement curve of fig. 5b) not common to both the specimens ($-\Delta U$) and (ΔW).

If we consider a real material, which exibits plasticity, and we unload the two specimens respectively at crack length a and $a + \Delta a$ we loose the energy ΔW that always is absorbed by the crack area $B \Delta a$.

So loading and unloading two specimens that differ each other of a crack length Δa we spend an energy $-\Delta U$ and we loose another ΔW.

The crack length has grown during the loading phase of $(a - a_o)$ that is a finite crack length.

If we divide now these energies by $B \Delta a$ and we do the limit for Δa approaching to 0 in the first case we have what is normally defined as J_R

$$J_R = \lim_{\Delta a \to 0} \frac{-\Delta U}{B \Delta a} = -\frac{1}{B} \int_{q_i}^{q} \frac{dP}{da} \, dq + J_{Ic} \qquad (1)$$

$$J_R = \lim_{\Delta a \to 0} -\frac{\Delta U}{B \Delta a} = -\frac{1}{B} \int_{P_i}^{P} \frac{dq}{da} \, dP + J_{Ic} \qquad (2)$$

In the second case we define:

$$J_{Pr} = \lim_{\Delta a \to 0} \frac{\Delta W}{B \Delta a} = \frac{1}{B} \int_{q_T}^{q} \frac{dP}{da} \, dq \qquad (3)$$

$$J_{Pr} = \lim_{\Delta a \to 0} \frac{\Delta W}{B \Delta a} = \frac{1}{B} \int_{0}^{P} \frac{dq}{da} \, dP \qquad (4)$$

J_{Pr} is formally defined as a J-integral value but the energy here considered has the aforementioned meaning.

$J_{Ic} = \sigma_Y q_i$ is an initiation parameter and J_R is something similar but it refers to the energy dissipated per unit of crack area before a certain point always situated after initiation. This energy is due to an infinitesimal drop in load and to a finite crack growth that occurred before this point.

J_{Pr} is an energy dissipated for propagation. This energy is due to a finite load for an infinitesimal (real) crack growth. In case of a plastic deformation J_R and J_{Pr} have different values and physical meaning while in elastic field the two have the same absolute value but opposite sign.

I CASE

Hypotheses : Negligible specimen elastic energy and

$$\frac{dq}{da} = \text{const. ;} \qquad P = P_L = \sigma_Y B (W-a)$$

$$\frac{dP_L}{da} = -\sigma_Y B \qquad q = \frac{dq}{da}(a-a_0) + q_i$$

$$\frac{dq}{da} = \frac{1}{2} tg \, \gamma_P$$
can be defined only after initiation and is a toughness parameter. q for a SENT specimen after initiation coincide with the COD at the fatigue crack tip.

From these hypotheses it follows:

$$J_R = \begin{cases} \sigma_Y(a-a_0)\dfrac{dq}{da} + J_{Ic} & (5) \\[2mm] \left(P_i - P_L\right)\dfrac{dq}{Bda} + J_{Ic} & (6) \\[2mm] \sigma_Y(q-q_i) + J_{Ic} & (7) \end{cases} \qquad J_{Pr} = \begin{cases} \sigma_Y(W-a)\dfrac{dq}{da} & (8) \\[2mm] P_L\dfrac{dq}{Bda} & (9) \\[2mm] \sigma_Y(q_T-q) & (10) \end{cases}$$

$$J_R + J_{Pr} = \sigma_Y(W-a_0)\frac{dq}{da} + J_{Ic} = \sigma_Y q_T$$

$$\frac{dJ_R}{da} + \frac{dJ_{Pr}}{da} = 0$$

where q_T is the total deformation of the specimen broken in plastic mode. If we now assume

$$U_i = \frac{P_L q_i}{2} \quad ; \quad U_R = \frac{P_L(q - q_i)}{2} \quad ; \quad U_{Pr} = \frac{P_L(q_T - q)}{2} \quad ; \quad U_T = \frac{P_L^o q_T}{2}$$

we get also :

$$J_{Ic} = \frac{2 U_i}{B(W - a_o)} \quad ; \quad J_R = \frac{2 U_R}{B(W - a)} + J_{Ic} \quad ; \quad J_{Pr} = \frac{2 U_{Pr}}{B(W - a)}$$

$$J_R + J_{Pr} = \frac{2 U_T}{B(W - a_o)}$$

$$\frac{dJ_R}{da} + \frac{dJ_{Pr}}{da} = 0$$

II CASE

Hypotheses : Negligible specimen elastic energy and

$$\frac{dq}{da} = \text{const.} \quad ; \quad q = \frac{dq}{da}(a - a_o) + q_i$$

$$P = P_L = \sigma_L B(W - a) \quad ; \quad \sigma_L = \sigma_Y + \frac{d\sigma_Y}{da}(a - a_o)$$

From these hypotheses it follows:

$$J_R = \left[\sigma_Y - \frac{d\sigma_Y}{da}(W - a) \right](a - a_o)\frac{dq}{da} + J_{Ic} \tag{11}$$

$$J_{Pr} = \left[\sigma_Y + \frac{d\sigma_Y}{da}(a - a_o) \right](W - a)\frac{dq}{da} \tag{12}$$

$$J_R + J_{Pr} = \sigma_Y(W - a_o)\frac{dq}{da} + J_{Ic} = \sigma_Y q_T$$

$$\frac{dJ_R}{da} + \frac{dJ_{Pr}}{da} = 0$$

III CASE

Hypotheses : Negligible specimen elastic energy and

$$\frac{dq}{da} = f(a) \quad ; \quad q = \left(q_T - q_i\right) \sqrt{\frac{P_i - P_L}{P_i}}$$

$$P = P_L = \sigma_Y B\left(W - a\right) \quad ; \quad \frac{dP_L}{da} = -\sigma_Y B$$

In this case we can use eqn. (7), (10) that are still valid and we get

$$J_R + J_{Pr} = \sigma_Y \left(q_T - q_i\right) + J_{Ic} = \sigma_Y q_T$$

J_{Pr} and J_R are plotted in Fig. 6 for the three cases considered.

2.2.   Analysis of the instability conditions

The instability condition derived by Paris [14] for a SENT specimen is:

$$\frac{L}{W} \geqslant T_{mat} = \frac{E}{\sigma_Y^2} \frac{dJ_R}{da} \tag{13}$$

In the previous enclosed report   an instability condition has been given:

$$\frac{dP_{LS}}{dq_{LS}} \geqslant \frac{dP_L}{dq} \tag{14}$$

where
$$C_{LS} = \frac{L}{BWE} \quad ; \quad dq = -dq_{LS}$$

Eq. (14) can be rewritten as follows:

$$\frac{L}{W} \geqslant -BE \frac{dq}{dP_L}$$

The two criteria coincide only if:

$$-BE \frac{dq}{dP_L} = T_{mat}$$

In the hypotheses of Case I the two criteria coincide.
Coincidence for remaining cases has not yet been checked.

2.3. Conclusions

In this report are given some analytical relationships regarding initiation and propagation of a crack. These make it possible to get a crack growth measurement all through the test by means of a two clip-gauges instrumentation.

A toughness propagation parameter J_{Pr} has been introduced. This parameter has the following properties: the algebric sum of J_R and J_{Pr} is a constant all through propagation and their "a" derivates are equal but of opposite sign.

The value of the constant is the total deformation of the specimens moltipled by the yield stress of the specimen.

The introduction of this parameter gives the possibility of computing J_{Pr} and dJ_R/da rather than J_R and dJ_R/da through the entire propagation by means of Fracture Kinematics concepts and offers a real alternative to traditional measurement of J_R that are restricted for their validity to a short crack extention.

The some conclusion can be extended to bending specimens for not too large plastic deformations.

ACKNOWLEDGMENTS

We would like to thank dr A.MARTINELLI (CISE) for his help in critically discussing the last part of this paper.

REFERENCES

/1/   Mirabile M., Venzi S. - CSM Internal Report 1763 R, 1973 Roma.

/2/   Venzi S. et al. - "Measurement of Fracture Initiation and Propagation
      Parameters From Fracture Kinematics" "Analytical and Experimental
      Fracture Mechanics", Sih G.C. and Mirabile M. Ed., Sijthoff and
      Noordhoff, Alphen aan den Rijn (Netherlands), 1981, pp. 737-756.

/3/   Venzi S., Re G., Sinigaglia D. - "Determination of Ductile Fracture
      Parameters on Small Specimens by Means of Fracture Kinematics" A.G.A.-
      EPRG Line Pipe Research Seminar IV, September 22-24, 1981, Duisburg
      (Germany), Vol. I, pp. 1-38.

/4/   Venzi S. - CSM Internal Report 1247 R , 1972, Roma.

/5/   Knauf A., Riedel H. - "A Comparative Study on Different Methods To
      Measure the Crack Opening Displacement" "Advances in Fracture Re-
      search" Francois D. Ed., Pergamon Press, Oxford, 1981, Vol.5,
      pp. 2547-2553.

/6/   Shih C.F. et al.-"Crack Initiation and Growth Under Fully Plastic
      Conditions. A Methodology For Plastic Fracture" EPRI Document NP
      701-SR, 1978, Palo Alto (Ca), pp. 6-1/6-57.

/7/   Willoughby A.A., Pratt P.L., Turner C.E. - Int. Journ. of Fracture,
      17, 449-466 (1981).

/8/   Towers O.L. - The Welding Institute Research Report 136, 1981,
      Cambridge.

/9/   Towers O.L., Garwood S.J. - The Welding Institute Research Report
      89, 1979, Cambridge.

/10/ Haigh J.R., Richards C.E. - Central Electricity Research Laboratories,
      Laboratory Memorandum RD/L/M 461, 1974, Leatherhead.

/11/ Walker E.F. - A.G.A. - EPRG Line Pipe Research, Seminar IV, September
      22-24, 1981, Duisburg (Germany), Vol. II, pp.1-30.

/12/ Mirabile M., Venzi S. - First International Conference on Crack
      Propagation, Lehigh University Bethleem, 1972.

/13/ Turner, C.E. - "Stable Crack Growth and Resistance Curves" "Develop-
      ments in Fracture Mechanics - 1" Chell G.C. Ed., Applied Science
      Publishers LTD., London, 1979, pp.107-144.

/14/ Paris P.C. et al. - "The Theory of Instability of the Tearing Mode
      of Elastic-Plastic Crack Growth", ASTM-STP 668, ASTM, Baltimore,
      1979, pp. 5-36.

/15/ Civallero M., Mirabile M., Sih G.C. - "Fracture Mechanics in Pipeline Technology " "Analytical and Experimental Fracture Mechanics",
Sih G.C. and Mirabile M. Ed., Sijthoff and Noordhoff, Alphen aan
den Rijn (Netherlands), 1981, pp.157-174; Kanninen M.F.-"Theoretical
Model of Propagating Fracture" A.G.A. - V Symposium on Line Pipe
Research, November 20-22, Houston, 1974; Maxey W.A. - "Fracture Initiation, Propagation and Arrest", ibidem.

/16/ Bramante M. et al. - "Pipe Deformation During Ductile Fracture Propagation in Full-Scale Tests" A.G.A. - EPRG Line Pipe Research,
Seminar IV September 22-24, 1981.

/17/ Venzi S., Martinelli A. - I.I.W. Colloquium su "Practical Applications
of Fracture Mechanics to the Prevention of Failure of Welded Structures"
Bratislava 10 Luglio 1979.

/18/ Venzi S. et al. - Giornate su "Sicurezza strutturale nella progettazione,
esercizio ed ispezioni degli impianti chimici" - Milano 6-7 Ottobre 1981.

/19/ Carpinteri A. - Eng.Fract. Mech. 16, 467-481 (1982)

SENB 3 SPECIMEN

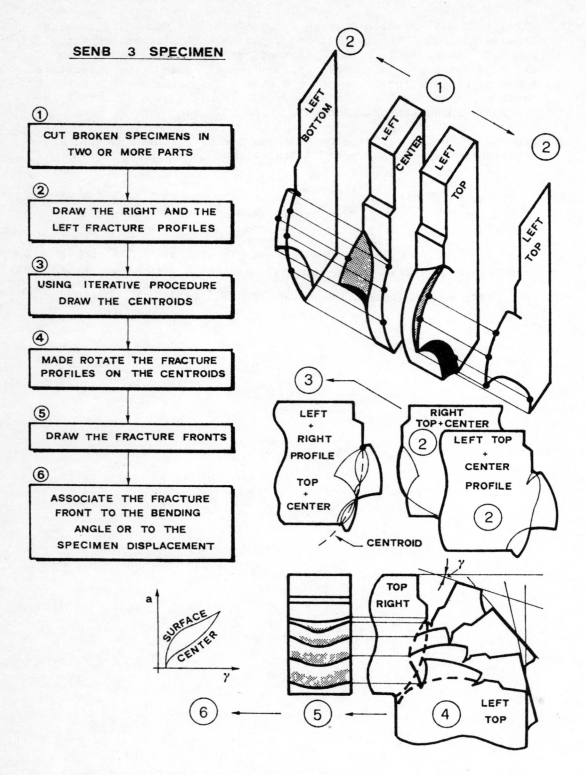

① CUT BROKEN SPECIMENS IN TWO OR MORE PARTS

② DRAW THE RIGHT AND THE LEFT FRACTURE PROFILES

③ USING ITERATIVE PROCEDURE DRAW THE CENTROIDS

④ MADE ROTATE THE FRACTURE PROFILES ON THE CENTROIDS

⑤ DRAW THE FRACTURE FRONTS

⑥ ASSOCIATE THE FRACTURE FRONT TO THE BENDING ANGLE OR TO THE SPECIMEN DISPLACEMENT

Fig. 1 - Reconstruction of the entire fracture process by examining the two broken parts of a test specimen.

- 422 -

SENB 3

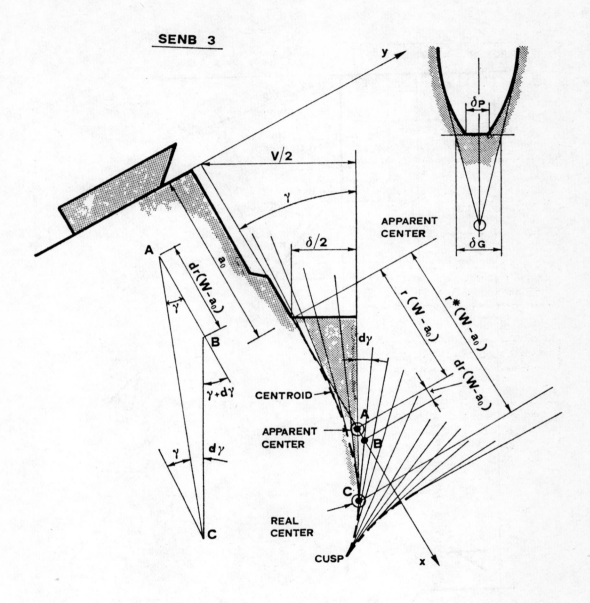

Fig. 2 – Definitions of the geometrical parameters used
in the equations describing fracture initiation.

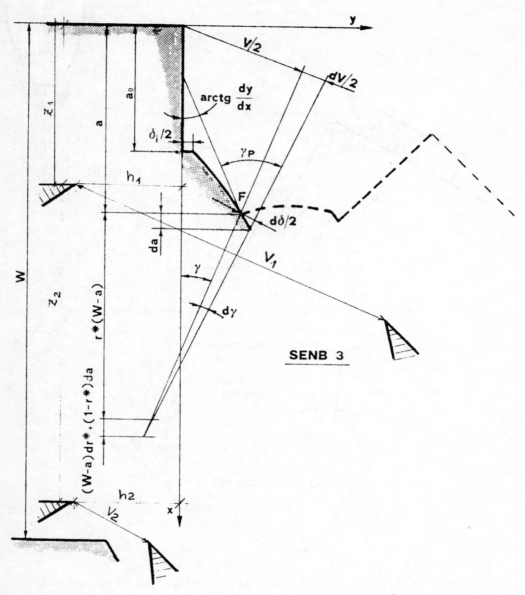

Fig. 3 — Definitions of the geometrical parameters used
in the equations describing fracture propagation.

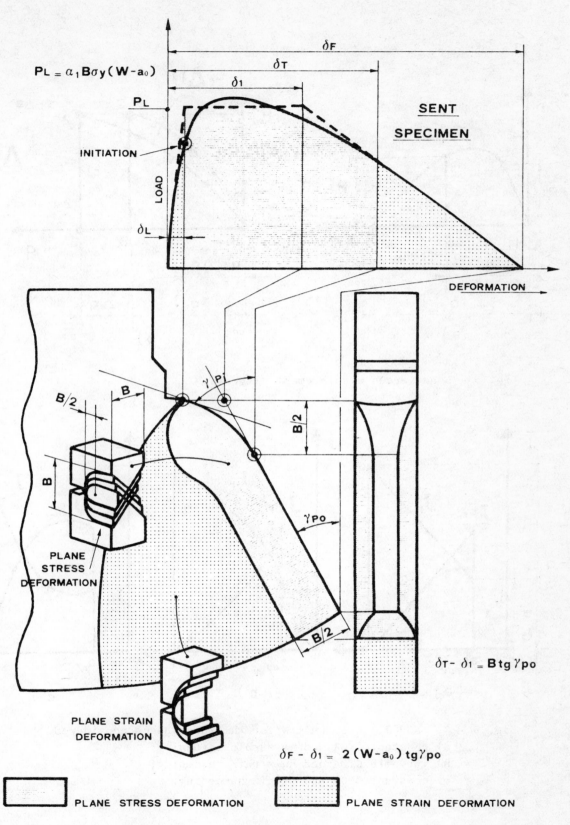

$P_L = \alpha_1 B \sigma y (W - a_0)$

SENT
SPECIMEN

δ_F

δ_T

δ_1

P_L

INITIATION

LOAD

δ_L

DEFORMATION

$B/2$

B

γ_{Pi}

$B/2$

B

γ_{PO}

PLANE
STRESS
DEFORMATION

$\delta_T - \delta_1 = B \, tg \, \gamma_{po}$

PLANE STRAIN
DEFORMATION

$\delta_F - \delta_1 = 2(W - a_0) \, tg \, \gamma_{po}$

PLANE STRESS DEFORMATION

PLANE STRAIN DEFORMATION

Fig. 4 – Schematic drawing of the load-displacement curve for a SENT
specimen, and its correlation with the different predominant
deformation modes.

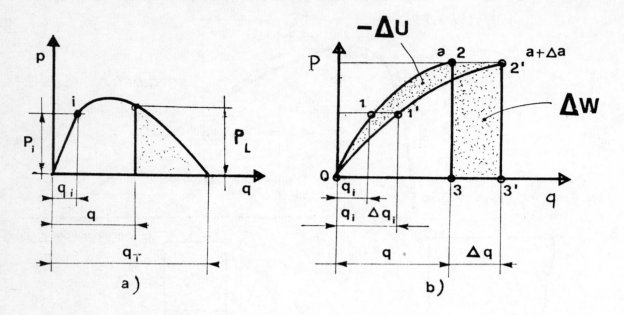

Fig. 5 - Load versus displacement diagrams.

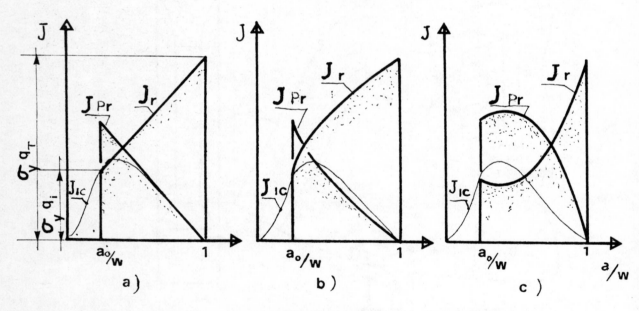

Fig. 6 - J_R and J_{Pr} versus a/W for the three cases considered.
  a) constant CTOA and no strain hardening;
  b) variable CTOA and no strain hardening;
  c) constant CTOA and strain hardening.

Photo 1 - Experimental assembling.

CHARACTERISATION OF DUCTILE CRACK GROWTH BEHAVIOUR

BASED ON ENERGY DISSIPATION WITHIN INTENSE

STRAIN REGION AT CRACK TIP

T Shoji and H Takahashi

Research Institute for Strength and Fracture of Materials
Faculty of Engineering
Tohoku University
Sendai 980
Japan

1. Recrystallisation-etch Technique

In order to analyse the formation and extension behaviour of intense strain region with stable crack growth, recrystallisation-etch technique[1,2] developed for delineating the intense strain region and for measurement of large plastic strain, is applied to a wide variety of elastic-plastic fracture toughness specimens (Fig 1). Figure 2 shows an example of intense strain region formed at the crack tip with initial bainitic microstructure changed to ferritic by recrystallisation heat treatment.

The heat treatment to recrystallise intense strain region for SA533B-1 is at 700°C for 3 hrs in vacuum followed by furnace cooling. For the purpose of quantitative evaluation, the energy dissipation[2] (or plastic work) within intense strain regions, W_p is measured at the mid-plane of specimens.

2. Influence of Specimen Size, Geometry and Loading Configuration upon W_p-R, J-R and δ-R Curves

The resistance curves based on J_p, δ and W_p showed various trends (Figs 3, 4, 5 and 6)[3]. Namely, J-R and δ-R curves essentially have dependence upon specimen geometry and size. Hence, as can be seen in

Figs 7 and 8, dJ/da (or T_j) and $d\delta/da$ (or T_δ) cannot be regarded as material properties, although dJ/da may be considered as a constant when an amount of crack extension is limited to about 2mm and a straight line is assumed for propagation data in J-R curves. The straight line for a value of Δa greater than 0.2mm can be approximately fitted by the expression $J = Ji (1 + \beta \Delta a)$ where β is the slope of J/J_i versus Δa curve. This normalised J-R curve can be used as a key curve for compact tension specimens within the limit of 30 percent scatter band, regardless of test temperature and specimen size as shown in Fig 9. An average value of $\beta = 857m^{-1}$ is obtained based on the experimental results for different alloy steels.[4]

On the other hand, as indicated in Fig 10, energy dissipation with crack growth, dW_p/da has a constant value irrespective of specimen size and goemetry during the steady propagation stage and is a material property in a wide range of Δa. This tendency can be seen clearly in Fig 11 where intense strain region developed along crack surfaces with almost constant thickness. Consequently, the crack tip resistance against crack growth in terms of the energy dissipation rate remains constant during crack growth in spite of the fact that the global resistance in terms of J and δ still increase with crack growth. The non-linear relationship between J and W_p is shown in Fig 12. All data on initiation and propagation are seen in Table 1. This difference can be considered from the viewpoint that the development of the intense strain region is not uniform with respect to specimen thickness and more large intense strain regions near the surface of a specimen leads to the formation of shear tip at the specimen surfaces (Figs 13, 14 and 15). In order to examine this effect on crack growth characteristics, a series of specimens with different specimen sizes and side groove depths were tested as shown in Fig 16. The introduction of a side groove of 25% significantly depressed the shear tip formation and maintained the specimen plane strain condition (Fig 17). Hence the J-R curve of 1T and 2T-CT specimens with 25% side groove coincide with each other as indicated in Fig 18. As can be expected easily, the W_p-R curve did not make any difference with the side groove depth (Fig 19).

3. A Criterion for Ductile Crack Growth Based on Energy Dissipation Within Intense Strain Region

A new tearing parameter[5] based on the rate of crack tip energy dissipation was proposed. That is

$$T_w = \frac{1}{R} \frac{E}{\sigma_o^2} \frac{d\bar{W}_p}{da} \tag{1}$$

where $\bar{W}_p$ is the plastic work done in the intense strain region with characteristic radius R at the growing crack tip. By using this parameter, a fracture criterion for stable crack growth can be expressed as

$$T_{\bar{w}} = (T_w)_c \tag{2}$$

where $(T_w)_c$ is the critical value of T_w which is considered as a material constant and can be measured by recrystallisation-etch technique as described in previous paragraphs in terms of dW_p/da. Saka et al[5] found that T_w can be directly related to the amplitude of the singularity field near the tip of a growing crack.

4. J-R Characterisation Based on dW_p/da (or T_w)-Criterion

A relationship between the local resistance and the global resistance J for crack growth is obtained experimentally in a wide range of crack growth, (Figs 12 and 20).

The J- integral can be related to W_p as

$$J/\sigma_o = \beta \sqrt{(E/\sigma_o^2)}W_p, \quad \beta = 0.11 \tag{3}$$

or
$$J = 50 \sqrt{W_p} \text{ for A533B-1} \tag{4}$$

The above equation was derived empirically, and the theoretical explanation of this expression has not been addressed. However, only for the deformation before the onset of crack growth, the similar expression has been given in the literature.[6]

Differentially, the equation by a

$$J \frac{dJ}{da} = \gamma \tag{5}$$

where $\quad \gamma = \dfrac{\beta^2 E}{2} \dfrac{dW_p}{da} = \dfrac{\beta^2 \sigma_o^2 R}{2} T_w \quad$ is a material constant.

Solving the above equation for the initial condition of $J = J_i$ at $\Delta a = \Delta a_o$, the J resistance curve based on the crack tip energy dissipation is obtained as

$$J = \sqrt{2\gamma (\Delta a - \Delta a_o) + J_i^2} \tag{6}$$

where J_i and Δa_o are the critical value of J and Δa at the onset of crack growth. This expression implies the non-linearity of the J-resistance curve as observed widely in experimental works. In order to evaluate the J-resistance curve for a wide range of crack extension, the non-linear expression is more suitable and accurate than a straight line approximation. It is worthwhile to mention that the non-linear J-R curve can be approximated by a straight line only when the crack extension is small.

In turn, eqn (6) suggests that the crack growth resistance γ based on energy dissipation can be estimated from the non-linear J-R curve. A conventional estimation procedure of γ and J_i in eqn (6) is based on a regression analysis for determination of two unknowns. Good agreement between the observed and estimated J-R curves as shwon in Figs 21 to 26 implies that the estimation procedure of γ would be feasible in general. Based on the J-R curves for different materials reported in the literature are reevaluated and the value of γ and J_i are estimated and listed in Table 2.

5. Concluding Remarks

The potential parameters to characterise the ductile crack growth behaviour are compared and evaluated from various points of view. Summarised results showed up the superiority of W_p-approach in all respects as shown in Table 3. The characterisation of ductile crack growth by use of small specimens as usually used in toughness assessment of irradiated materials can be performed more widely by means of W_p-approach.

REFERENCES

1. T Shoji, Metal Science, Vol 10, No 4, 1976, pp 165.

2. T Shoji, M Takahashi and M Suzuki, ibid, Vol 12, No 12, 1976, pp 579.

3. T Shoji, J of Testing and Evaluation, Vol 9, No 6, 1981, pp 324.

4. M A Kahn, T Shoji, H Takahashi and M Suzuki, Trans, ASME, J of
    Engineering Materials and Technology, Vol 103, No 10, 1981,
    p 276.

5. M Saka, T Shoji, H Takahashi and H Abé, presented at the 2nd
    International Symposium on Elastic-Plastic Fracture Mechanics,
    6-9 Oct, 1981, Philadelphia, USA, to appear in ASTM STP.

6. M Saka, T Shoji, H Takashasi and H Abé, J Mech Phys Solids, Vol 30,
    No 4, 1982, p 209.

7. W R Andrew and C F Shih, in Elastic-Plastic Fracture, ASTM STP 668,
    ASTM, 1979, p 426.

8. M G Vassilarous, J A Joyce and J P Gudas, in Fracture Mechanics,
    ASTM STP 700, ASTM, 1980, p 251.

9. F J Loss, Quarterly Progress Report, Naval Research Laboratory,
    Washington, DC, Apr - June, 1980.

10. F J Loss, Quarterly Progress Report, Naval Research Laboratory,
    Washington, DC, Apr - June, 1979.

TABLE 1. A comparison of candidate elastic-plastic fracture mechanics parameters characterising the ductile crack initiation and propagation.

Specimen Size and Geometry	a/W	Initiation						Propagation		
		J_i [a]	J_i [b]	J_i kJ/m² [c]	δ_i, mm [c]	w_p^i, j/m [c]	$(dw_p/da)i$ kJ/M² [c]	(dJ/da), MPa [d]	$d\delta/da$	(dw_p/da), kJ/m²
Three-point-bend specimen										
Charpy size	0.2	-	300	200	0.2	16	160	-	0.94	120
Charpy size	0.4	-	300	200	0.2	16	160	-	0.94	120
Charpy size	0.6	360	360	200	0.2	16	160	280	0.94	120
1T square	0.6	600	400	200	0.2	16	160	520	0.94	-
2T sqaure	0.6	600	400	200	0.2	16	160	520	0.96	170
Compact bend specimen										
0.4T	0.6	560	360	200	0.2	16	160	540	0.54	140
1T	0.6	500	330	200	0.2	16	160	470	0.30	140
2T	0.6	500	330	200	0.2	16	160	470	0.30	170

a Measured by ASTM E813

b Experimental blunting line

c initiation was defined as the stage of ductile crack growth of $\Delta s = 0.2$mm (actual crack tip separation).

d Averaged value during limited crack extension of $\Delta a = \Delta a_{initiation} \sim 2.0$mm assuming a straight line on J-R curve.

TABLE 2. Summarised data of ductile crack initiation toughness, J_i and propagation toughness, γ estimated by Eqn 6.

$$1 MJ^2 m^{-5} = 10.4 kgf^2 mm^{-2}$$

| Material | | Specimen (CT) | | | | Radiation (i) | Temp (°C) | Δa_o (mm) | J_i (kJm^{-2}) | (ii) γ (MH^{2}m^{-2}) | ω_{fin} |
		Orientation	Size	S.G (%)	a_o/W						
Andrews-Shih	A 533 B	Base metal L-T	4T	12.5	0.570	U	93	0.31	300	133(129)	6.2
"	"	" L-T	4T	12.5	0.615	U	93	0.31	300	185(181)	9.3
"	"	" L-T	4T	12.5	0.659	U	93	0.20	200	124(99.3)	3.4
"	"	" L-T	4T	12.5	0.718	U	93	0.41	400	215(233)	4.3
"	"	" L-T	1.2T	25	0.6	U	R.T.	0.18	200	267(238)	3.3
Shoji(ii)	"	" L-T	1T	25	0.6	U	288	0.094	100	88.0(-)	6.6
Vassilaros	"	" T-L	1T	20	0.72	U	150	0.16	150	42.8(56.0)	2.6
"	"	" T-L	2T	20	0.63	U	150	0.14	160	27.9(30.6)	2.3
Loss	"	Weldment T-L	1T	20	0.62	U	20	0.035	35	32.6(32.9)	3.3
"	"	" T-L	1T	20	0.62	U	200	0.075	75	17.5(19.6)	3.1
"	"	" T-L	1T	20	0.61	I	200	0.017	20	6.50(8.20)	3.6
"	"	" T-L	1T	20	0.61	I	200	0.017	20	8.66(12.0)	2.1
"	"	" T-L	1T	20	0.61	IAR	200	0.031	40	8.64(10.7)	2.5
"	"	" T-L	1T	20	0.62	IAR	200	0.015	20	7.64(10.0)	2.2
"	"	" T-L	1T	20	0.61	IARAR	200	0.015	20	8.87(10.2)	3.4
"	"	" T-L	1.6T	20	0.519	1*	200	0.021	25	12.3(14.9)	1.4
"	"	" T-L	1.6T	20	0.535	1	200	0.021	25	5.88(7.72)	1.3
Gudas	HY130	Base metal T-L	1T	12.5	0.67	U	R.T.	0.078	150	13.6(-)	1.0
"	A508Cl.3	" T-L	1T	25	0.52∿0.73	U	100	0.098	100	87.9(102)	1.6

(i) U = Unirradiated; I = Irradiated (*1.6 x 10^{19} n/cm^2 > 1MeV. Others 1.2 x 10^{19} n/cm^2 > 1MeV).
A = Annealed (399°C 168 hr); R = Reirradiated (0.7 x 10^{19} n/cm^2 > 1MeV).

(ii) Value in bracket are ω > 10   (iii) Shoji: tested in BWR condition

TABLE 3. Potential approaches and candidate parameter from characterising stable crack growth.

		J-approach dJ/da or T_J	COD-approach dδ/da or T_δ	W_p-approach dW_p/da or T_w
Controlling parameter of crack tip stress/strain-field (theoretical background)	crack extension/small	yes[a]	yes[a]	yes[a]
	crack extension/large	unknown	yes	yes
Constant during crack growth	crack extension/small	yes	yes	yes
	crack extension/large	no	yes	yes
Independent of specimen size		yes[b]	yes	yes
Independent of side groove		no	yes	yes
Independent of loading configuration		no	no	yes
Direct measurement at crack tip		no	yes[c]	yes
Global measurement		yes	no	no[d]
Applicability to instability analysis		yes	yes	yes[e]
Applicability to microfracture mechanism analysis		difficult	possible	possible

(a) Perfect plastic material

(b) Limited within the J-control crack growth

(c) Difficult as a CTOA value

(d) Possible by use of γ value

(e) M Saka and H Abe, Proc JSME, No 821-3, Dec 1982

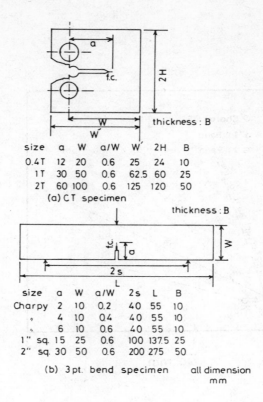

size	a	W	a/W	W'	2H	B
0.4T	12	20	0.6	25	24	10
1T	30	50	0.6	62.5	60	25
2T	60	100	0.6	125	120	50

(a) C T specimen

thickness : B

size	a	W	a/W	2s	L	B
Charpy	2	10	0.2	40	55	10
"	4	10	0.4	40	55	10
"	6	10	0.6	40	55	10
1" sq.	15	25	0.6	100	137.5	25
2" sq.	30	50	0.6	200	275	50

(b) 3 pt. bend specimen   all dimension mm

Fig 1.   Specimen geometries and dimensions (A533B-1).

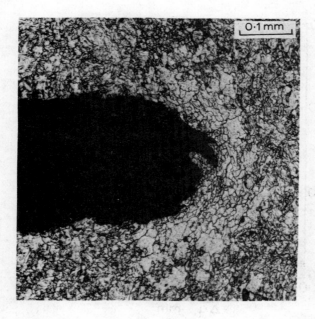

Fig 2.   An example of intense strain region formed at crack tip, with initial bainitic microstructure changed to ferritic by recrystallisation heat-treatment.

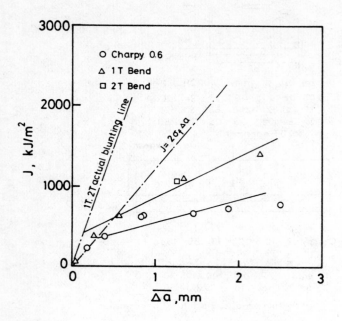

Fig 3.   J-R curves for three-point bend specimens.

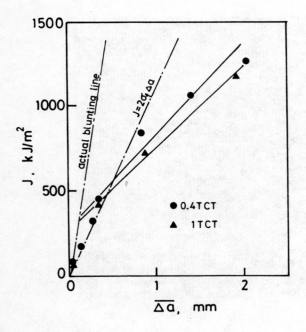

Fig 4.   J-R curves for compact tension specimens.

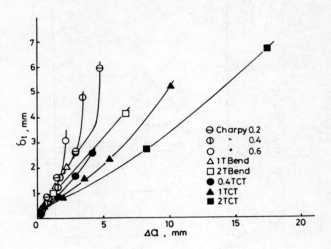

Fig 5. δt-R curves for three-point bend and compact tension specimens.

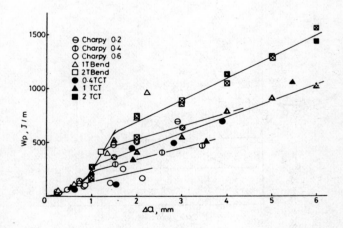

Fig 6. Wp-R curves for three-point bend and compact tension specimens.

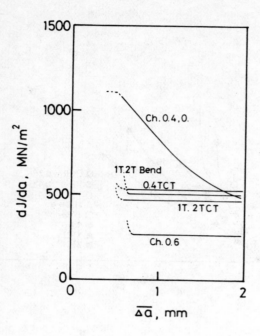

Fig 7. Variation of tearing parameter, dJ/da with stable crack growth.

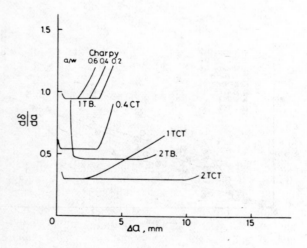

Fig 8. Variation of tearing parameter, dδ/da with stable crack growth.

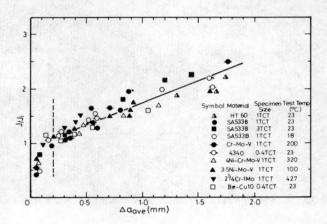

Fig 9. Experimentally determined J/Ji versus Δa key curve for several
alloy steels. Data point marked as x are invalid per specimen
size criteria, B, b > 25J/σy

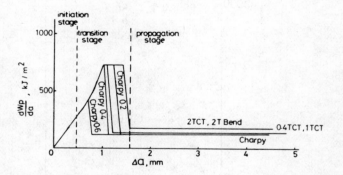

Fig 10. Variation of energy dissipation, dWp/da with ductile crack growth.

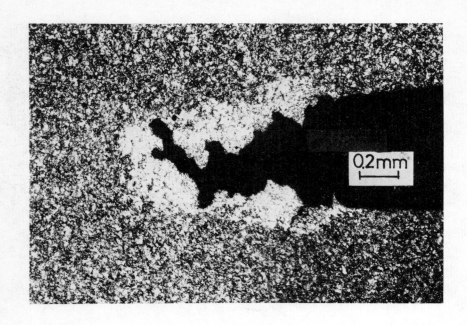

Fig 11. An example of intense strain region formed along ductile crack
surfaces.

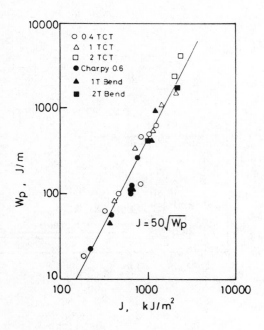

Fig 12. Relationship between crack tip energy dissipation, Wp and global
toughness, J.

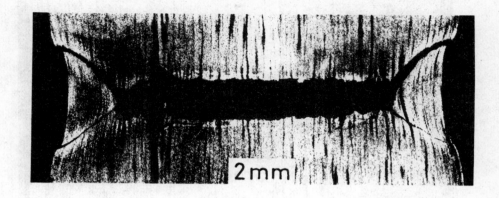

Fig 13.   Through-thickness distribution of intense strain region at
section normal to plate surface and crack plane obtained at
section 2mm from original crack tip.

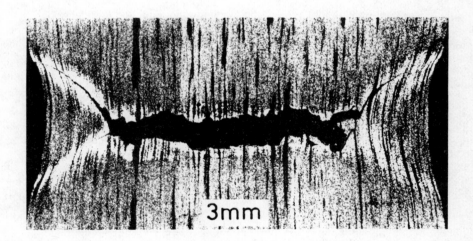

Fig 14.   Through-thickness distribution of intense strain region at
section normal to plate surface and crack plane obtained at
section 3mm from original crack tip.

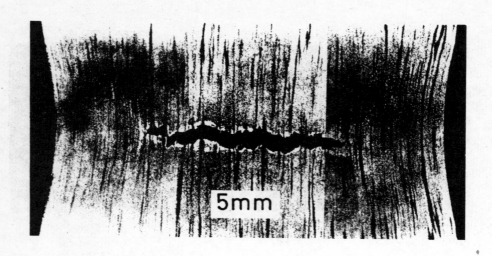

Fig 15. Through-thickness distribution of intense strain region at section normal to plate surface and crack plane obtained at section 5mm from original crack tip.

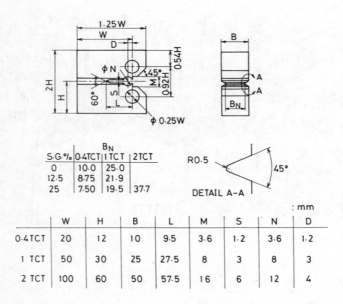

B_N			
S.G.%	0.4TCT	1TCT	2TCT
0	10.0	25.0	
12.5	8.75	21.9	
25	7.50	19.5	37.7

: mm

	W	H	B	L	M	S	N	D
0.4TCT	20	12	10	9.5	3.6	1.2	3.6	1.2
1 TCT	50	30	25	27.5	8	3	8	3
2 TCT	100	60	50	57.5	16	6	12	4

Fig 16. Dimensions of compact tension specimens with various side groove depth (A533B-1).

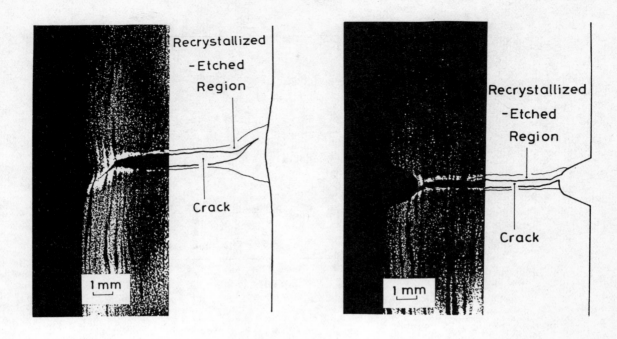

Fig 17. Through-thickness distribution of recrystallised-etched region
at section normal to plate surface and crack plane in 0.4T
compact tension specimens with (a) 0% and (b) 25% side groove.

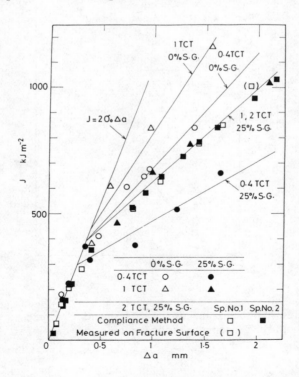

Fig 18. Effects of side groove depth and specimen size on J-R curves.

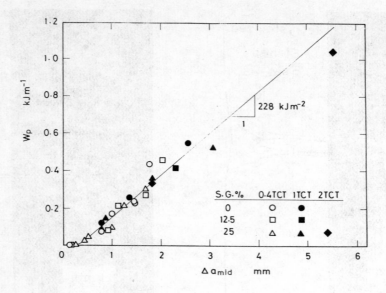

Fig 19. Wp-R curves for side grooved compact tension specimens with various sizes.

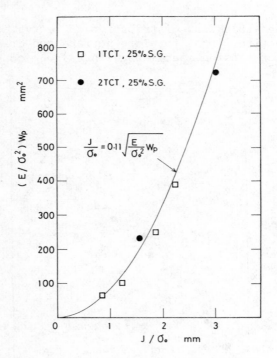

Fig 20. Relationship between J-integral and plastic work done, Wp for 1T and 2T compact tensions specimens with 25% side groove.

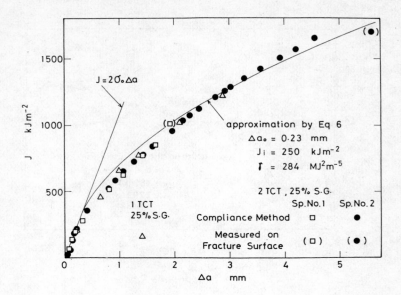

Fig 21.   Non-linear J-R curve for a wide range of crack extension
          (1T and 2T compact tension specimens and 25% side groove)
          and a comparison with an estimated curve by Eqn 6.

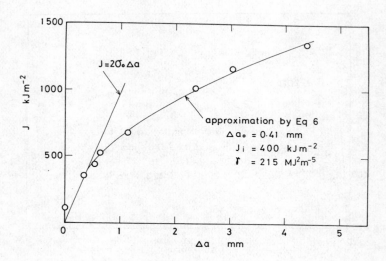

Fig 22.   J-R curve for 4T compact tension specimen with 12.5% side
          groove, A533B tested at $93^{\circ}C^{(7)}$ and a estimated curve by Eqn 6.

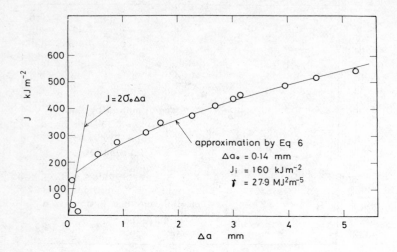

Fig 23. J-R curve for 2T compact tension specimen with 20% side groove, A533B tested at $150^{\circ}C^{(8)}$ and a estimated curve by Eqn 6.

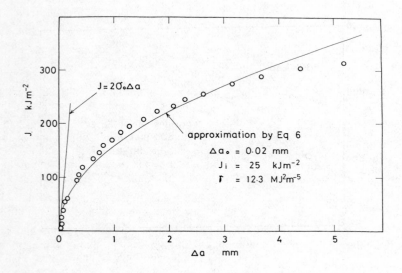

Fig 24. J-R curve for 1.6T compact tension specimen with 20% side groove, A533B weld deposit irradiated with neutron influence: $1.6 \times 10^{19}n/cm^2 > 1MeV$, tested at $200^{\circ}C^{(9)}$ and an estimated curve by Eqn 6.

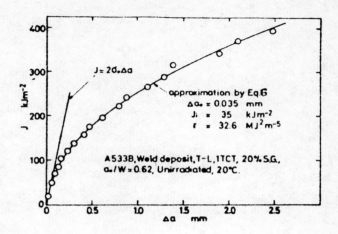

Fig 25. J-R curve for 1T compact tension specimen with 20% side groove,
A533B weld deposit tested at 20°C [10] and an estimated curve
by Eqn 6.

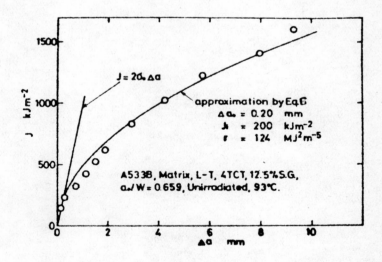

Fig 26. J-R curve for 4T compact tension specimen with 12.5% side groove,
A533B tested at 93°C [7] and an estimated curve by Eqn 6.

Session V

NEW AND IMPROVED TEST TECHNIQUES

Chairman - Président

M.I. DeVries
(Netherlands)

Séance V

METHODES NOUVELLES ET AMELIOREES

SESSION NO. 5:   NEW AND IMPROVED TECHNIQUES

Summary by the Chairman: M. I. de Vries

The work presented in this session deals with
non-conventional experimental techniques for nuclear surveillance
programmes and high-rate tests, when the standard test methods are not
applicable. The major experimental restrictions with respect to
surveillance testing are the dimensional limitations of shielding and
remote manipulation of the irradiated material. Another disadvantage
of the irradiation programmes is the long period between the start of
the irradiations and the final testing of the specimens.

In the past 15 years failure assessment methods and
consequently the needs for materials data have changed considerably.
With respect to surveillance testing it is hard to anticipate what
safety analyses will be required in the future (period of about 30
years). Therefore, adaptations of both the experimental techniques
and the theoretical analyses are necessary to measure the appropriate
materials data for the actual fracture mechanics analyses. The
existing surveillance programmes use notched specimens, being
originally aimed at the qualitative safety assessment based on Charpy
tests. In the contribution of Neale a method is presented to generate
relevant data from Charpy specimens for J-based crack growth
resistance analyses.

To overcome the technical limitations in testing irradiated
small specimens, alternative displacement measuring techniques are
necessary. This topical point is discussed for CT-specimens by
Landes, Loss and Varga. Loss shows equivalent J, R-curves by using
displacement ratios to relate different displacement measurements.
Landes obtained identical results for lOT-CT specimens and adapted
IX-WOL specimens originally intended for K_{1c}-testing, using the
modified J-analysis. Varga presented alternative procedures to cope
with the request of the Swiss Nuclear Safety Authorities for
J-experiments in surveillance programmes.

With the key curve method, crack length data are derived from
a comparison of the actual normalized load-displacement curves with
the normalized curves from specimens having different orginal notch
depths but without crack growth (key-curve). Joyce used this
technique to derive J, R-curves at high crack growth rate (2.5 m/s)
when the conventional crack monitoring techniques cannot be applied.

The work on the validity of the alternative displacement
procedures offers the perspective for the development of identical
test methods for LEFM and EPFM. Further, the key curve method
promises to be an alternative for crack length measurements. Both are
very useful for the application of EPFM testing techniques to
specimens which have already been irradiated as well as to new
surveillance programmes.

MODIFICATION OF THE J_{Ic} TEST PROCEDURE FOR NUCLEAR SURVEILLANCE SPECIMEN TESTING

J. D. Landes
Materials Engineering Department

SUMMARY

Westinghouse has fracture mechanics specimens placed in its nuclear surveillance capsules along with the other conventional specimens. Among these are specimens of the 1X-WOL geometry, Fig. 1. These were placed with the hope of conducting LEFM K_{Ic} tests for fracture toughness and predated the development of EPFM. The toughness of the irradiated material is high enough that J_R curves must be developed to measure toughness. This leads to testing problems in that the specimen capacity for ductile crack extension is small and the specimen design results in arm bending, Fig. 2, when they are deformed plastically.

To test the 1X-WOL specimen several modifications are necessary including modifications in test fixturing, specimen design and method of analysis, Fig. 3. The modification in the test fixturing provides large flat regions in the clevis to accommodate large rotations in the specimen arms, Fig. 4. The specimen itself must be modified to eliminate arm bending. This can be done by side grooving or reducing the uncracked ligament size. A diagram was developed, Fig. 5, to determine the amount of side grooving or ligament reduction needed to eliminate arm bending. This was done by equating the limit load on the uncracked ligament to the load needed for plastic deformation in the arms. Several options are marked on Fig. 5.

The analysis of the J_R curves was made using the modified J developed by Ernst, Fig. 6, which has the effect of extending the useful range of crack extension in specimens with small remaining ligaments, Fig. 7.

A test matrix was developed, Fig. 8, to test the various specimen modifications illustrated in Fig. 5. These test specimens were taken from the broken half of a 10T-CT specimen of A508 Cl 2a steel for which the J_R curve behavior had been documented. J_R curves were developed for specimens in this matrix at 477°K (400°F) using the unloading compliance test method. The success of the specimen modification was judged by whether the arm bending had been eliminated and by how well the small specimen R curve matched the 10T R curve.

Results from the test matrix are given in Figs. 9, 10, and 11. The standard 1X WOL specimen, Fig. 9, must be side grooved to 50% to eliminate arm bending. This specimen then gives an R curve identical to the 10T R curve if a modified J analysis is used. A modification of reducing the specimen width to 0.95 inch and side grooving 20% also gives the desired result, Fig. 10. Specimens of the 1/2T-CT geometry are also included in some surveillance capsules. Figure 11 shows that these specimens can be used to successfully develop R curves if they are side grooved 20% and a modified J analysis used.

The conclusion to this work is given in the form of testing guidelines in Fig. 12. The test fixture modification and use of the Ernst modified J are recommended. Several modified specimen designs could be used for the 1X-WOL. The recommended one is the 20% side grooving and shortening the specimen width to 0.95 inch.

This work was partially supported by EPRI under Project 1238-2.

REFERENCES

1. Standard Test for J_{Ic}, a Measure of Fracture Toughness, ASTM E813-81, 1981 Annual Book of ASTM Standards, Part 10.

2. Albrecht, P., et al., "Tentative Test Procedure for Determining the Plane Strain J_1-R Curve" to be published in the Journal of Testing and Evaluation, JTEVA, November, 1982.

3. Standard Test Method for Plane-Strain Fracture Toughness of Metallic Materials, ASTM E399-81, 1981 Annual Book of ASTM Standards, Part 10.

4. Landes, J. D., "J Calculation from Front Face Displacement Measurement on a Compact Specimen", Int. Journal of Fracture, Vol. 16, 1980, pp. 183-186.

5. Paris, P. C., Ernst, H. A. and Turner, C. E., "A J Integral Approach to Development of η Factors", Fracture Mechanics: Twelfth Conference, ASTM STP 700, American Society for Testing and Materials, 1980, pp. 338-351.

6. McCabe, D. E., Landes, J. D. and Ernst, H. A., "An Evaluation of the J_R-Curve Method for Fracture Toughness Characterization" presented at the Second International Symposium for Elastic Plastic Fracture Mechanics, Philadelphia, PA, October, 1981.

7. Ernst, H. A., Paris, P. C. and Landes, J. D., "Estimations on J-Integral and Tearing Modulus T from Single Specimen Test Record", ASTM STP 743, 1981.

8. Rice, J. R., Drugan, W. J., and Sham, T. L., "Elastic Plastic Analysis of Growing Cracks", Fracture Mechanics: Twelfth Conference, ASTM STP 700, American Society for Testing and Materials, 1980, pp. 189-221.

9. Ernst, H. A., "Material Resistance and Instability Beyond J Controlled Crack Growth", presented at the Second International Symposium on Elastic Plastic Fracture Mechanics, October 1981, Philadelphia, PA.

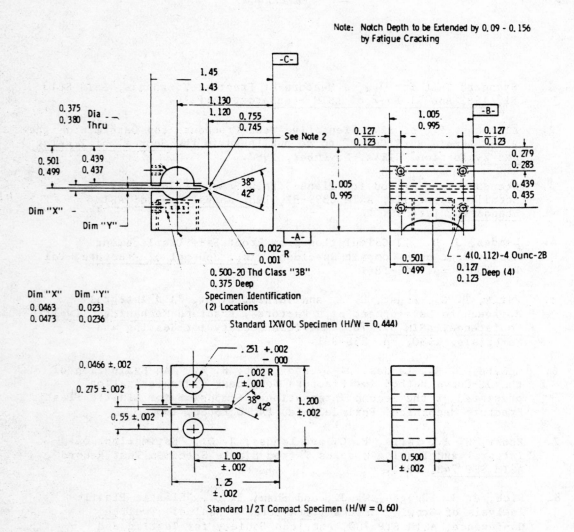

Note: Notch Depth to be Extended by 0.09 - 0.156 by Fatigue Cracking

Standard 1XWOL Specimen (H/W = 0.444)

Dim "X"	Dim "Y"
0.0463	0.0231
0.0473	0.0236

Standard 1/2T Compact Specimen (H/W = 0.60)

Figure   1   Surveillance capsule fracture mechanics specimens.

PRE-TEST   **POST-TEST**

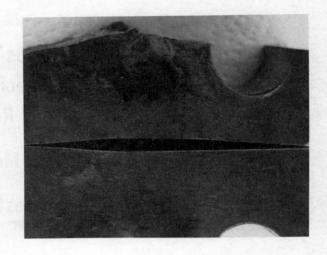

1X WOL SPECIMEN

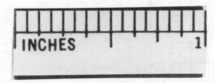

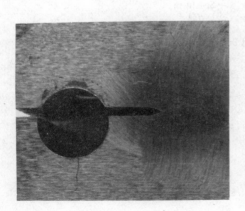
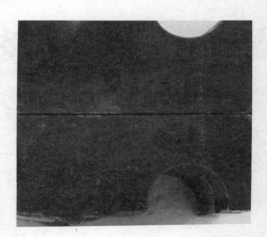

MODIFIED 1X WOL SPECIMEN

Fig. 2 — Pre and post-test pictures of 1X WOL and modified 1X WOL specimens tested at shelf temperature

Contents Of Lecture

A. Modifications

    1. Test Fixtures

    2. 1X-WOL Specimen

    3. Analysis Of Results

B. Program To Test Modifications

C. Guidelines For Testing And Analysis

FIGURE 3

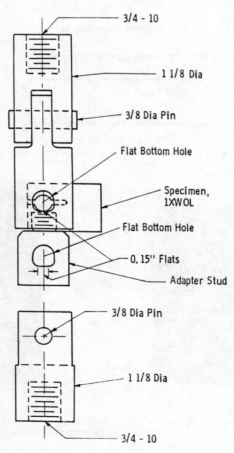

Fig. 4 —Fixtures for 1XWOL testing

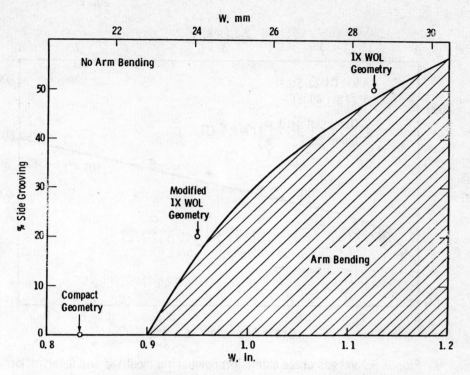

Fig. 5 –Percent side grooving versus W for the limit on arm bending

FIGURE 6

Deformation J

$$J_{(i + 1)} = \left[J_i + \left(\frac{\eta}{b} \right)_i \frac{A_{i, i + 1}}{B_N} \right]\left[1 - \left(\frac{\gamma}{b} \right)_i \left(a_{i + 1} - a_i \right) \right]$$
$$= J_D$$

Modified J

$$J_M = J_D - \int_{a_0}^{a} \frac{\partial (J - G)}{\partial a} \bigg|_{\delta_{pl}} da$$

For The Compact Specimen

$$J_M = J_D + \int_{a_0}^{a} \gamma \frac{J_{pl}}{b} da$$

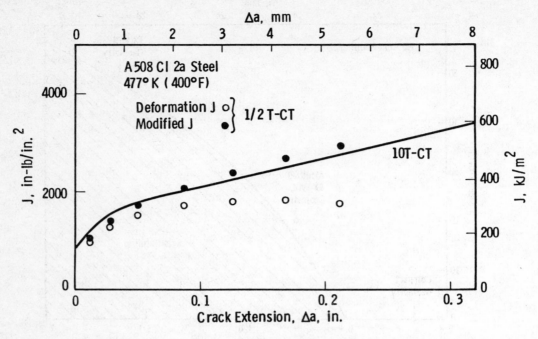

Fig. 7 —J versus crack extension comparing modified and deformation J for a 1/2 T - CT specimen

S-89-26

FIGURE 8

Matrix A

Specimens Taken From The A508 C12a 10T-CT Specimen

% Side Groove	Standard 1XWOL (H/W = .444)	Modified 1XWOL (H/W = .526)	1/2 T-Compact (H/W = .60)
0	X	X	X*
20	X	X*	X
50	X*		X

* Minimum Side Groove For Zero Arm Bending

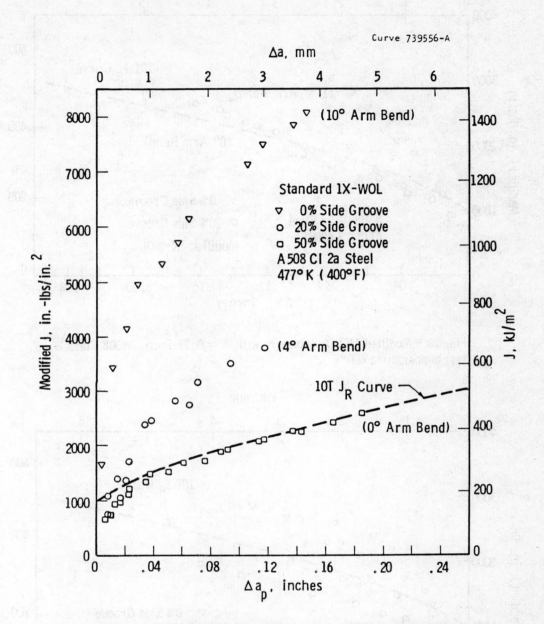

Curve 739556-A

Fig. 9 – Standard 1XWOL geometry. Three side groove levels. A508 Class 2A. Test temperature 400°F

- 461 -

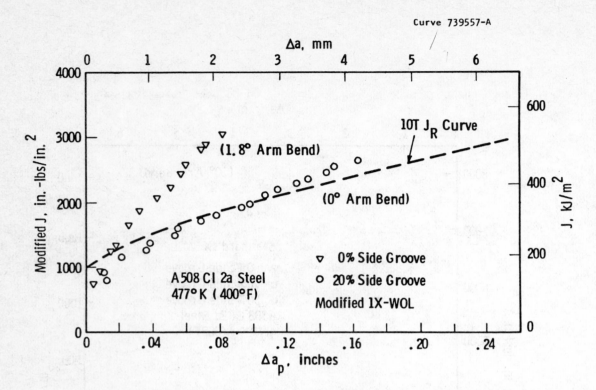

Fig. 10 – Modified 1XWOL geometry with W = 0.95-inch. A508 Class 2A, test temperature 400°F

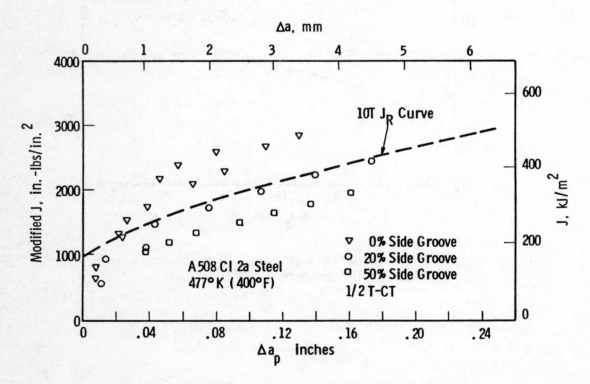

Fig. 11 – 1/2T compact specimens. Standard thickness A508 Class 2A, test temperature 400°F

FIGURE 12

S-89-28

Surveillance Specimen Testing Guidelines

A. 1X-WOL

1. Modify Loading Fixtures To Provide Rotation Capacity

2. Modify Specimen To W = 0.95 In, 20% Side Groove

3. Modify Analysis
   a) Front Face Displacement To Load Line
   b) Use Modified J Analysis For R Curve

B. 1/2 T - CT

1. Optional 20% Side Grooves Recommended

2. Use Modified J Analysis For Side Grooves

The fracture toughness of irradiated pressure vessel steels determined directly from pre-cracked Charpy specimens

by

B.K. Neale & R.H. Priest

CEGB, Berkeley Nuclear Laboratories, Berkeley, Glos., GL13 9PB

Introduction

In order to fully assess the structural integrity of pressure vessels in nuclear reactors, a measure of the fracture toughness of the component steels is required. The established method of assessing the effect of irradiation on the fracture properties of pressure vessel steels is to measure the absorbed impact energy levels using Charpy V-notch surveillance specimens. Although several empirical correlations exist between fracture toughness and impact energy levels, their accuracy for irradiated steels is uncertain unless the fracture toughness can also be measured. This note describes CEGB experience gained using the single specimen unloading compliance method to measure the fracture toughness of irradiated pre-cracked Charpy specimens.

Specimen behaviour

Figure 1 shows a standard V-notch Charpy specimen containing a crack nominally 3 mm deep below the root of the 2 mm deep V-notch giving an effective crack length to specimen width ratio of 0.5. The specimen is loaded in three-point bend at a span to width ratio $^s/w$ of 4. Side-grooves, usually of Charpy V-notch profile, are machined into the specimen to promote straight fronted crack growth.

Figures 2 and 3 show the effect of side-groove depth determined from elastic three-dimensional finite element analyses of a pre-cracked Charpy specimen, Neale (1980), on the stress intensity factor along the crack front and the stress state ahead of the crack, respectively. These results suggest that the stress intensity factor will be constant along the crack front if the side-groove depth is between 0-1 mm deep and that a near uniform plane strain stress state exists ahead of the crack for 1 mm deep side-grooves.

Figure 4 shows the effect of side-groove depth on the elastic compliance, $E\Delta/P$, and compares the values with analytical and two-dimensional finite element results. Figure 5 compares the compliance of non-sidegrooved Charpy specimens measured from load point and clip gauge displacements with the two-dimensional analytical solution. Extraneous displacements were removed from the load point measurements and the finite element results were used to relate clip gauge measurements to equivalent load point displacements. The results are in reasonably close agreement providing $0.5 \leq {}^a/w \leq 0.65$.

Practical considerations

Load point displacements are measured with LVDT gauges mounted across the loading shackles of the testing machine in the irradiated test facility described by Neale and Priest (1981). This approach while avoiding the difficulty of attaching clip gauges to irradiated test specimens does introduce extraneous displacements into the measurements. From a series of tests performed on blunt notched Charpy specimens, Neale and Priest (1982) found that the extraneous displacement was a linear function of both load and displacement. The load term was associated with elastic deformation in the testing fixture and was independent of test specimen material properties. The displacement term, which showed a dependence on test specimen material properties that scaled approximately with yield stress, was associated with indentation effects. Figures 6 and 7 compare the predicted and measured extraneous displacements for EN3A and A508 steel, respectively. The negative extraneous displacement arises from material pile-up, Figure 8, which gives rise to a force opposing the applied load with indentation and motion around the roller support.

Unloading compliance

A typical load displacement record for a pre-cracked Charpy specimen tested in three-point bend is given in Figure 9a. Prior to each unloading cycle the specimen is held at constant displacement until the load has relaxed. The compliance is evaluated from the unloading portions of the unloading cycle, Figure 9b. Removing the extraneous displacements produces the corrected compliance shown in Figure 9c.

The crack length a is evaluated at each unloading cycle using the plane strain analytical compliance function modified to account for side-groove depth, Neale (1982a), and the effect of temperature on Young's modulus, Neale (1982b).

The fracture resistance J is evaluated at each unloading cycle using the side-grooved thickness B_n in the Rice, Paris and Merkle (1973) relationship. This relationship is compatible in the elastic regime with the analytical compliance function and the stress intensity function, and in the plastic regime to the J-integral function calculated from the limit load function, Neale (1982a).

Figure 10 shows the fracture resistance versus crack length corresponding to the data given in Figures 9a and 9c.

Interpretation

A best fit blunting line of slope of either $2\sigma_f$ ot $4\sigma_f$ is currently drawn by eye through the J-a data points where σ_f is the flow stress, Figure 11. The predicted initial crack length a_i is given by the intersection of the blunting line and the abscissa. Providing the final unloading compliance data point corresponds to the end of the test, a_f can be used to estimate the final crack length. Figure 12 compares for a number of steels at different test temperatures the predicted crack lengths with the corresponding measured values obtained using the 7-point averaging procedure, BS5762. Crack lengths are predicted to within $^+_-4$ percent and crack growths to within $^+_-8$ percent. The accuracy is independent of side-groove depth.

The initiation fracture resistance J_q is given by the intersection of the blunting line with a least squares fit line of slope $\partial J/\partial a$ drawn through the J-a data points lying between the blunting and exclusion line, Figure 13. The exclusion line is drawn parallel to the blunting line at a distance corresponding to 0.06 (w - a_o) where a_o is the measured initial crack length, Neale (1982a).

The initiation fracture toughness K_q is given by $\sqrt{(EJ_q/(1 - \nu^2))}$ where E is Young's modulus at the test temperature and ν is Poisson's ratio.

Results

Fracture resistance curves for sidegrooved and non-sidegrooved Charpy specimens are compared in Figure 14 with the results obtained from 25 mm non-sidegrooved compact tension specimens using the multispecimen method. The non-sidegrooved specimen results, which were obtained for C-Mn steel as part of an accelerated irradiation programme, are in reasonably close agreement.

Figure 15 compares the fracture toughness of A508 Class II steel determined from pre-cracked Charpy specimens containing 1 mm deep side-grooves with values measured from large sized compact tension specimens, Neale (1982b). The Charpy specimen results fall on the J_{Ic} validity limit beyond which the K_q values cannot be used as estimates of K_{Ic}. The fracture resistance curves of the Charpy specimens at 65° are compared in Figure 16 with the results for the compact tension specimens. Although J-controlled crack growth in a Charpy specimen is limited to 0.3 mm at a nominal $^a o/w$ of 0.5, it is interesting to note the close agreement with the compact tension specimen data at larger crack growths.

References

Anon, 1979, ASTM Standards, Part 10, E399-78a, 580-601.

Krägeloh, E., Issler, L. and Zirn, R., 1980, 6 MPA-Seminar, University of Stuttgart, West Germany, 9-10 October.

Kussmaul, K. and Issler, L., 1981, Second International Symposium on Elastic-Plastic Fracture Mechanics, Philadelphia, 6-9 October.

Neale, B.K., 1980, CEGB Report RD/B/N4927 and Int.J. of Pressure Vessels and Piping, Vol. 10, No. 5, 1982, 375-398.

Neale, B.K., 1982a, CEGB Report TPRD/B/0012/N82 and to be published in Int. J. of Pressure Vessels and Piping.

Neale, B.K., 1982b, Draft CEGB Report.

Neale, B.K. and Priest, R.H., 1981, CEGB Report RD/B5208N81 and Int. Conf. on Fracture Toughness Testing, The Welding Institute, London, 1982, Paper 2.

Neale, B.K. and Priest, R.H., 1982, CEGB Report TPRD/B0085N82 submitted for publication.

Rice et al, 1973, ASTM STP536, 231-245.

Sinz, R., Doll, W., Zirn, R., Denster, G., Jöst, H. and Lottermoser, J., 1979, 5 MPA-Seminar, University of Stuttgart, West Germany, 11-12 October.

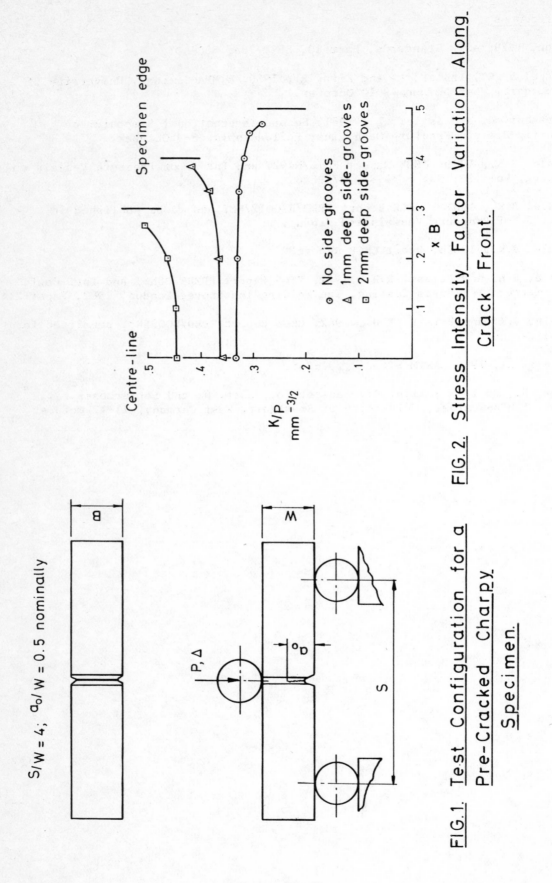

FIG.2. Stress Intensity Factor Variation Along. Crack Front.

FIG.1. Test Configuration for a Pre-Cracked Charpy. Specimen.

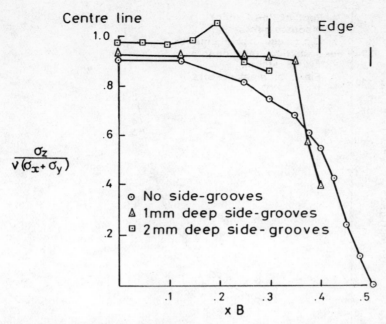

FIG.3. The Effect of Side-Groove Depth on Stress State at a Distance r/a = 0.06 from the Crack Front in the Plane of the Crack.

a/w	Compliance E (Δ/p) mm^{-1}	Finite element idealisation / analysis.
0.25	2.83	2-D, plane stress
	2.70	Analytical relationship, plane stress.
	2.64	Analytical relationship, plane strain.
0.50	5.66	2-D, plane stress
	5.73	Analytical relationship, plane stress.
	5.39	Analytical relationship, plane strain
	5.50	3-D, no side-grooves
	5.89	3-D, 1mm deep side-grooves.
	6.90	3-D, 2mm deep side-grooves.
0.75	22.7	2-D, plane stress.
	22.7	Analytical relationship, plane stress.
	20.8	Analytical relationship, plane strain.

E is Young's modulus, $\frac{S}{W}$ = 4.

FIG.4. Compliance of a Charpy Specimen in Three-Point Bend.

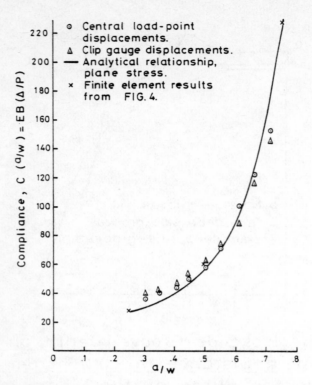

FIG.5. Compliance of a Charpy Specimen
without Side-Grooves.

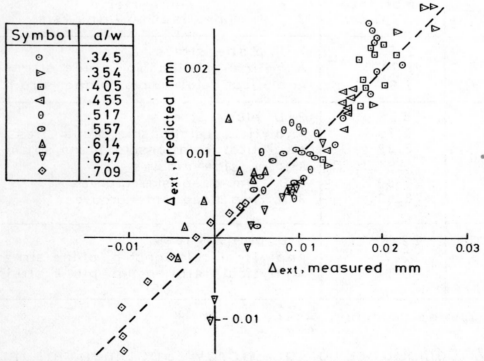

FIG.6. Predicted Extraneous Displacement versus Measured
Extraneous Displacement for EN3A Steel.

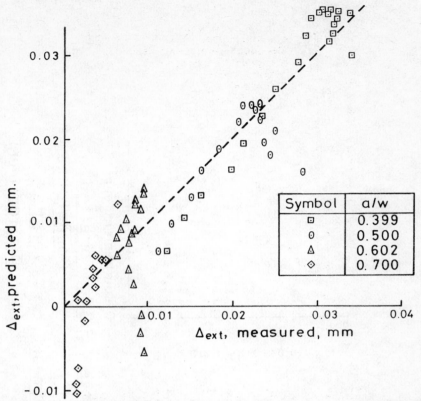

<u>FIG. 7.</u> <u>Predicted Extraneous Displacement versus Measured</u>
<u>Extraneous Displacement ·for A 508 Steel.</u>

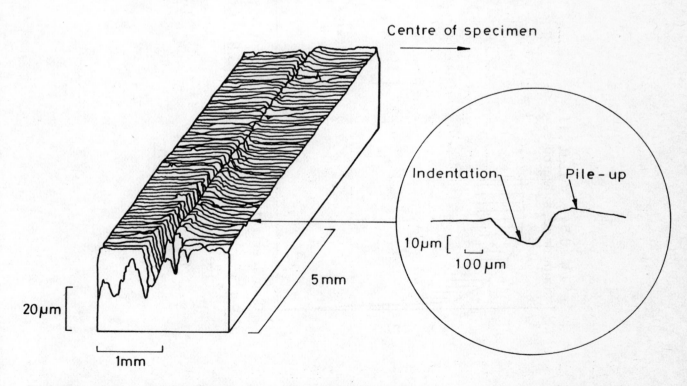

<u>FIG.8.</u> <u>Talysurf Profiles of Roller Support Contact Area.</u>

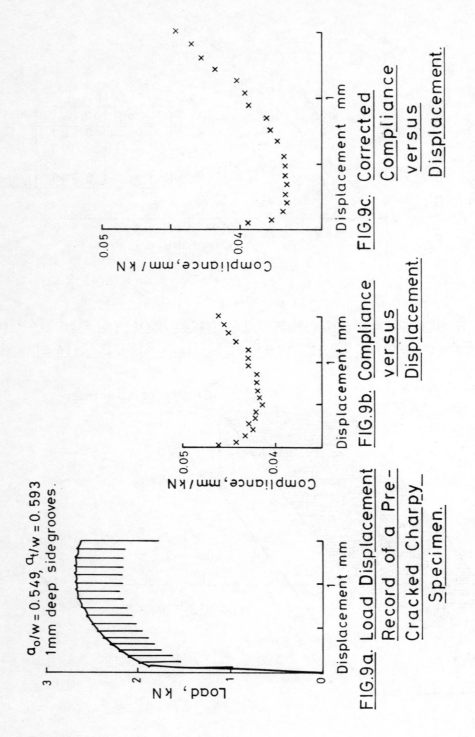

FIG.9a. Load Displacement Record of a Pre-Cracked Charpy Specimen.

FIG.9b. Compliance versus Displacement.

FIG.9c. Corrected Compliance versus Displacement.

$a_o/w = 0.549$, $a_{t/w} = 0.593$
1mm deep sidegrooves.

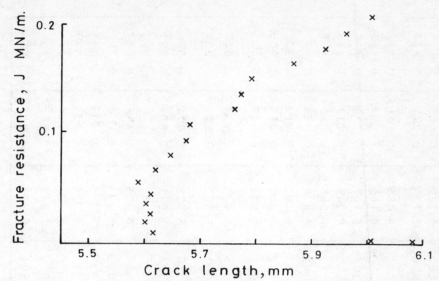

FIG.10. Fracture Resistance versus Unloading Compliance Estimated Crack Length of a Pre-Cracked Charpy Specimen.

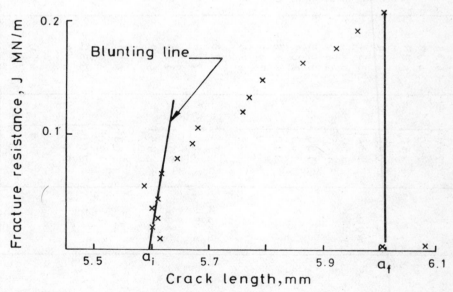

FIG.11. Crack Length Estimation Procedure.

Material	Temperature °C	Yield stress MPa	Initial crack length Predicted (a_i) mm	Initial crack length Measured (a_o) mm	Final crack length Predicted (a_f) mm	Final crack length Measured (a_t) mm	Crack growth Predicted (a_f−a_i) mm	Crack growth Measured (a_t−a_o) mm
Weld metal	250 }1	438	4.830	4.770	5.04	4.974	0.210	0.204
	250	438	5.330	5.190	5.451	5.309	0.121	0.119
	250	438	5.517	5.490	5.808	5.763	0.291	0.273
	280	438	5.610	5.503	6.191	6.065	0.581	0.562
	250	337	5.745	5.540	----*	6.100	----*	0.560
	190	340	5.590	5.400	6.309	6.147	0.719	0.747
	170	342	5.506	5.410	6.317	6.245	0.811	0.835
	150 }2	353	5.656	5.600	6.115	6.065	0.459	0.465
	140	362	5.588	5.357	6.069	5.836	0.481	0.479
	110	428	6.225	6.119	6.611	6.492	0.386	0.373
	90	391	5.561	5.465	6.118	6.071	0.557	0.606
	26	421	5.585	5.490	6.012	5.934	0.427	0.444
EN 3 A steel	250 }1	198	5.100	5.000	----*	5.623	----*	0.623
	55 }3	256	5.660	5.530	6.011	5.869	0.351	0.339
A508 steel	200 }2	390	5.532	5.340	6.136	5.928	0.604	0.588
	65	420	5.524	5.420	6.125	6.011	0.601	0.591
	65	420	5.500	5.418	5.983	5.905	0.483	0.487
	130 }2	404	5.250	5.337	6.063	6.139	0.813	0.802
	200	390	5.422	5.335	5.938	5.876	0.516	0.541
	300	382	5.468	5.406	6.347	6.315	0.879	0.909
	65 }1	420	5.516	5.576	7.072	7.212	1.556	1.636
	65 }2	420	5.484	5.447	6.500	6.431	1.016	0.984

1. Non-sidegrooved
2. 1mm deep sidegrooves
3. 2mm deep sidegrooves
* Last unloading compliance measurement not coincident with end of test.

FIG.12. Predicted and Measured Crack Lengths for Various Steels Tested over a Range of Temperatures.

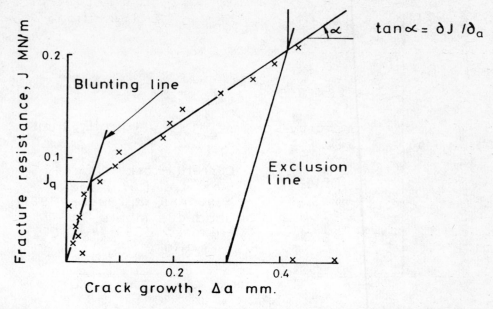

$tan \alpha = \partial J / \partial_a$

FIG.13. $\underline{J_q}$ and $\partial J / \partial_a$ Interpretation Procedure.

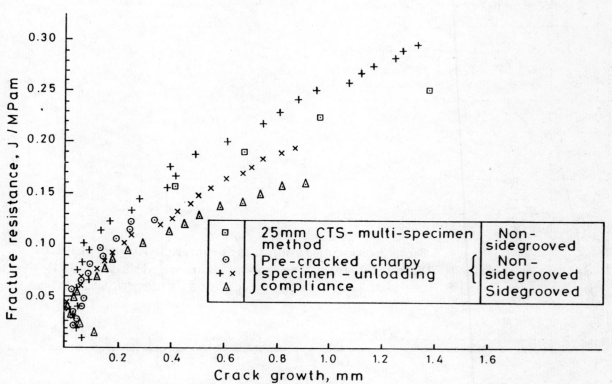

FIG.14. Fracture Resistance of C-Mn Steel.

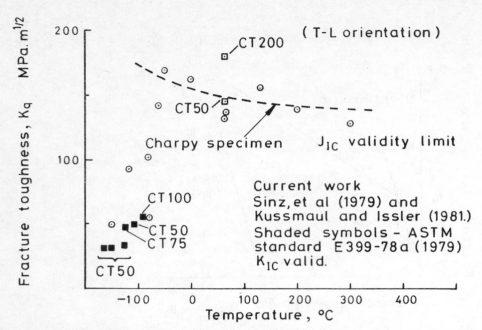

FIG.15. Comparison of the Fracture Toughness of A508 Class II Steel Determined using Pre-Cracked Charpy and Compact Tension Specimens.

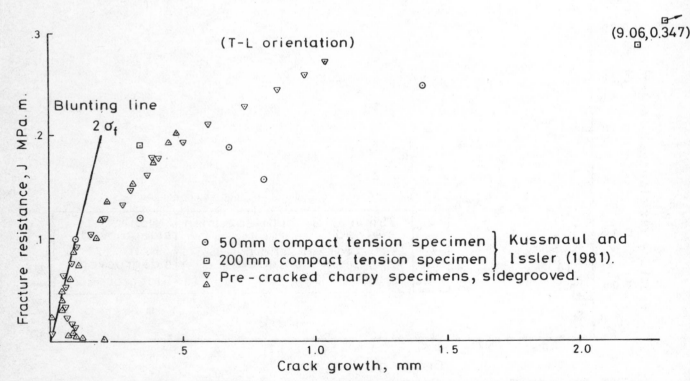

FIG.16. Comparison of the Fracture Resistance of A508 Class II Steel Determined using Pre-Cracked Charpy and Compact Tension Specimens at 65°C.

ALTERNATIVE DISPLACEMENT PROCEDURES FOR DETERMINATION OF THE J-R CURVE

F.J. LOSS and A.L. HISER
Materials Engineering Associates, Lanham, MD 20706, USA

BACKGROUND

Experimental determination of the J-R curve requires displacement measurement at the points of load application (V_{LP}). Because of the practical difficulty of this requirement, displacements are usually measured at the load line (V_{LL}) under the assumption that the latter is an adequate approximation to V_{LP}. With compact toughness (CT) specimens of the ASTM E-399 standard design, V_{LL} measurements have required that the notch region of the specimen be enlarged and the spacing of the loading pin holes increased to permit the placement of a displacement gage on the load line (Fig. 1). It has long been recognized that it would be advantageous if the standard CT specimen, in which only crack mouth displacement (V_M) is measured, could be used for both K_{Ic} and J-R curve determinations. A single specimen design would also benefit the testing of standard CT specimens which have already been irradiated in test reactors or in reactor surveillance capsules.

Studies of elastic loading in CT specimens by Newman [1] and Orange [2] have suggested that it is not necessary to modify the CT specimens for V_{LL} determination. Instead, they have proposed that V_M provides an approximation to V_{LP} that is even better than V_{LL}. However, verification of this hypothesis for elastic-plastic loading is required.

The need to establish J-R curves on the basis of displacements other than V_{LL} became apparent in the testing of 0.5T- and 0.8T-CT specimens that were irradiated in the Heavy Section Steel Technology (HSST) program. While these specimens were of the standard design, the notch had been modified for V_{LL} measurements (integral knife edge design). However, this design proved to be unsuitable for reliable measurements of V_{LL} with irradiated specimens, and the authors devised a "double clip gage" procedure to compute the J-R curve on the basis of displacements at two locations on the specimen: V_M and V_{LL}', where the latter is the load line displacement at the outside specimen surface (Fig. 1). With this method J is computed with the V_{LL}' displacements and V_M is used to infer crack extension by the compliance procedure.

This report discusses the use of V_M as an alternative to load line displacement measurements for J in terms of the double clip gage procedure.

SUMMARY OF RESULTS

In order to establish a relationship between V_{LL}' and V_M for elastic-plastic loading, these quantities were compared for several specimens (irradiated and unirradiated) for A508 weld deposit from the HSST program (Fig. 2, 3). While the relationship is of a quadratic form, a linear fit to the data also may be considered acceptable. Although the specimens in Figs. 2 and 3 were geometrically similar, a small difference in the correlation between V_{LL}' and V_M is apparent.

The nonlinear relationship between V_{LL}' and V_M as a function of relative crack length (a/W) is more clearly illustrated in Figs. 4 and 5. Also shown in these figures is the elastic displacement ratio (V_{LL}'/V_M) computed by Newman [1] for the standard CT specimen. In this case the enlarged notch (Fig. 1 - bottom) was not modelled. The average values of the displacement ratio from experiment (i.e., 0.756 and 0.732 for 0.5T- and 0.8T-CT specimens, respectively) correspond well with the values computed by Newman.

Figure 6 illustrates various elastic displacement ratios for the standard CT specimen [1]. With several of these ratios the value at a/W = 0.5 was chosen as the constant to relate the displacements in the CT specimen. In this way it is possible to compare R curves based on displacements measured at other than the load point. (Chosing a fixed displacement ratio at a/W = 0.5 will produce lower bound or conservative R curves.) Figure 7 illustrates this procedure for a specimen tested by the double clip gage technique (i.e., points denoted as "Combination"). The curve labelled "Orange L-P Data" was computed from V_M displacements expressed in terms of V_{LP}. The curve "Front Face Data" was computed from V_M displacements expressed in terms of V_{LL}'. Finally the curve "Load Line Prime Data" was computed from V_{LL}' where both J and crack extension were computed from this quantity. Another example of these comparisons is given in Fig. 8. An example of the variation in the J_{Ic} and tearing modulus (average between 0.15 and 1.5 mm of crack extension) is illustrated in Table 1. From these comparisons it is clear that both V_M and V_{LL}' displacements can be used to establish an R curve.

Tests with the double clip gage procedure did not permit an assessment to be made of the V_{LL} vs. V_{LP} displacements for R curve determination. From Fig. 6, it is apparent that displacement ratios based upon V_{LL} exhibit somewhat more variation with a/W than do V_{LL}' or V_M. Consequently, if the displacement ratio is assumed to be constant over a range in a/W, better R curve results would be obtained with V_{LL}' and V_M.

A comparison of R curves determined by the double clip gage technique (V_{LL}' and V_M) with R curves determined by the more common single clip gage (compliance) technique using V_{LL} is illustrated in Fig. 9. The data from both techniques compare quite well. This method of presenting the data assumes that V_{LL}/V_{LL}' is unity; the computed values in Fig. 6 suggest that this is very near the case. It should be noted, however, that the double clip gage data in the initial portion of the R curve ($\Delta a < 0.15$ mm) lie toward the high side of the band produced by the single clip gage tests. If, on the other hand, a ratio of $V_{LL}/V_{LL}' = 0.95$ is used for this region (e.g. Fig. 6) the data at small crack extension are moved to the center of the single clip gage band (Fig. 10).

In summary, it has been suggested by Newman that the conventional practice of measuring V_{LL} as an approximation to V_{LP} could lead to differences in J. Through the use of displacement ratios, a way exists to determine the R curve without the practical difficulty of direct load-point measurements. Experimental results have been presented for a limited case of nuclear pressure vessel steels. On the basis of experimentally determined displacement ratios, confirmed by finite element calculations, it appears that R curves obtained from V_M or V_{LL}' are equivalent to R curves obtained from the commonly used V_{LL} displacements. This suggests that the ASTM E-399 specimen, using V_M, can be used for both K_{Ic} and J-R curve determinations.

ACKNOWLEDGEMENT

This research was sponsored by the U. S. Nuclear Regulatory Commission, Office of Nuclear Regulatory Research (M. Vagins, Project Manager), under contract to ENSA, Inc. and subcontract to Materials Engineering Associates, Inc.

REFERENCES

1. J. C. Newman, Jr., "Stress Intensity Factors and Crack-Opening Displacements for Round Compact Specimens", Technical Memorandum TM-80174, National Aeronautics and Space Administration, 1979.

2. T. W. Orange, "Crack Displacements for J_I Testing with Compact Specimens", Int. Journal of Fracture, 19, (1982), pp. R59-R61.

TABLE 1

	Jic (kJ/m^2)	% CHANGE	T average	% CHANGE
61W—42				
COMBINATION	68.8	--	45	--
FRONT FACE	79.5	16	46	2
ORANGE L-P	82.3	20	48	7
63W—41				
COMBINATION	77.9	--	31	--
FRONT FACE	81.8	5	32	3
ORANGE L-P	85.5	10	33	6
61W—45				
COMBINATION	90.0	--	50	--
FRONT FACE	90.8	1	52	4
ORANGE L-P	97.1	8	52	4
LL PRIME	90.5	1	51	2

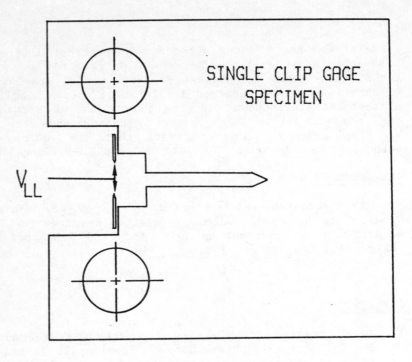

SINGLE CLIP GAGE
SPECIMEN

V_{LL}

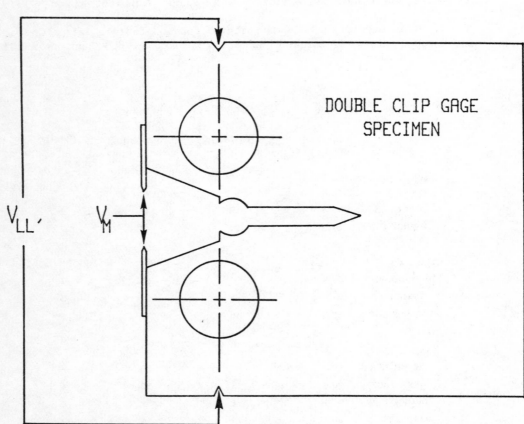

DOUBLE CLIP GAGE
SPECIMEN

V_{LL}'　　V_M

Fig. 1   Modified versions of ASTM E-399 compact specimens. In both
designs the notch area has been enlarged for measurement of
load line displacement ($_{LL}$).   In addition, the pin hole spacing
has been increased in the top specimen.

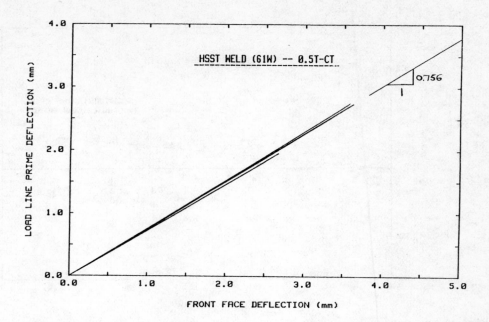

Fig. 2   Comparison of V_{LL}' and V_M for several specimens tested over a range in a/W (0.5 - 0.75) and temperature (75 - 200°C).

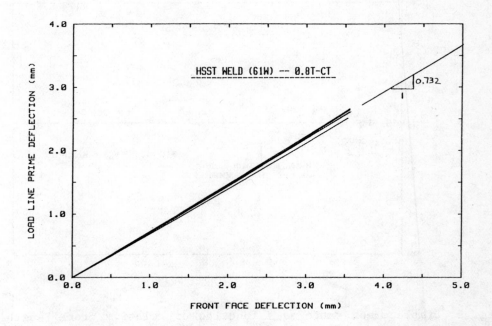

Fig. 3   Comparison of V_{LL}' and V_M for several specimens tested over a range in a/W (0.5 - 0.75) and temperature (75 - 200°C).

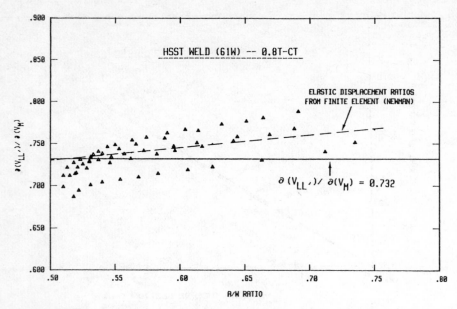

Fig. 4 Displacement ratio as a function of relative crack length (a/W). An average value of this ratio (solid line) is compared with that computed from a finite element analysis [1].

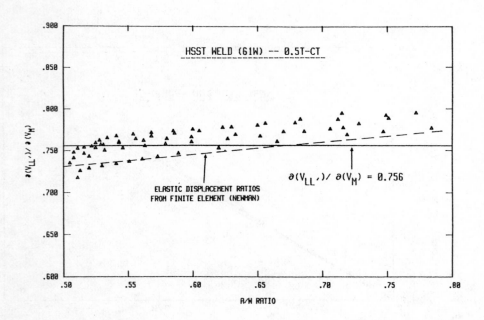

Fig. 5 Displacement ratio as a function of relative crack length (a/W). An average value of this ratio (solid line) is compared with that computed from a finite element analysis [1].

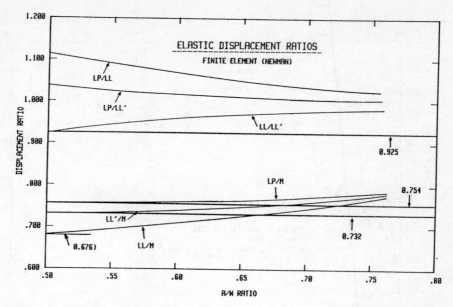

Fig. 6   Elastic displacement ratios for the standard CT specimen computed by Newman [1]. The values at a/W = 0.5 were used to compute R curves from displacements measured at different locations on the specimen.

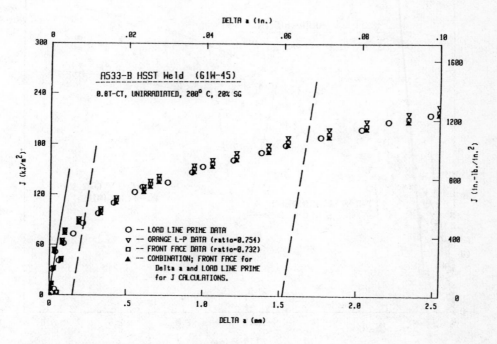

Fig. 7   J-R curves computed from V_M or V_{LL}' displacements using the displacement ratios of Newman [1] at a/W = 0.5 (Fig. 6).

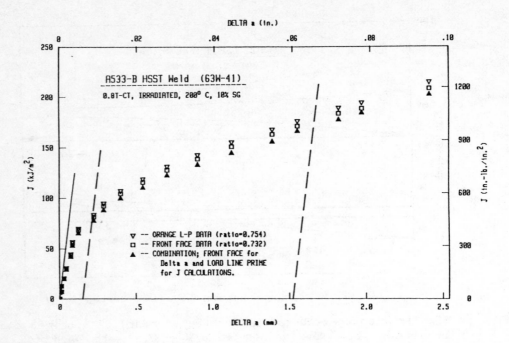

Fig. 8 J-R curves computed from V_M or V_{LL}' displacements using the displacement ratios of Newman [1] at a/W = 0.5 (Fig. 6).

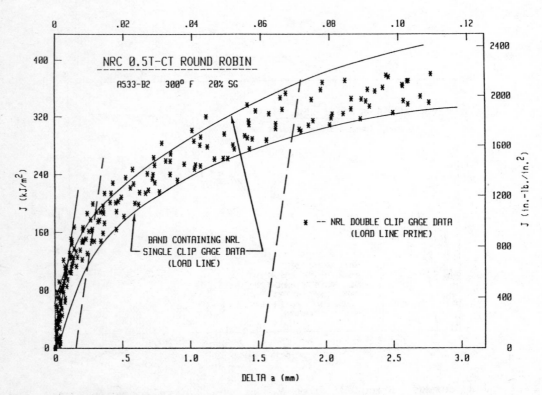

Fig. 9 Comparison of R curve data from compliance measurements determined by the double- and single clip gage techniques.

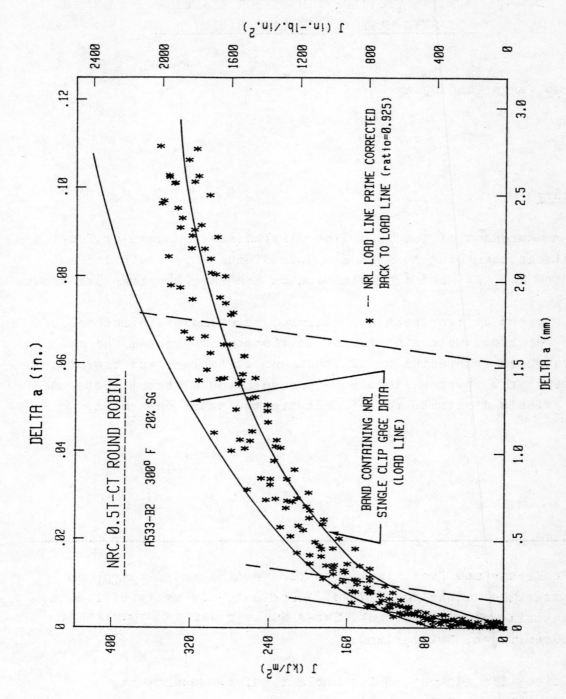

Fig. 10 Comparison of the single- and double clip gage data of Fig. 9 using a V_{LL}/V_{LL}' ratio of 0.95 (Fig. 6).

SPECIAL TESTING METHODS FOR ELASTIC-
PLASTIC MATERIAL CHARACTERIZATION

Thomas Varga[I] and Djing Han Njo.[II]

Summary

The measurement of the load line displacement on Compact Tension Specimens according to ASTM E 399 is discussed. Possibilities for the K_{Ic}, J_{Ic} and δ_c evaluation on the same specimen are shown.

Measurement by extrapolation compared to calculation methods of δ is reported. Determination of initiation in general and by ultrasonics in special on CT specimens at ambient and elevated temperatures is presented. Some pressure vessel burst tests and the effects of stress relief heat treatment are commented.

I   Professor and Head, Institute for Testing and Research in
    Materials Technology (TVFA), Technical University of Vienna,
    Austria; also Consultant, Swiss Nuclear Safety Department,
    Würenlingen, Switzerland.

II Scientific Adjunct, Swiss Nuclear Safety Department,
    Würenlingen, Switzerland.

Introduction

Experimental methods to measure elastoplastic fracture mechanics
characteristics of materials are currently under development in
many laboratories all over the world. In consequence, many
variations of test specimens, test methods, measuring devices
and the corresponding evaluation procedures are applied.

However, for fracture safety assessments of nuclear components,
a clear separation of research and development activities on one side
and sufficiently qualified materials testing and evaluation methods on
the
/other side is needed; otherwise "Continuing Uncertainty" on the applicability will result. Of course, a periodical updating of the applied
practices is absolutely necessary. For this latter purpose
specialists meetings could offer a platform, if a representative
participation of national authorities, utilities and system ven-
dors can be secured.

Some illustrations to the abovementioned will be presented in the
following:

Linear Elastic and Elastoplastic Materials Characteristics
Measurement by Using the same Specimen

Compact tension, CT specimens have been used according to
ASTM E 399 and other national standards and recommended practices
for many years. For J_{Ic} testing, the shape of the CT specimen has
been changed according to ASTM E 813-81, (Fig. 1), to accomodate
load line opening COD measurement. Furthermore, the range of a/w
values has been changed from 0.45/0.55 to 0.50/0.75.

Even if one neglects cantilever beam stiffnes and angular changes
during testing of ductile materials, the large difference allowed
in a/w, i.e. 0.45 to 0.75 may result in different ductile to
brittle relations in the transitional region. Therefore, different,

i.e. incompatible data may be generated on the same material under otherwise identical testing conditions.

From the testing point of view, the use of the same specimen geometry for the measurement of all fracture mechanics material characteristics eliminates several problems. There is no pre-evaluation needed to anticipate test results, i.e. how many specimens will behave in linear elastic respectively in elasto-plastic manner. The same group and gages may be used and a cross-check of the measured values is possible.

Load line opening may be measured by internal clip gages also in the slot of the ASTM E 399 CT specimen beginning with the 1 T size. The 3 mm-wide slot permits the insertion of two small gages. If the force of the spring arms is sufficient, no grooves are needed for keeping the gage in position. The 2 T size even accomodates a multiple point clip-in gage "Drawer" like the one in Fig. 2.

Force-displacement diagrams recorded from the output of a six-point clip-in gage are shown in Fig. 3.

For CT specimens up to 10 T single point clip-in gages have been used and results reported [1].

Whereas three-point bend specimens already have been used for simultaneous measurement of J_{Ic} and δ_c (according to BS 5762 or of K_{Ic}, if appropriate), the use of CT specimen as a universal or multipurpose specimen (type) seems to be unknown (even has not been reported yet). A code of practice has been published in Switzerland [2].

Measurement of Representative Fracture Mechanics Material Characteristics

J_{Ic} and δ_i represent safe lower limit values of the elasto-plastic fracture mechanics materials characteristics at a given termperature, specimen size and geometry.

Minimum specimen sizes given in ASTM E 813-81 by

$$B \gtrsim \alpha \frac{J_{Ic}}{\sigma_y} , \qquad \alpha = 25$$

seem also to be material and microstructure dependent. It has been reported several times in the open literature that α-values up to 100 were necessary to obtain size-independent (i.e. valid for higher thickness) J_{Ic} values.

Mildner, Varga and Schneeweiss [3]
found α varying with microstructure. From a Mn Si alloyed, Al Mn QT weldable steel plate several pieces were taken, of which some were normalized, some left in the delivered condition or overheated again. The necessary α-values to obtain the full section (50.5mm thickness) J_{Ic} are given in Fig. 4 in function of the testing temperature.

There is, however, another aspect to be considered if the lowest loading and consequently the testing temperature fall into the brittle-ductile transition range. The author observed repeatedly that small specimens exhibit much more ductility than thick (large) one do under otherwise identical testing conditions. An impressive example was presented recently by Marandet et al [4], (see Fig. 5).

Because of this effect of size on constraint and hence on the brittle-ductile transition the author was not able to agree with a draft ASTM Standard of Committee E 24.04.03, which tended to use small specimens to obtain lower bound fracture mechanics material characteristics.

As it was shown by Carlsson several years ago
the CT geometry with ratio a/w = 0.5 yields the smallest plastic
zone size for a given K_I compared to all other specimen geometries
commonly in use. In consequence the CT specimen exerts the
highest constraint for a given thickness and K_I. These findings
justify the choice of CT specimens for consecutive safety
assessments of nuclear components by the author back in the early
seventies.

Three point bend specimens broken in a ductile manner, however, failed
to predict partly brittle behavior of test models [5,6]. Therefore,
too optimistic, a value of critical (and tolerable) defect sizes
may result.

When using EPFM material characteristics originating from three-
point bend specimens in a limit test approach for a component it
should, therefore, be checked whether the material of the component
will have a ductile fracture behavior at the lowest material temperature of interest, (usually lowest loading temperature, i.e.
hydrotest or thermal shock case).

On the other hand CT specimen may yield unconservative high material
characteristics in the fully ductile condition of tough material.
The high degree of constraint which remains up to high specific
loads inhibits yielding. Values, measured under such conditions,
may be unconservative if applied to components which are subjected
to less constraint, notably also to less superinposed bending when
failing.

A MnNiV alloyed QT containment vessel steel showed much less
ductile deformation and strain hardening in surface fatigue
precracked wide plate test specimens than full thickness CT
specimens at the same temperature. All specimens were taken from
the same plate, a/w were similar [6].

Generally, if there is little bending and strain hardening no
normal stress (with large flaws: not net section stress) should
be predicted above the flow stress (still safer: above yield stress)
even in the fully ductile condition (see also: R6 procedure of the
CEGB).

Determining Initiation

The **corresponding** load/opening displacement values /which indicate
the beginning of the extension of a separation from a pre-existing
flow,represents the so-called "Initiation Point". The point of
initiation represents the lowest safe limit below which the onetime
loaded, partly or fully ductile material does not undergo any change
which impairs its integrity. With brittle material behavior no
similar statement of the initiation point is possible. Safety
expressed in load bearing and plastic deformation capacity increases with increasing difference in load and a difference in
COD at initiation and fracture (or plastic collapse).

Whether initiation has to be detected and taken into account at
separations on the microscopic scale or whether initiation becomes
realistic in the order of millimeters (difference three or more
orders of magnitude) is depending on the material (extent of
ductility, strain hardening, recristallisation temperature, creep
properties), geometry size and stress gradients, loading sequence,
temperature, environmental effects, accuracy and reliability
of the testing equipment. Because of these different conditions
which have to be taken into account, no rule for the fixing of a
generally valid initiation point can probably be defined.
To define the point of initiation according to ASTM E 813-81 heat
tinting may be used. If there is no sufficient material to conduct
heat tinting series for every testing temperature, single probe
techniques may be used. According to the author's experience, partial
unloading compliance technique may be the solution in some cases
but not in other cases, e.g. with certain steels and specimen
gemometries where this method was not successful.

Besides the potential drop technique an ultrasonic method was developed in the TVFA. A probe is fixed perpendicular to the slot in such a way that partial reflection from the pre-crack takes place. When the pre-crack begins to open the echo signal starts to decrease. As soon as the pre-crack starts to extent (i.e. formation of a new sharp crack) a remarkable increase of the echo signal (reflection) from the pre-crack front is observed (see also Fig. 6 for the single and Fig. 7 for multiple loading steps; in Fig. 8 a heat tintet fracture surface of a specimen according to Fig. 1 is shown).

Remarks on Stress Relief Heat Treatment

Cylindrical vessels of 2 m diameter with a wall thickness from about 30 to 60 mm have been manufactured of Mn Si alloyed, Al-killed QT steels. From three pairs of vessels three vessels were subjëcted to burst tests in the welded, three vessels in the stress relieved condition [6].

Whereas there was very little difference in burst pressure, the COD measured in about the wall thickness distance /ahead of the 0.3 mm through saw cut notch the stress relieved vessels showed the relatively greater local plastic deformation capacity of the stress relieved vessels (see Fig. 9). Burst stresses were best approximated by the simple

K_{Ij} approach and the Folias correction. Also the COD design curve (with a correction factor of two, taking into account the built-in safety margin in the design-curve), gave good approximations.

Literature

1. Sahgal, S., Varga, T., Junker, M. and Shakeshaft,
   M: Pressure Vessel Fracture Safety Investigations
   using as Criterion the Onset of Crack Propagation.
   Conference on Structural Mechanics in Reactor Technology. Vol. G. I/4, San Francisco, 1977.

2. Prüfvorschrift für das COD (Crack Opening Displacement,
   Rißöffnungsverschiebungs-) Verfahren, Eidgenössisches
   Amt für Energiewirtschaft ASK-AN-220, Rev. 1.

3. Mildner, H., Varga, T., and Schneeweiss, G.,
   Influence of Microstructure and other Parameter on
   Fracture-Mechanics Values of a Structural Steel.
   Proceedings of the 4th European Conference on
   Fracture, September 22-24, 1982, Leoben, Austria.

4. Marandet B., Phelippeau G., De Roo P. and Rousselier G.,
   Effect of Specimen Dimensions on J_{Ic} at Initiation of Crack
   Growth by Ductile Tearing.
   15th Symposium on Fracture Mechanics, ASTM, 7-9 Juillet,
   1982, College Park, Maryland, U.S.A.

5. Personal communication of Mr. I. Rak.

6. Mildner, H., Beitrag zur Ermittlung von J - Integral und
   Rißaufweitung als Werkstoffkennwerte und zu deren Anwendung
   bei der Sicherheitsabschätzung von Druckbehältern.
   Dissertation TU - Wien, 1981.

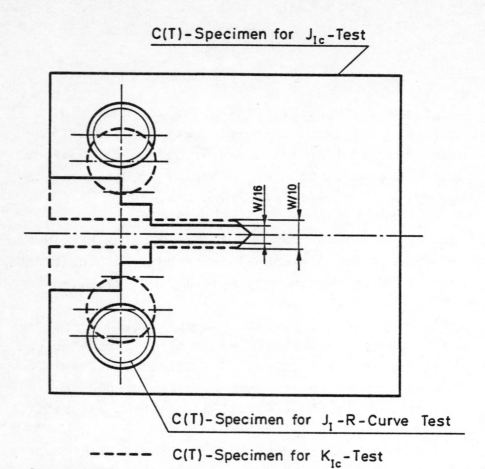

C(T)-Specimen for J_{Ic}-Test

W/16   W/10

C(T)-Specimen for J_I-R-Curve Test

— — — —   C(T)-Specimen for K_{Ic}-Test

FIG. 1.   Compact Tension Specimen, changed to accomodate load line
clip gages, fixed on razor blades.

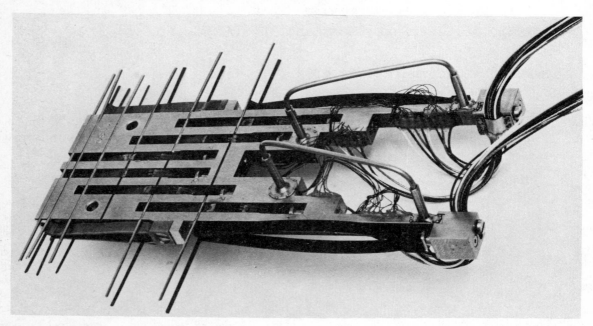

FIG. 2.   Multiple clip-in gages, applicable to 2T Compact Tension
Specimens according to ASTM E 399.

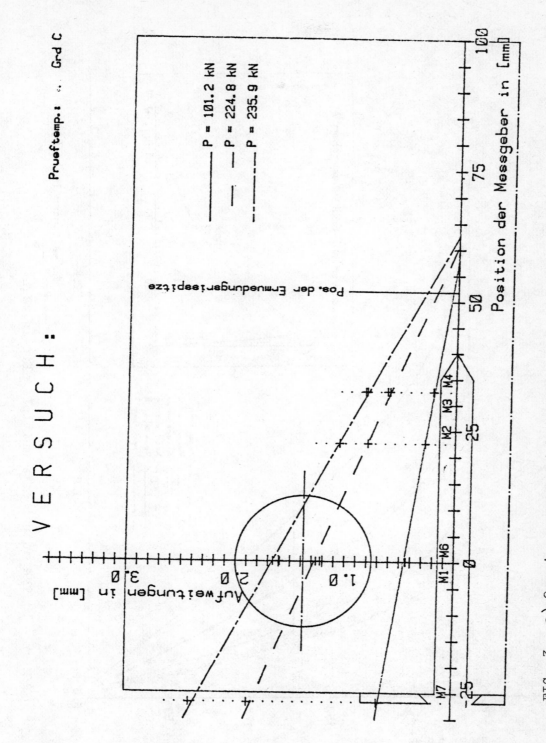

FIG. 3. a) Openings measured by the multiple clip-in gage and their extrapolation to the original front of the fatigue pre-crack.

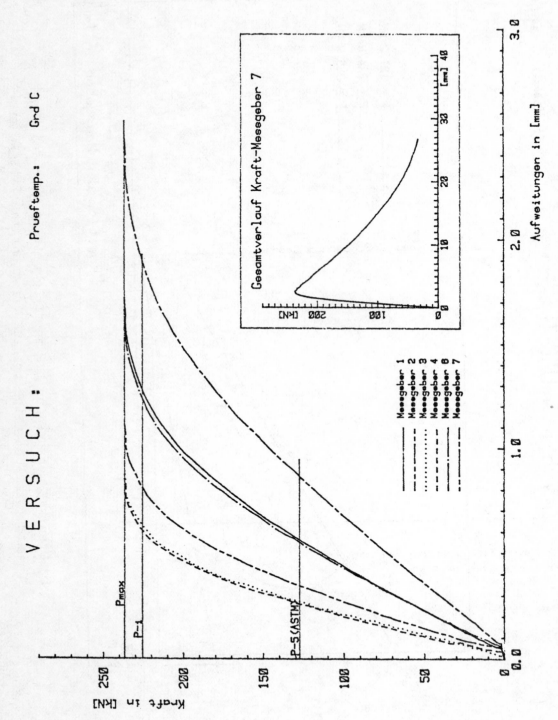

FIG. 3. b) The output of a multiple clip-in gage.

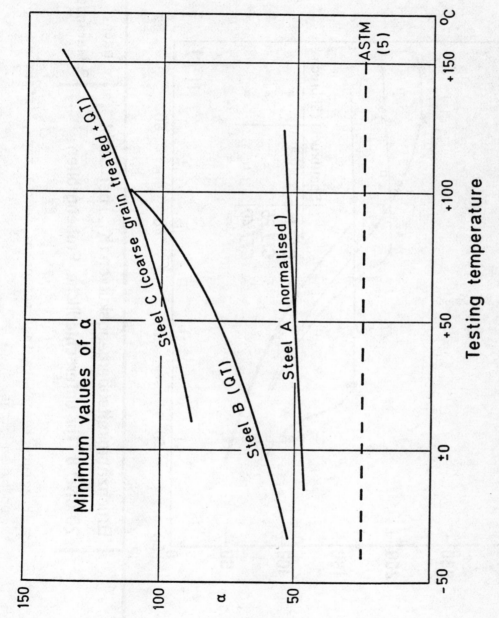

FIG. 4. α -values, necessary to obtain the full plate thickness J_{Ic} - values, normalized (A), quenched and tempered (B), coarse grain heated and quenched and tempered (C) heat treatment.

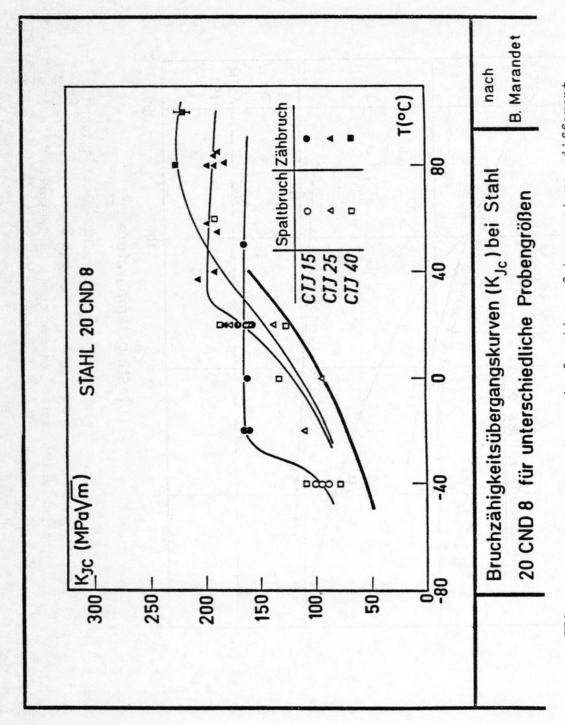

FIG. 5. Fracture toughness in function of temperature, different specimen sizes.

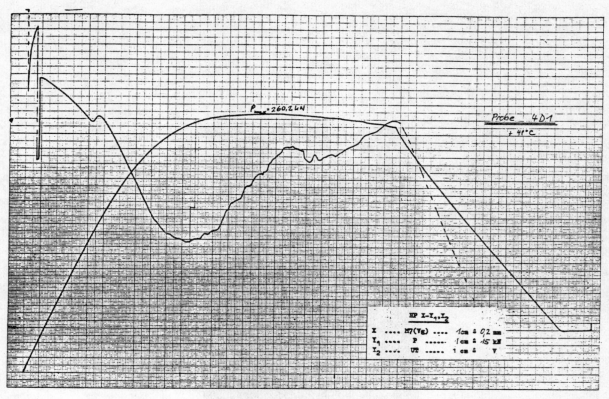

FIG. 6. Load and ultrasonic signal during a single loading sequence.

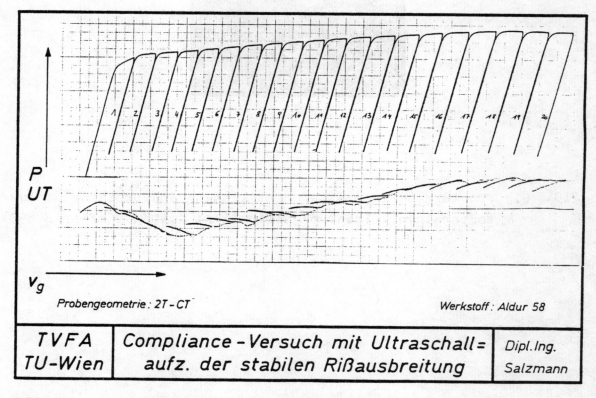

FIG. 7. Load and ultrasonic signal during repeated loading-
unloading steps.

FIG. 8. Fracture surface after heat tinting and breaking at low
temperature. Linear magnification: 3,4:1 (CT after E 399).

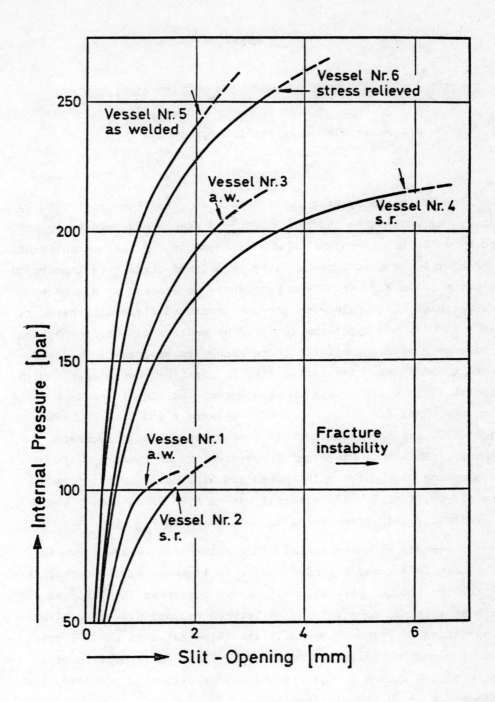

FIG. 9. The stress relieved vessels 2, 4 and 6 show significantly higher COD compared to the as welded vessels 1, 3, 5.

HINGE POINT MEASUREMENT OF CRACK TIP OPENING DISPLACEMENT

J. D. Landes
Materials Engineering Department

SUMMARY

The crack tip opening displacement, δ_t, has been analytically related to J by expressions such as $\delta_t = \dfrac{dn}{\sigma_o} J$ when δ_t is defined by and measured by a finite element calculation such as in Fig. 1c. Experimental measurement of δ_t is most often made by an assumed rigid rotation around a hinge point as in Fig. 1b. Test procedures to measure δ_t assume a constant ratio of hinge point distance to uncracked ligament length, r; for ASTM r is 0.4. An apparatus designed by Andrews of G.E. for an EPRI sponsored elastic-plastic fracture mechanics program can be used to measure r and δ_t experimentally, Fig. 2. Specimens were tested with this apparatus; five gages were used along the load line to measure displacements. Any two could be used to measure r and δ_t. The load line gage (δ_{LL}) and back face gage (δ_b) provided the most accurate measurement capability. Hinge point measurements of r and δ_t were made for CT specimens with varying thickness and ligament in sizes from 1T-CT to 12T-CT, Fig. 3. The material was an A508 Cl 2a steel; the R curve results for this steel based on J is shown in Fig. 4.

A summary of experimental hinge point results are shown in Fig. 5. There is a large dependance of r on ligament size and thickness. However for traditional proportionally sized specimens (W = 2B) the variation in r is not so great and is between 0.3 and 0.4. The effects of the variation of r shown in Fig. 5 are illustrated in the following results. Figure 6 compares a δ_t defined by the ASTM assumption of a constant r of 0.4 with a δ_t measured by the experimental procedure for CT specimens of proportional dimensions.

The two correlate within 15%. The R curve, δ_t versus crack extension, Δa_p, for the two definitions of δ_t are shown in Fig. 7 for 1T-CT specimens. The R curves are similar for the a/W = 0.5 (proportional specimen); however, for a/W = 0.75 the two definitions of

δ_t give different R curves. Also R curves for 10T, 4T and 1T-CT specimens as characterized by the experimental δ_t are shown to be different in Fig. 8. These results show that the variation in r due to specimen size variation results in a δ_t which does not uniformly correlate R curve behavior. The conditions needed to make the 1T and 10T specimens have identical R curves as characterized by δ_t the illustrated in Fig. 9.

Some analytical comparisons between a hinge point measurement of δ_t and J are taken from the Shih-Hutchinson estimation procedure using the results from the EPRI-G.E. Handbook developed in EPRI project RP-1237-1. Handbook values of J were calculated as well as δ_t inferred from a handbook calculation of δ_{LL} and assuming r = 0.4 for proportional specimens (a/W = 0.5) ranging from 1/2T to 10T. The results compared in Fig. 10 show that the hinge point δ_t versus J plots are specimen size dependent. The same comparison is made in Fig. 11 for specimens with vastly different in-plane dimensions but identical uncracked ligament sizes. The result is that J and δ_t are identical. These calculations are confirmed by experimental results in Fig. 12.

The conclusion from this work is that a hinge point measurement of δ_t does not correlate with J independently of specimen size as the finite element definition of δ_t does. When J correlates R curve behavior δ_t should not. The hinge point δ_t and finite element δ_t are defined differently and therefore should not be interchanged indiscriminately.

This work was partially supported by EPRI under Project 1238-2.

REFERENCES

1. Merkle, J.G., "An Approximate Method of Elastic-Plastic Fracture Analysis for Nozzel Corner Cracks" Elastic-Plastic Fracture, ASTM STP 668, 1979, pp 674-702.

2. Fearnehough, G.D., Lees, G.M., Lowes, J.M., and Weiner, R.T., "The Role of Stable Ductile Crack Growth in Failure of Structures", Applied Mechanics Group Conference, Dec., 1970.

3. Dawes, M.G., "The COD Design Curve Approach", Book, Advances in Elastic Plastic Fracture Mechanics, Applied Science Publishers LTD, 1979.

4. Hutchinson, J.W. and Paris, P.C., "Stability Analysis of J-Controlled Crack Growth", Elastic-Plastic Fracture, ASTM STP 668, 1979, pp 37-64.

5. Tanaka, K., and Harrison, J.P., "An R-Curve Approach to COD and J
    for an Austenitic Steel", Int. J. Pres Ves. and Piping, 1978, 6, pp
    177-202.

6. Turner, C.E., "Stable Crack Growth and Resistance Curves", Book,
    Developments in Fracture Mechanics, Applied Science Publishers,
    LTD., 1979 pp 107-144.

7. Garwood, S.J., Turner, C.E., "Slow-Stable Crack Growth in Structural
    Steel", Int. J. of Fracture, 1978, 14, pp 195-198.

8. Methods of Crack Opening Displacement (COD) Testing, British
    Standard BS5762, British Standards Institute, London 1972.

9. DeKoning, A.U., "A Contribution to the Analysis of Slow Crack
    Growth" The Netherlands National Aerospace Lab Report. NLR MP
    75035U, 1975.

10. Kanninen, M.F., and others, "Elastic-Plastic Fracture Mechanics for
    Two-Dimensional Stable Crack Growth and Instability Problems",
    Elastic-Plastic Fracture, ASTM STP 668, ASTM 1979, pp 121-150.

11. Draft Test Method for Crack Tip Opening Displacement (CTOD) Testing,
    Dec. 1, 1981 Draft, ASTM E24.08.05.

12. Shih, C.F., "Relationships between the J-Integral and the Crack
    Opening Displacement for Stationary and Extending Cracks, J. Mech.
    Phys. Solids. Vol 29., No. 4, pp 305-326, 1981.

13. Sorensen, E.P., "A Numerical Investigation of Plane Strain Stable
    Crack Growth Under Small Scale Yielding Conditions," Elastic Plastic
    Fracture, ASTM STP 668, 1979, pp 151-174.

14. Shih, C.F., German, M.D., and Kumar, V., "An Engineering Approach
    for Examining Crack Growth and Stability in Flawed Structures," G.E.
    Report No. 80CRD205. September 1980.

15. McCabe, D.E., Landes, J.D., Ernst, H.A., "An Evaluation of the J_R-
    Curve Method for Fracture Toughness Characterization", 2nd
    International Symposium on Elastic-Plastic Fracture Mechanics,
    Philadelphia, PA., October 6-9, 1981. To be published.

16. Andrews, W.R., and Shih, C.F., "Thickness and Side Groove Effects on
    J and δ-Resistance Curves for A533B Steel at 93°C," Elastic-Plastic
    Fracture, ASTM STP 668, 1979, pp. 426-450.

17. "Standard Test for J_{Ic}, A Measure of Fracture Toughness", ASTM Book
    of Standards, Part 10, 1981.

18. Kumar V., German, M.D., Shih, C.F., "An Engineering Approach for
    Elastic-Plastic Fracture Analysis", Handbook, NP-1931, July, 1981.

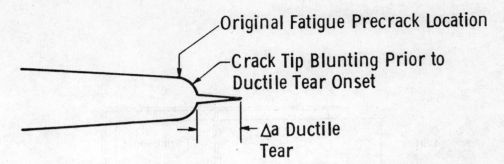

A) Schematic of Crack Shape at Midthickness After Stable Crack Growth.

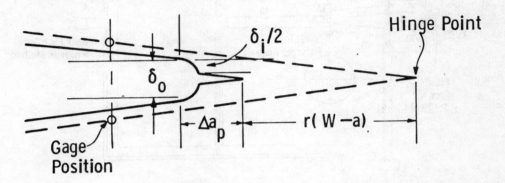

B) Inferred Crack Opening Displacement From Hinge Rotation Assumption.

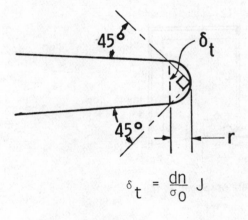

$$\delta_t = \frac{dn}{\sigma_0} J$$

C) Crack Tip Opening Displacement Described by 45° Projection

Fig. 1 – Crack opening displacement models under ductile tear conditions.

Dwg. 4253B76

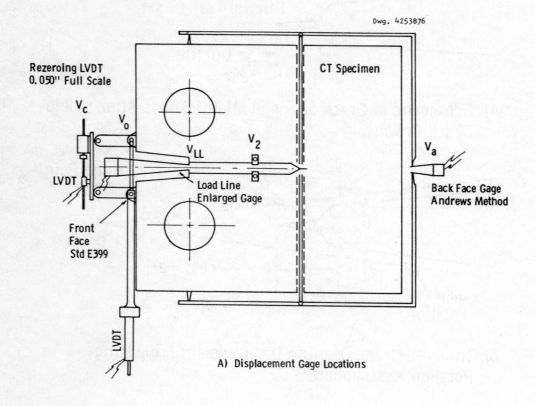

Rezeroing LVDT
0.050" Full Scale

V_c

V_o

V_{LL}

V_2

V_a

LVDT

Load Line
Enlarged Gage

CT Specimen

Back Face Gage
Andrews Method

Front
Face
Std E399

LVDT

A) Displacement Gage Locations

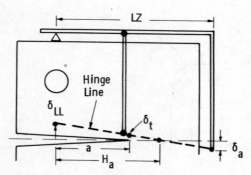

LZ

Hinge
Line

δ_{LL}

δ_t

δ_a

a

H_a

B) δ Quantities Sketched Here Are Shown as 1/2
the Measured Values.

Fig. 2 — Instrumentation of specimens with five measurements
points and schematic of imaginary hinge line

Figure 3

Table 1
Elastic-Plastic Methodology To Establish δ_t - Δa_p R-Curves

Thickness Inches Gross/net	Size	Ligament Size - Inches				
		1/2	1	2	4	10
0.5/0.4	1T	W = 2	W = 2			
1.0/0.8	1T	W = 2	W = 2			
	4T		W = 8	W = 8	W = 8	W = 20
2.0/1.6	2T	W = 4		W = 4		
	4T		W = 8	W = 8		
4.0/3.2	4T		W = 8		W = 8	
10.0/8.0	10T					W = 20

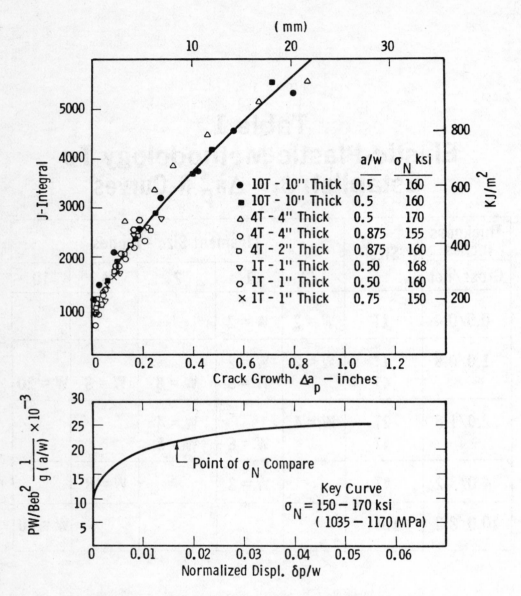

Fig. 4 · J$_R$-curves for material in the 150-170 ksi nominal strength category.

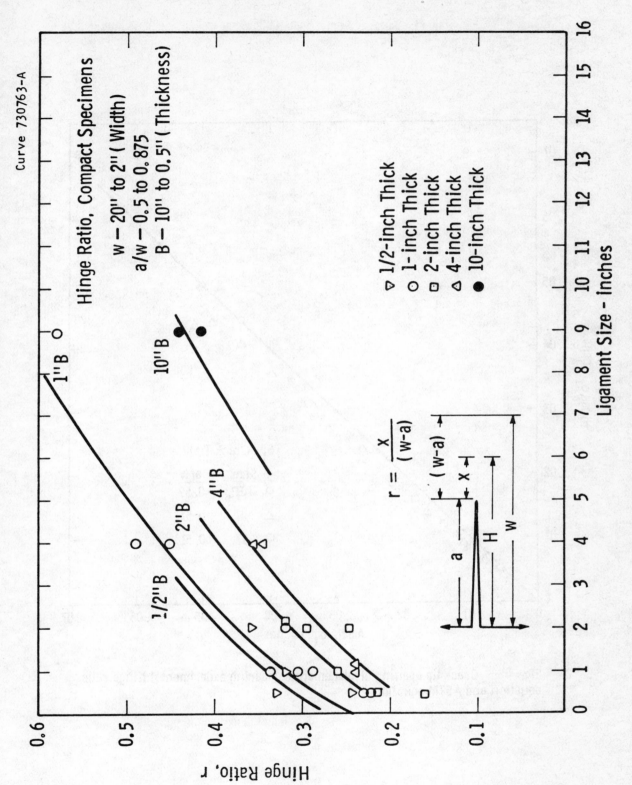

Fig. 5 – Hinge ratio versus thickness, B, and ligament size.

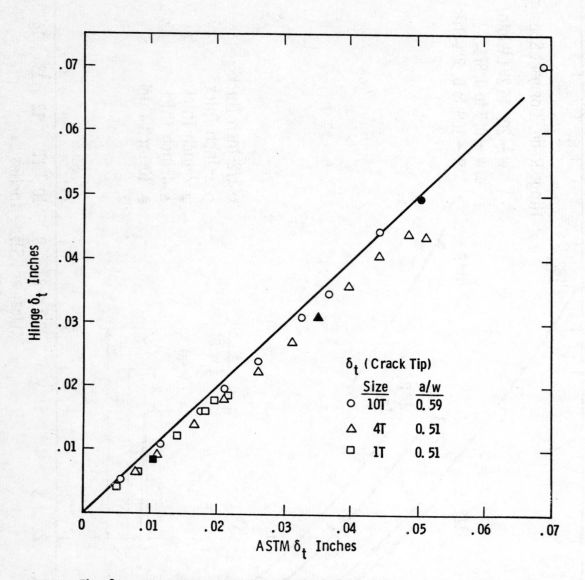

Fig. 6 — Crack tip opening displacement comparing experimental hinge ratio equation and ASTM equation

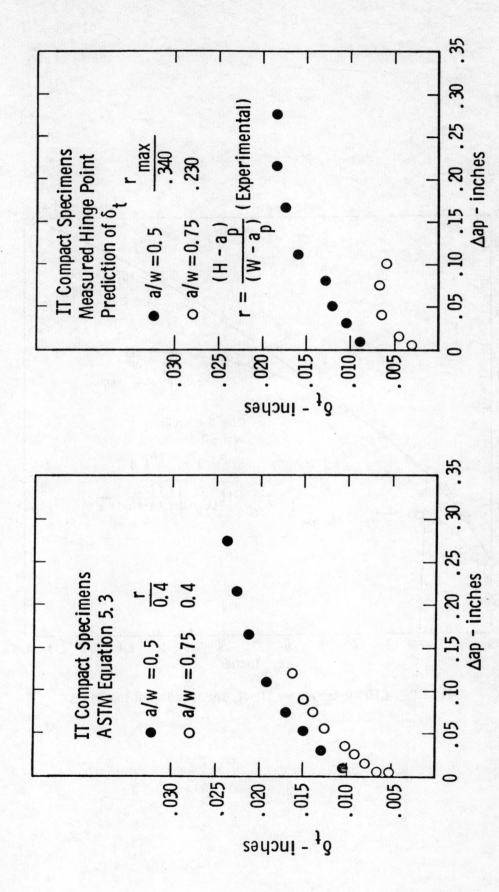

Fig. 7 — Prediction of δ using ASTM equation (left) and experimental hinge (right). IT size compact specimens

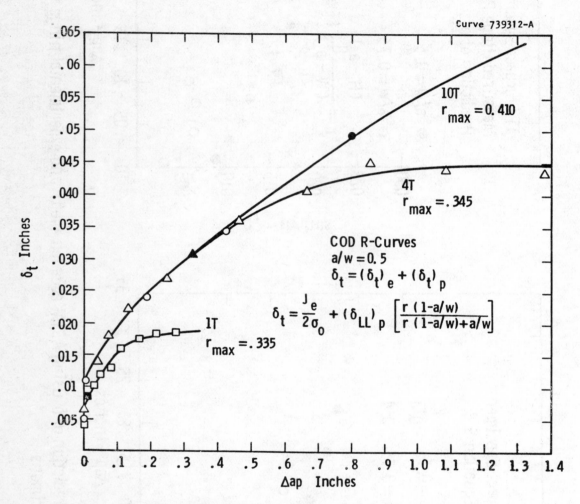

Fig. 8 — CTOD R-curves for 1T, 4T, and 10T compact specimens

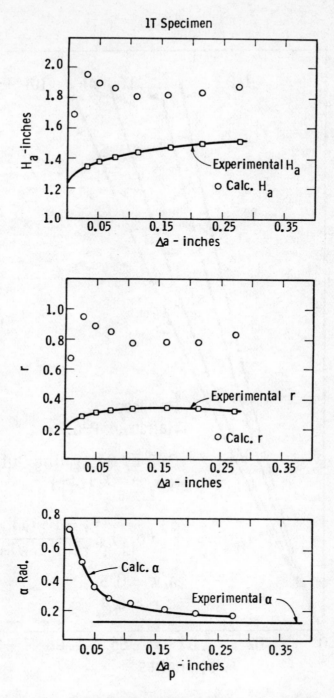

Fig. 9 — Comparison of hinge distance, H_a, ratio, r, and COA, α. Calculated values needed to match 10T $\delta_t - \Delta a_p$ R-curve compared to actual experimental values for a 1T compact specimen

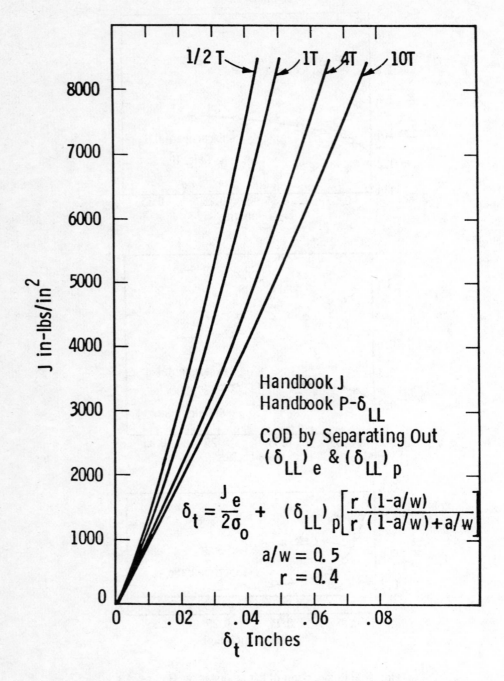

Fig. 10 Handbook predicted J and CTOD. Separate J elastic and plastic hinge contributions

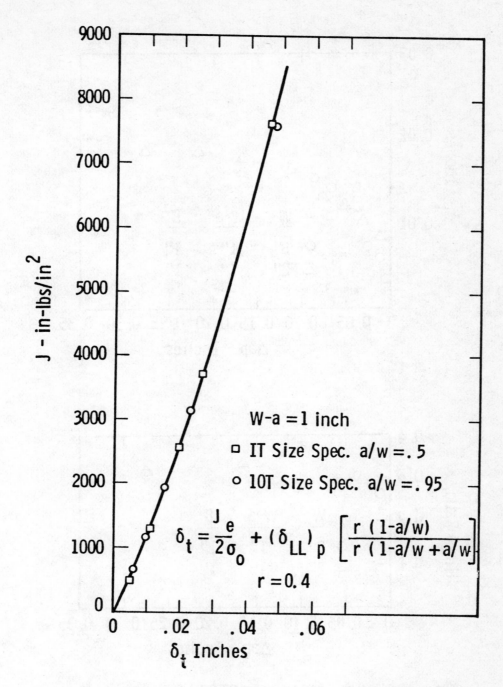

Fig. 11 – Handbook predicted J and CTOD.
Separate J elastic and plastic hinge contributions.
Fixed ligament size

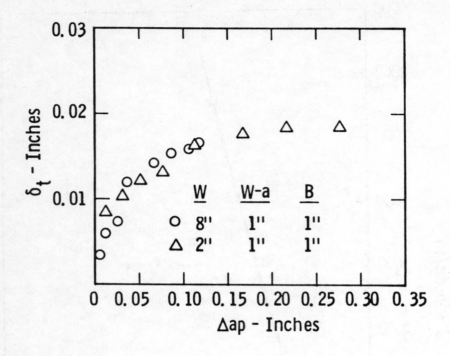

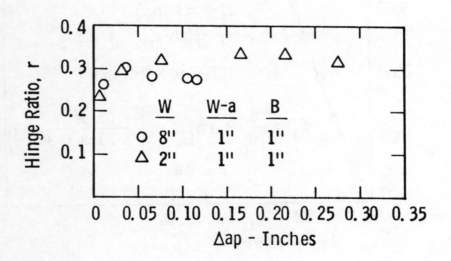

Fig. 12 · Trend in CTOD and hinge ratio, **r**, with crack growth. 4T and 1T plan view size but fixed ligament size

Drop Weight J-R Curve Testing
Using the Key Curve Method
Dr. James A. Joyce

Introduction

The Key Curve Method was developed by Ernst et.al.[1] as a method for
determination of a material J-R curves directly from fracture specimen load
displacement records without the use of additional crack length measurement
techniques. The method has been used previously by the present author on
static specimens of A533B steel [2] and on high rate loading tests on HY130 [3]
and A533B [4] steel alloys. In the static cases the key curve method was
shown to give J-R curves which agreed closely with J-R curves obtained by
more conventional unloading compliance methods. For the high rate tests the
resulting J-R curves showed elevated toughness as expected for those materials.
The key curve method demonstrated an ability in all cases to accurately predict
the extent of crack growth.

Other advantages of the key curve method are as follows:

(1) J values incorporating crack extension effects are evaluated
directly.

(2) Crack extensions are available directly from the load displacement
record without additional instrumentation.

(3) Additional corrections like those for specimen rotation are included
in the calculation of J and crack extension.

(4) A J-R curve point is evaluated for each point on the load displacement record allowing accurate measure of the J-R curve slope and
the T material tearing modulus quantity introduced by Paris et al.[5]

(5) J-R curves can be evaluated for high speed tests - tests done
at any rate at which an accurate load displacement record can
be recorded - which is possible today at load point velocities up to
100 in/sec.

For static tests the complexity of the key curve method and the necessity of additional specimens in general negates many positive features, but for high rate tests the key curve method shows great promise for J-R curve determination.

Objective

The objective of the present test program has been to develop a capability to evaluate J-R curves from fracture mechanics specimens at test rates in the range of 100 in/sec. Methods being utilized are the multi specimen - stop block approach utilized by [6]; a potential drop crack length measurement technique; and the key curve method. In this report only the key curve method will be discussed.

Experimental Description

The test machine utilized for these tests was a standard drop weight machine with an instrumented tup. A schematic of the specimen, transducers, and associated electronics is shown in Fig 1. The light probe displacement transducer is offset from the crack and load line by about .02L where L is the specimen length. Both displacement and load transducers had frequency responses in excess of 10KHz. The digital oscilliscope was connected via an instrumentation interface to a microcomputer for data transfer, storage, and analysis as described in the following section.

As shown in Fig 2 the light probe was in fact inserted inside of a stop block apparatus of variable height which straddled the specimen in the center region as shown in Fig 2 to obtain a quick and distinct stop of the load tup. To keep the stop block load at reasonable levels additional stop blocks were placed at the sides of the apparatus shown which were hit by the drop weight at approximately the same instant as the load tup impacted its stop block fixture. This procedure required some experience but it was found possible to keep the load tup stopping loads to about 100000 lbs which was within the tup calibrated capacity.

The material used in this program was a medium strength alloy steel with tensile properties shown in Fig 3 and the nominal chemistry shown in Fig 4. In these tests the large scale specimens were 1T bend bars in accordance with the ASTM E399 and E561T fracture test methods. The subsize specimens were half scale or $\frac{1}{2}$T specimens except that the notch tip radii were machined to <.003 inch radius rather than being fatigue cracks. The 1T specimens were tested with a drop weight velocity of 100 inches/sec while the $\frac{1}{2}$T specimens were tested at 50 inches/sec.

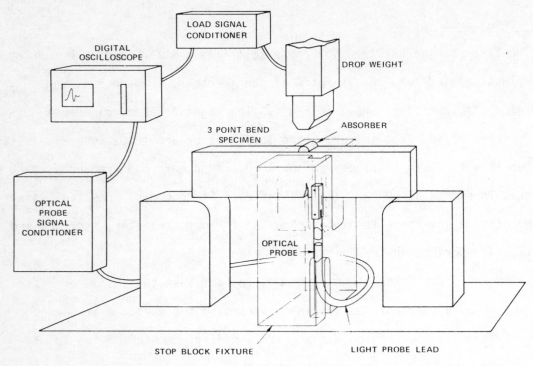

Figure 1. Schematic of drop weight specimen, transducers, and associated electronics.

EXPERIMENTAL SETUP

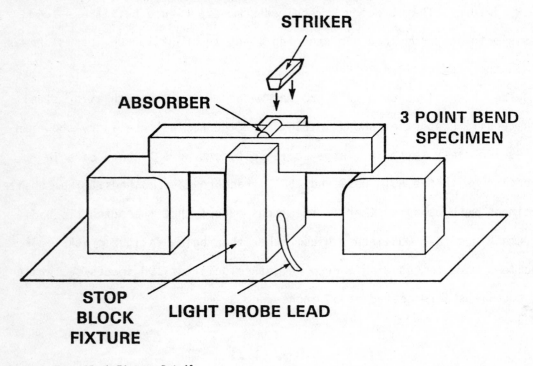

Figure 2.   Stop Block Fixture Detail.

yield strength MPa (psi)	Tensile strength MPa (psi)	Elongation in 2 in	RA %
941.0 (136,600)	994.0 (144,200)	19.5%	62%

Fig 3 Tensile mechanical properties of test steel

Center Code	Chemical Composition (Wt %)												
	C	Mn	P	Si	Ni	Cr	Mo	V	S	Cu	Al	Co	Ti
FTF	0.11	0.76	0.005	0.03	5.00	0.42	0.53	0.043	0.004	0.022	0.021	0.02	0.008

Fig. 4 Nominal Chemistry of test steel.

Analysis

In the Ernst, et al [1] analysis, dimensional analysis is used to show that for simple geometries in which the plasticity is confined to the uncracked ligament region, the load sisplacement relationship must have the form:

$$\frac{PW}{Bb^2} = F1 \left(\frac{\Delta}{W}, \frac{a}{W}, \frac{H}{W}, \frac{B}{W}, \text{ material properties}\right) \tag{1}$$

Where: P = applied load

Δ = total load line crack opening displacement

a = crack length

b = uncracked ligament

B = specimen thickness

W = a + b = specimen width

H = specimen height

Starting from the deformation plasticity theory formula for the path independent J integral that [7]

$$J = \frac{-1}{W} \int_0^\Delta \left(\frac{\partial P}{\partial(a/W)}\right)_\Delta d\Delta \tag{2}$$

an incremental formula for J can be obtained in terms of F1 giving

$$dJ = \left[\frac{2b}{W} F1 - \frac{b^2}{W^2} \frac{\partial F1}{\partial(a/W)}\right] d\Delta$$

$$+ \left[\int_0^\Delta - \frac{2}{W} F1 d\Delta \quad +\int_0^\Delta \frac{4b}{W^2} \frac{\partial F1}{\partial(a/W)} d\Delta\right.$$

$$\left.+\int_0^\Delta \frac{b^2}{W^3} \frac{\partial^2 F1}{\partial(a/W)^2} d\Delta\right] da \tag{3}$$

To evaluate J from an integral of Eqn 3 requires knowledge of the instantaneous crack length. This is obtained by Ernst in a differential form as:

$$da = \frac{\frac{b^2}{W^2} \frac{\partial F1}{\partial(\Delta/W)} \; d\Delta - dP}{\frac{2b}{W} F1 - \frac{b^2}{W^2} \frac{\partial F1}{\partial(a/W)}} \qquad (4)$$

Essentially if the load displacement record of the specimen under analysis is coincident with the key curve surface when evaluated using the initial flaw size no crack extension is predicted. If the load increment dP falls below that predicted by the key curve Eqn 4 estimates the corresponding increment of crack extension.

An alternating evaluation of Eqns (3) and (4) can thus be utilized to evaluate a J-R curve if the F1 function of Eqn (1) is available through analytical or experimental techniques.

For the special case that the key curve of Eqn 1 is independent of crack length Eqns (3) and (4) reduce to the following forms

$$dJ = \frac{2b}{W} F1 \; d\Delta \quad + \left[\int_0^\Delta \frac{-2}{W} F1 \; d\Delta \right] da$$

$$\qquad (5)$$

$$da = \frac{\frac{b^2}{W^2} \frac{\partial F1}{\partial(\Delta/W)} \; d\Delta - dP}{\frac{2b}{W} F1} \qquad (6)$$

These forms in fact are useful for the bend bar geometry as demonstrated in the following section.

Discussion

In the previous work [2] [3] [4] the key curve function of Eqn 1 was
evaluated for the particular steel by testing a series of subsized or blunt
notched compact specimens with various crack lenghts. A typical result for A533B
is shown in Fig 5. Using this surface and Eqns (3) and (4) J-R curves were
developed from load displacement records and compared with results obtained by
unloading compliance methods. The correspondence between key curve results
and those of other methods was found to be excellent and the method was demonstrated to accurately predict crack extensions even in regions beyond which J
was applicable.

Load - time and displacement- time records for a typical ½T specimen are
shown in Fig 6. Clearly inertial oscillations are a problem for this geometry
tested at a 50 in/sec loading rate. As shown in Fig 7 a smoothing operation was
utilized which produced an adequately smooth and accurate load time trace. The
resulting load displacement curve was now reasonably smooth but needed a further
correction to remove a machine compliance component corresponding to theoretical
elastic handbook [2] predictions and ascribing the difference to machine compliance,
then applying the correction to the complete load displacement record.

This process of smoothing and compliance correcting was applied in turn
to each of six ½T bend specimens with a/W values of .4,.5,.55,.6,.65,.7. The
resulting load displacement curves plotted in the key curve normalization are
shown in Fig 8. A band of curves results with oscillations present which are
apparently residuals of the smoothing process, material variability, and
system noise. No apparent crack length dependence was found in these six results,
in agreement with the predictions of Ernst et.al. [1] for the bend geometry. The
one curve marked by the arrow corresponding to the deepest crack length was considered
to be too irregular and was excluded from the subsequent analysis. The key curve
was thus taken to be the average of the remaining five curves of Fig 8 which is
shown in Fig 9.

The 1T speciments loaded at 100 in/sec also showed oscillatory behavior due

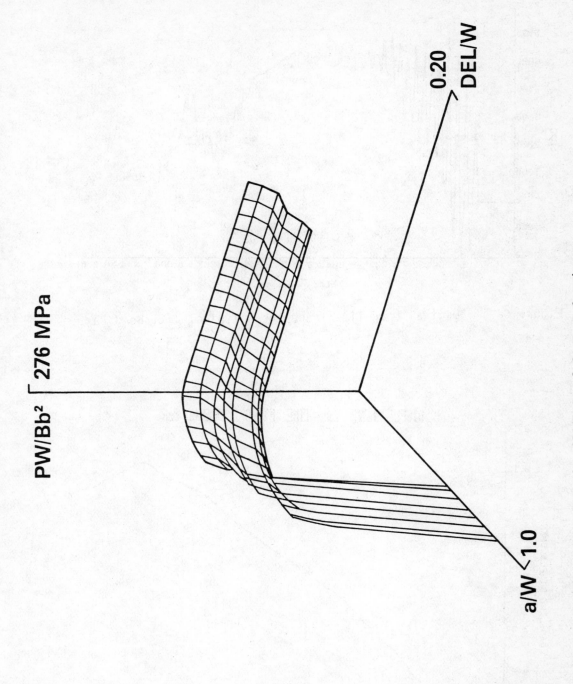

Figure 5. Key Curve File for A533B Steel Compact Specimens.

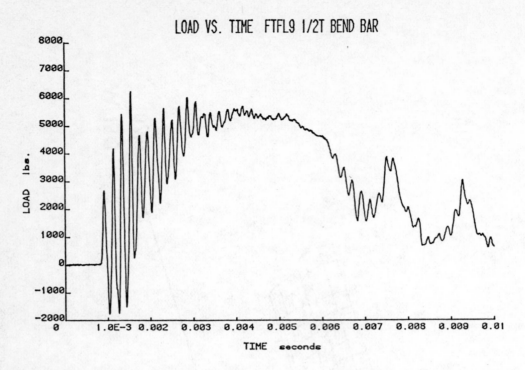

Figure 6a. Digital Load-Time Trace for a Typical 1/2T Bend Bar.

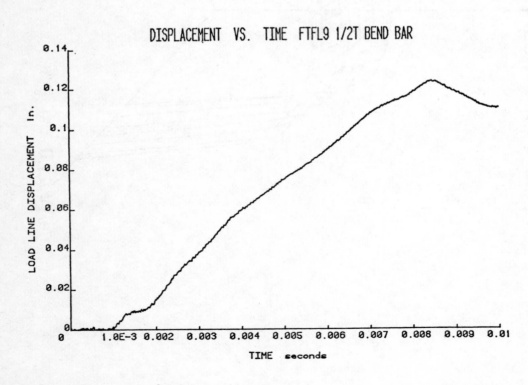

Figure 6b. Digital Displacement-Time Trace for a Typical 1/2T Bend Bar.

- 526 -

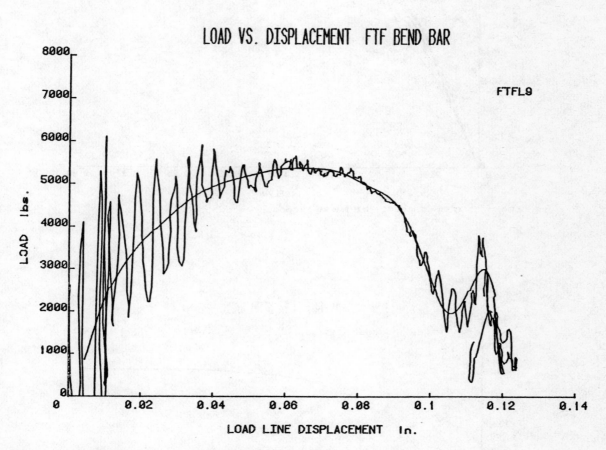

Figure 7.   Smoothed Load-Time Trace Compared with Original Unsmoothed Data.

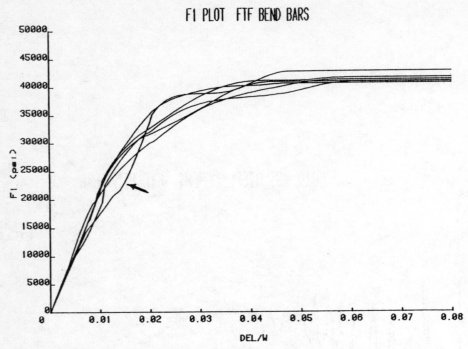

Figure 8.   Key Curve Plots of Six 1/2T Bend Bar Specimens.

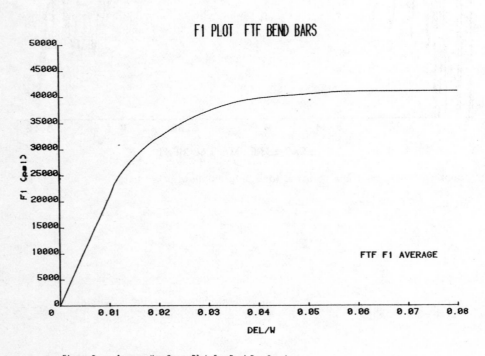

Figure 9.   Average Key Curve Plot for Bend Bar Specimens

apparently to inertial elastic vibrations, and since the frequency of the
transients was closer to the overall test frequency these oscillations were
somewhat harder to smooth. Nonetheless it was done here in a fashion similar to
that used on the subsize specimens and a typical result including a compliance
correction is shown in Fig 10. Plotting this result after key curve normalization
using the initial remaining ligament dimension gives the result shown in Fig 11.
Deviation between the two curves corresponds to a prediction of crack extension
which can be evaluated by using Eqn 6 for each increment and accumulating a total
crack extension. This process as well as that of calculating J by Eqn (5) were
done by microprocessor and the resulting J-R curve is shown in Fig 11. A
comparison is made on Fig 12 of the drop weight key curve result, and
three static results obtained on compact and bend specimens using an unloading
compliance method[8].

Clearly these results are in close agreement and imply that this material
is not rate sensitive within the range of rates used here. When this material
was chosen it was anticipated that a rate dependence would be demonstrated, but
the low J_{IC} and J-R curves obtained are consistent with the result of Fig 12
that for this lower toughness material no upper shelf toughness increase results
from the test rate increase applied here.

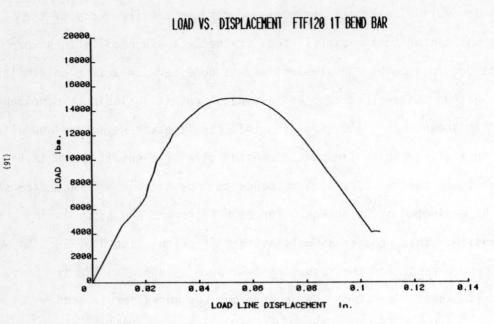

Figure 10.   Smoothed Load Displacement Curve for a 1T Bend Bar with a/W = 0.5.

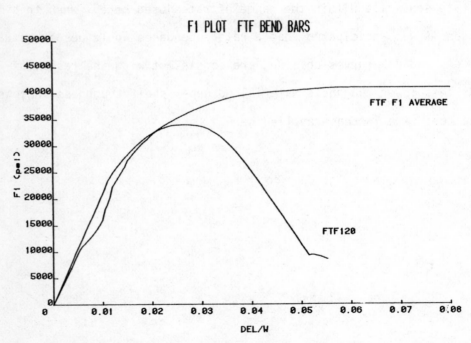

Figure 11.   Key Curve Plot Comparing the Average Bend Bar Key Curve with the 1T Bend Bar.

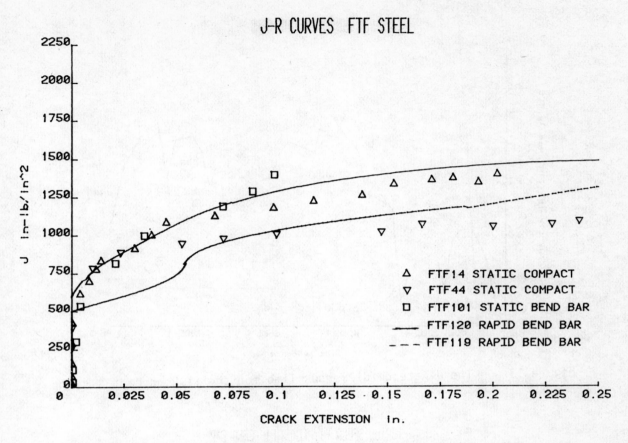

Figure 12.   J-R Curve for the 1T Bend Bar at 100 in./sec. Loading Rate

Crack Velocity Estimates

The final J-R curves, as shown for instance in Fig 12, consist of J-Δa pairs estimated at 2×10^{-6} second intervals. Using this fact both dJ/dt and da/dt can easily be estimated by a numerical differentiation routine giving the results shown in Fig 13 and 14. The maximum crack velocity achieved in these tests was then about 1500 in/second which is impressive but well below a true dynamic crack velocity. This is consistant with fractographic observations that all fracture surfaces in these tests remained fully ductile without even a hint of cleavage fracture.

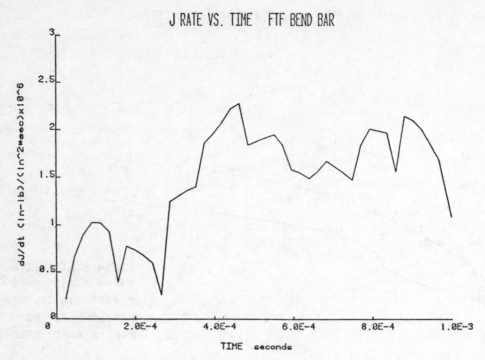

Figure 13.   The J-Rate dJ/dt Versus Time for the 1T Bend Bar.

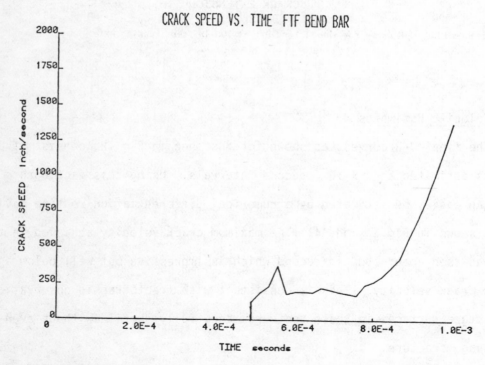

Figure 14   The Crack Velocity, da/dt, Versus Time for the 1T Bend Bar.

Drop Weight Tearing Instability

As described under the experimental method the center stop block was used to hold the optical displacement transducer but also to stop the drop weight after specified specimen displacements had been accomplished. This was done so that a multispecimen type J-R curve could be obtained for comparison with the key curve result discussed above. This did not turn out to be possible for this material. Various stop block heights were utilized but only two results were obtained, the first with the specimen load as shown in Fig 15, the second after the load had fallen to ≅ 25% of the maximum specimen load. Comparing the observed load displacement records of these specimens with the machine stiffness estimated in the previous section (giving values of approximately 500,000 lbs/in) showed that for this material a ductile tearing instability was occuring as described and utilized earlier in static test situations [9] [10]. Figure 16 shows a comparison of the test machine unloading relaxation versus the specimen load displacement curve which, following the lead of Ernst [11], predicts instability with this machine - specimen combination. Further work is needed to verify this observation but the tearing instability prediction of Fig 16 is consistent with the experimental observation that even slight extensions of the allowed bend displacement beyond that which produced the result of Fig 15 (with a measured crack extension of ≅ .005 in.) gave crack extensions of ≅ 0.5 inches.

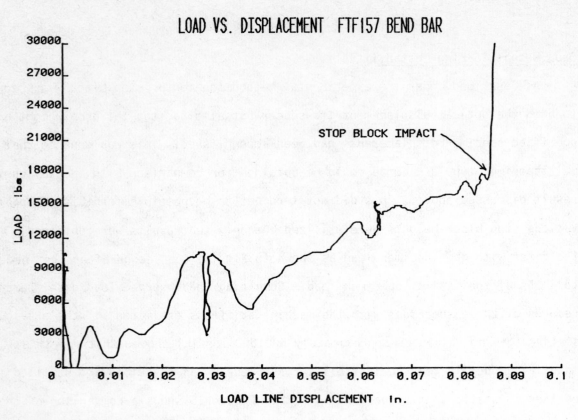

Figure 15   Specimen Load Displacement Curve Stopped by Stop Blocks Near Maximum Load.

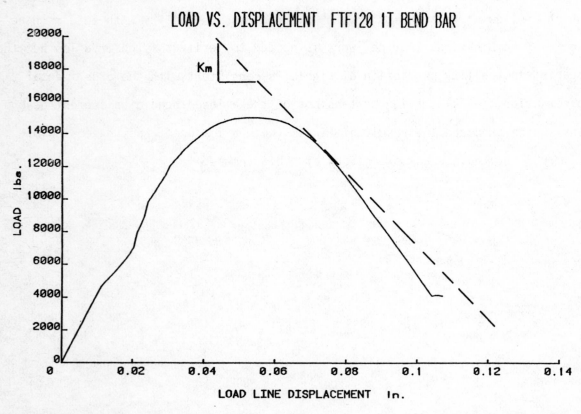

Figure 16.   Comparison of Test Fixture Relaxation Stiffness to the 1T Specimen Load Displacement Record.

Conclusions

Since all data taken has not as yet been analyzed it might be somewhat premature to draw too many conclusions but the following points are now clear:

1) The key curve technique can be applied to drop weight fracture tests though smoothing and compliance corrections necessitated by the present apparatus cause both difficulty and concern for overall accuaracy.

2) The key curve obtained for the three point bend geometry is apparently crack length independent. This result has been predicted but has not previously been demonstrated by static tests.

3) The particular medium strength alloy steel tested here demonstrates no measurable toughness increase thru the range from static tests (.010 in/min) to 100 inch/second drop weight tests.

4) All specimen tests were conducted on the ductile upper shelf with fully ductile fracture surfaces resulting. No cleavage fracture was observed on the fracture surfaces.

5) Crack velocities of up to 1500 in/sec were obtained in these tests. an impressive value but far below that usually observed in dynamic elastic fractures (on the order of 10,000 in/sec).

6) Ductile tearing instability occurs for the alloy tested in a drop weight machine and this made impossible attempts to obtain a multiple specimen J-R curve from the tests for comparison with the key curve result.

References

1. Ernst, H. et.al., "Analysis of Load - Displacement Relationships to Determine J-R Curve and Tearing Instability Material Properties," Fracture Mechanics, ASTM STP 677, pp 581-599, (1979).

2. Joyce, J.A., "Application of the Key Curve Method to Determining J_I-R Curves for A533B Steel," NUREG/CR - 1290, U.S. Nuclear Regulatory Commission, Washington, D.C. (Jan 1980).

3. Joyce, J.A., et.al., "Dynamic J_I-R Curve Testing of HY-130 Steel," DTNSRDC/SME-81/57, David Taylor Naval Ship Research and Development Center.

4. Joyce, J.A., "Static and Dynamic J-R Curve Testing of A533B Steel Using the Key Curve Analysis Technique," NUREG/CR-2274, U.S. Nuclear Regulatory Commission, Washington, D.C. (July 1981).

5. Paris, et.al., "The Theory of Instability of the Tearing Mode of Elastic Plastic Crack Growth," Elastic - Plastic Fracture, ASTM STP-668, pp. 5-36, (1979).

6. Logsdon, W.A., "Dynamic Fracture Toughness of ASME-SA508 Class 2a Base and Heat - Affected - Zone Material," Elastic Plastic Fracture, ASTM STP-668, pp. 515-536.

7. Rice, J.R., "A Path Independent Integral and the Approximate Analysis of Strain Concentration by Notches and Cracks," Journal of Applied Mechancis, pp. 379-386, (June 1968).

8. Joyce, J.A. et.al. "Computer Interactive J_{IC} Testing of Naval Alloys," Elastic - Plastic Fracture, ASTM STP 668, pp. 451-468, (1979).

9. Paris, P.C., et.al., "An Initial Experimental Investigation of the Tearing Instability Theory," Elastic Plastic Fracture, ASTM STP 668, pp. 251-265, (1979).

10. Joyce, J.A., et.al. "An Experimental Evaluation of Tearing Instability Using the Compact Specimen," Fracture Mechanics, ASTM STP 743, pp. 525-543, (1981).

LIST OF PARTICIPANTS
LISTE DES PARTICIPANTS

BELGIUM - BELGIQUE

DENYS, R.M., Rijksuniversiteit Gent, Labo Soete, St. Pietersnieuwstraat, 41,
B-9000 Gent

CANADA

SIMPSON, L.A., Atomic Energy of Canada Ltd., Whiteshell Nuclear Research
Establishment, Pinawa, Manitoba ROE 1LO

DENMARK - DANEMARK

DEBEL, C.P., Metallurgy Department, Risø National Laboratory,
DK-4000 Roskilde

FINLAND - FINLANDE

WALLIN, T., VTT/Metals Laboratory, SF-02150 Espoo 15

FRANCE

BALLADON, P., Creusot-Loire, Centre de Recherches d'Unieux, B.P. 34,
F-42701 Firminy Cedex

BETHMONT, M., EDF Renardières, RNE-EMA, Route de Sens, F-77250 Ecuelles

DE ROO, P., IRSID, 185 Ave Président Roosevelt, F-78105 St. Germain-en-Laye

HOUSSIN, B., Framatome Materials Department, Tour Fiat, Cedex 16,
F-92084 Paris La Défense

MARANDET, B., IRSID, 185 Rue Président Roosevelt,
F-78105 St. Germain-en-Laye

MIANNAY, D., C.E.A., Centre d'Etudes de Bruyères le Chatel, Service
Métallurgie, B.P. no. 561, F-92542 Montrouge Cedex

ROUSSELIER, G., EDF, Département Etude des Matériaux, Les Renard...,
F-77250 Moret/Loing

SOULAT, P., C.E.A., D. Tech - SRMA, Centre d'Etudes Nucléaires de Saclay,
    B.P. no. 2, F-91191 Gif-sur-Yvette

TIBI, H., C.E.A., B.P. no. 6, F-92260 Fontenay-aux-Roses

FEDERAL REPUBLIC OF GERMANY - REPUBLIQUE FEDERALE D'ALLEMAGNE

AZODI, D.J., c/o G.R.S., Glockengasse 2, D-5000 Köln

BERGER, C., Kraftwerk Union A.G., Technik Werkstoffe, Postfach 011420,
    D-4330 Mülheim

DORMAGEN, D., Inst. F. Eisenkunde, RWTH Aachen, Intzerstr. 1,
    D-5100 Aachen

KLAUSNITZER, E., Kraftwerk Union A.G., Hammerbacherstr. 12+14,
    D-8520 Erlangen

MUNZ, D., Kernforschungszentrum Karlsruhe, Postfach 3640,
    D-7500 Karlsruhe

KOCKELMANN, H., Staatl. Materialprüfungsanstalt Stuttgart,
    Pfaffenwaldring 32, D-7000 Stuttgart 80

ROOS, E., Staatl. Materialprüfungsanstalt Stuttgart, Pfaffenwaldring 32,
    D-7000 Stuttgart 80

SCHWALBE, K.H., GKSS-Forschungszentrum, Max Planck Str., D-2054 Geesthacht

TERLINDE, G., GKSS-Forschungszentrum, Max Planck Str., D-2054 Geesthacht

VOSS, B., Fraunhofer-Institut für Werkstoffmechanik, Wöhlerstr. 11,
    D-7800 Freiburg

ITALY - ITALIE

ANGELINO, G., C.I.S.E., Via Reggio Emilia, Segrate (Milano)

BERNARD, J.A., C.E.C., Joint Research Centre, Ispra, I-21020 Varese

MILELLA, P.P., E.N.E.A., Viale Regina Margherita 125, I-00198 Roma

REALE, S., Instituto di Ingegneria Meccanica, Facolta di Ingegneria,
    Via de S. Marta n. 3, I-50139 Firenze

VENZI, S., S.N.A.M. s.p.a. Metal, S. Donato Milanese, I-20097 Milano

NETHERLANDS - PAYS BAS

DE VRIES, M., Energy Research Foundation of the Netherlands, Westerduinweg 3, Postbus 1, NL-1755 Petten ZG

PRIJ, J., Energy Research Foundation of the Netherlands, Department of Energy Technology, Westerduinweg 3, Postbus 1, NL-1755 Petten ZG

SCHAAP, B., Energy Research Foundation of the Netherlands, Materials Department, Westerduinweg 3, Postbus 1, NL-1755 Petten ZG

STEENKAMP, P., Laboratory for Thermal Power Engineering, Delft University of Technology, P.O. Box 5055, NL-2600 GB Delft

SWEDEN - SUÈDE

KAISER, S., Royal Institute of Technology, Dept. of Strength of Materials, S-10044 Stockholm

SWITZERLAND - SUISSE

PRANTL, G., Gebr. Sulzer A.G., Abt. 1502, Winterthur

UGGOWITZER, P., Eidgenössische Technische Hochschule, ETH, Inst. für Metallurgie, ETH-Zentrum, CH-8092 Zürich

VARGA, T., c/o Hauptabteilung für die Sicherheit der Kernanlagen, Bundesamt für Energiewirtschaft, CH-5303 Würenlingen

UNITED KINGDOM - ROYAUME UNI

CRESWELL, S.L., H.M. Nuclear Installations Inspectorate, Thames House North, Millbank, London SW1P 4QZ

DRUCE, S.G., A.E.R.E. Harwell, Metallurgy Division, B 388, Oxon OX11 ORA

GARWOOD, S.J., The Welding Institute, Abington, Cambridge CB1 6AL

INGHAM, T., R.N.P.D.L., U.K.A.E.A. (Northern Division), Risley, Warrington WA3 6AT, Cheshire

NEALE, B.K., C.E.G.B., Berkeley Nuclear Laboratories, Berkeley, Glos. GL13 9PB

PRIEST, A.H., British Steel Corporation, Sheffield Laboratories, Swindon House, Moorgate, Rotherham, S. Yorks.

QUIRK, A., Safety & Reliability Directorate, U.K.A.E.A., Wigshaw Lane, Culcheth, Warrington, Cheshire

SHOJI, T., Department of Metallurgy and Engineering Materials, University of Newcastle upon Tyne, Haymarket Lane, Newcastle upon Tyne NE1 7RU

UNITED STATES - ETATS-UNIS

JOYCE, J.A., Engr. & Weapons Division, U.S. Naval Academy, Annapolis, MD 21658

LANDES, J.D., Westinghouse R&D Center, 1310 Beulah Road, Pittsburgh, PA 15235

LOSS, F.J., Materials Engineering Associates, 9700-B Palmer Highway, Lanham, MD 20706

READ, D.T., Mail Stop 2-1601, National Bureau of Standards, 325 Broadway, Boulder CO 80303

VAN DER SLUYS, W.A., Babcock & Wilcox, R&D Division, 1562 Beeson Street, Alliance, Ohio 44621

WILKOWSKI, G., Battelle Columbus Laboratory, 505 King Ave, Columbus, Ohio 43201

COMMISSION OF THE EUROPEAN COMMUNITIES

COMMISSION DES COMMUNAUTES EUROPEENNES

MAURER, H.A., Commission of the European Communities, 200 rue de la Loi, B-1049 Brussels, Belgium

OECD NUCLEAR ENERGY AGENCY

AGENCE DE L'OCDE POUR L'ENERGIE NUCLEAIRE

OLIVER, P., OECD Nuclear Energy Agency, 38 boulevard Suchet, F-75016 Paris, France

OECD SALES AGENTS
DÉPOSITAIRES DES PUBLICATIONS DE L'OCDE

ARGENTINA – ARGENTINE
Carlos Hirsch S.R.L., Florida 165, 4° Piso (Galería Guemes)
1333 BUENOS AIRES, Tel. 33.1787.2391 y 30.7122
AUSTRALIA – AUSTRALIE
Australia and New Zealand Book Company Pty, Ltd.,
10 Aquatic Drive, Frenchs Forest, N.S.W. 2086
P.O. Box 459, BROOKVALE, N.S.W. 2100
AUSTRIA – AUTRICHE
OECD Publications and Information Center
4 Simrockstrasse 5300 BONN. Tel. (0228) 21.60.45
Local Agent/Agent local :
Gerold and Co., Graben 31, WIEN 1. Tel. 52.22.35
BELGIUM – BELGIQUE
CCLS – LCLS
19, rue Plantin, 1070 BRUXELLES. Tel. 02.521.04.73
BRAZIL – BRÉSIL
Mestre Jou S.A., Rua Guaipa 518,
Caixa Postal 24090, 05089 SAO PAULO 10. Tel. 261.1920
Rua Senador Dantas 19 s/205-6, RIO DE JANEIRO GB.
Tel. 232.07.32
CANADA
Renouf Publishing Company Limited,
2182 St. Catherine Street West,
MONTRÉAL, Que. H3H 1M7. Tel. (514)937.3519
OTTAWA, Ont. K1P 5A6, 61 Sparks Street
DENMARK – DANEMARK
Munksgaard Export and Subscription Service
35, Nørre Søgade
DK 1370 KØBENHAVN K. Tel. +45.1.12.85.70
FINLAND – FINLANDE
Akateeminen Kirjakauppa
Keskuskatu 1, 00100 HELSINKI 10. Tel. 65.11.22
FRANCE
Bureau des Publications de l'OCDE,
2 rue André-Pascal, 75775 PARIS CEDEX 16. Tel. (1) 524.81.67
Principal correspondant :
13602 AIX-EN-PROVENCE : Librairie de l'Université.
Tel. 26.18.08
GERMANY – ALLEMAGNE
OECD Publications and Information Center
4 Simrockstrasse 5300 BONN Tel. (0228) 21.60.45
GREECE – GRÈCE
Librairie Kauffmann, 28 rue du Stade,
ATHÈNES 132. Tel. 322.21.60
HONG-KONG
Government Information Services,
Publications/Sales Section, Baskerville House,
2/F., 22 Ice House Street
ICELAND – ISLANDE
Snaebjörn Jönsson and Co., h.f.,
Hafnarstraeti 4 and 9, P.O.B. 1131, REYKJAVIK.
Tel. 13133/14281/11936
INDIA – INDE
Oxford Book and Stationery Co. :
NEW DELHI-1, Scindia House. Tel. 45896
CALCUTTA 700016, 17 Park Street. Tel. 240832
INDONESIA – INDONÉSIE
PDIN-LIPI, P.O. Box 3065/JKT., JAKARTA, Tel. 583467
IRELAND – IRLANDE
TDC Publishers – Library Suppliers
12 North Frederick Street, DUBLIN 1 Tel. 744835-749677
ITALY – ITALIE
Libreria Commissionaria Sansoni :
Via Lamarmora 45, 50121 FIRENZE. Tel. 579751/584468
Via Bartolini 29, 20155 MILANO. Tel. 365083
Sub-depositari :
Ugo Tassi
Via A. Farnese 28, 00192 ROMA. Tel. 310590
Editrice e Libreria Herder,
Piazza Montecitorio 120, 00186 ROMA. Tel. 6794628
Costantino Ercolano, Via Generale Orsini 46, 80132 NAPOLI. Tel.
405210
Libreria Hoepli, Via Hoepli 5, 20121 MILANO. Tel. 865446
Libreria Scientifica, Dott. Lucio de Biasio "Aeiou"
Via Meravigli 16, 20123 MILANO Tel. 807679
Libreria Zanichelli
Piazza Galvani 1/A, 40124 Bologna Tel. 237389
Libreria Lattes, Via Garibaldi 3, 10122 TORINO. Tel. 519274
La diffusione delle edizioni OCSE è inoltre assicurata dalle migliori
librerie nelle città più importanti.
JAPAN – JAPON
OECD Publications and Information Center,
Landic Akasaka Bldg., 2-3-4 Akasaka,
Minato-ku, TOKYO 107 Tel. 586.2016
KOREA – CORÉE
Pan Korea Book Corporation,
P.O. Box n° 101 Kwangwhamun, SÉOUL. Tel. 72.7369

LEBANON – LIBAN
Documenta Scientifica/Redico,
Edison Building, Bliss Street, P.O. Box 5641, BEIRUT.
Tel. 354429 – 344425
MALAYSIA – MALAISIE
and/et SINGAPORE - SINGAPOUR
University of Malaya Co-operative Bookshop Ltd.
P.O. Box 1127, Jalan Pantai Baru
KUALA LUMPUR. Tel. 51425, 54058, 54361
THE NETHERLANDS – PAYS-BAS
Staatsuitgeverij
Verzendboekhandel Chr. Plantijnstraat 1
Postbus 20014
2500 EA S-GRAVENHAGE. Tel. nr. 070.789911
Voor bestellingen: Tel. 070.789208
NEW ZEALAND – NOUVELLE-ZÉLANDE
Publications Section,
Government Printing Office Bookshops:
AUCKLAND: Retail Bookshop: 25 Rutland Street,
Mail Orders: 85 Beach Road, Private Bag C.P.O.
HAMILTON: Retail Ward Street,
Mail Orders, P.O. Box 857
WELLINGTON: Retail: Mulgrave Street (Head Office),
Cubacade World Trade Centre
Mail Orders: Private Bag
CHRISTCHURCH: Retail: 159 Hereford Street,
Mail Orders: Private Bag
DUNEDIN: Retail: Princes Street
Mail Order: P.O. Box 1104
NORWAY – NORVÈGE
J.G. TANUM A/S Karl Johansgate 43
P.O. Box 1177 Sentrum OSLO 1. Tel. (02) 80.12.60
PAKISTAN
Mirza Book Agency, 65 Shahrah Quaid-E-Azam, LAHORE 3.
Tel. 66839
PHILIPPINES
National Book Store, Inc.
Library Services Division, P.O. Box 1934, MANILA.
Tel. Nos. 49.43.06 to 09, 40.53.45, 49.45.12
PORTUGAL
Livraria Portugal, Rua do Carmo 70-74,
1117 LISBOA CODEX. Tel. 360582/3
SPAIN – ESPAGNE
Mundi-Prensa Libros, S.A.
Castelló 37, Apartado 1223, MADRID-1. Tel. 275.46.55
Libreria Bosch, Ronda Universidad 11, BARCELONA 7.
Tel. 317.53.08, 317.53.58
SWEDEN – SUÈDE
AB CE Fritzes Kungl Hovbokhandel,
Box 16 356, S 103 27 STH, Regeringsgatan 12,
DS STOCKHOLM. Tel. 08/23.89.00
SWITZERLAND – SUISSE
OECD Publications and Information Center
4 Simrockstrasse 5300 BONN. Tel. (0228) 21.60.45
Local Agents/Agents locaux
Librairie Payot, 6 rue Grenus, 1211 GENÈVE 11. Tel. 022.31.89.50
TAIWAN – FORMOSE
Good Faith Worldwide Int'l Co., Ltd.
9th floor, No. 118, Sec. 2
Chung Hsiao E. Road
TAIPEI. Tel. 391.7396/391.7397
THAILAND – THAILANDE
Suksit Siam Co., Ltd., 1715 Rama IV Rd,
Samyan, BANGKOK 5. Tel. 2511630
TURKEY – TURQUIE
Kültur Yayinlari Is-Türk Ltd. Sti.
Atatürk Bulvari No : 77/B
KIZILAY/ANKARA. Tel. 17 02 66
Dolmabahce Cad. No : 29
BESIKTAS/ISTANBUL. Tel. 60 71 88
UNITED KINGDOM – ROYAUME-UNI
H.M. Stationery Office, P.O.B. 276,
LONDON SW8   5DT. Tel. (01) 622.3316, or
49 High Holborn, LONDON WC1V 6 HB (personal callers)
Branches at: EDINBURGH, BIRMINGHAM, BRISTOL,
MANCHESTER, BELFAST.
UNITED STATES OF AMERICA – ÉTATS-UNIS
OECD Publications and Information Center, Suite 1207,
1750 Pennsylvania Ave., N.W. WASHINGTON, D.C.20006 – 4582
Tel. (202) 724.1857
VENEZUELA
Libreria del Este, Avda. F. Miranda 52, Edificio Galipan,
CARACAS 106. Tel. 32.23.01/33.26.04/31.58.38
YUGOSLAVIA – YOUGOSLAVIE
Jugoslovenska Knjiga, Terazije 27, P.O.B. 36, BEOGRAD.
Tel. 621.992

Les commandes provenant de pays où l'OCDE n'a pas encore désigné de dépositaire peuvent être adressées à :
OCDE, Bureau des Publications, 2, rue André-Pascal, 75775 PARIS CEDEX 16.

Orders and inquiries from countries where sales agents have not yet been appointed may be sent to:
OECD, Publications Office, 2 rue André-Pascal, 75775 PARIS CEDEX 16.

66628-6-1983

PUBLICATIONS DE L'OCDE, 2, rue André-Pascal, 75775 PARIS CEDEX 16 - N° 42690 1983
IMPRIMÉ EN FRANCE
(66 83 04 3) ISBN 92-64-02492-1